AF581851

Neuroendocrine Control of Energy Metabolism

Neuroendocrine Control of Energy Metabolism

Editors

Giancarlo Panzica
Stefano Gotti
Paloma Collado Guirao

MDPI • Basel • Beijing • Wuhan • Barcelona • Belgrade • Manchester • Tokyo • Cluj • Tianjin

Editors
Giancarlo Panzica
Department of Neuroscience
Rita Levi Montalcini
University of Torino
Torino
Italy

Stefano Gotti
Department of Neuroscience
Rita Levi Montalcini
University of Torino
Torino
Italy

Paloma Collado Guirao
Departamento de Psicobiología
National University of Distance
Education (UNED)
Madrid
Spain

Editorial Office
MDPI
St. Alban-Anlage 66
4052 Basel, Switzerland

This is a reprint of articles from the Special Issue published online in the open access journal *Metabolites* (ISSN 2218-1989) (available at: www.mdpi.com/journal/metabolites/special_issues/Neuroendocrine_Control_Energy_Metabolism).

For citation purposes, cite each article independently as indicated on the article page online and as indicated below:

LastName, A.A.; LastName, B.B.; LastName, C.C. Article Title. *Journal Name* **Year**, *Volume Number*, Page Range.

ISBN 978-3-0365-3494-7 (Hbk)
ISBN 978-3-0365-3493-0 (PDF)

Contents

About the Editors

Giancarlo Panzica

Prof. Giancarlo Panzica, currently Honorary Professor at the University of Torino and Principal Investigator at the Neuroscience Institute Cavalieri Ottolenghi (NICO), started his research career as an electron microscopist, performing studies on synaptic developments of the chicken optic tectum. Later, he became interested in the neuroanatomical organization of the hypothalamus, which he studied both through Golgi impregnation and electron microscopy. Following these studies, he began to explore the neural circuits controlling sexual behaviour and gonadal hormone dependence through interdisciplinary methods; in particular, the vasopressin and the NO-producing systems, as well as the roles that gonadal hormones play in modulating them

In the last 15 years, his main research topic has been the behavioural and neural effects of the precocious exposure to environmental endocrine disruptors, covering the effects of several endocrine disruptors.

Prof. Panzica is also the organizer of a series of conferences which are held in Torino, dedicated to the effects of steroids on the nervous system.

He has published about 200 papers in international, peer-reviewed journals.

Stefano Gotti

Stefano Gotti is an Associate Professor of Human Anatomy at the University of Torino, and Co-PI in the Neuroendocrinology Lab at the Neuroscience Institute Cavalieri Ottolenghi (NICO). He obtained his PhD in Neuroscience in 2004. His research originally focused on morphological studies of the distribution of several neurotransmitters in rodent brains. Later, he became interested in behavioural and morphological analysis regarding the effect of a "bad" environment on the development of neural circuits in rodents. Work in recent years on endocrine disruptors and on disease models such as anorexia nervosa fall into this vein. His research has been published in numerous international peer-reviewed journals.

Paloma Collado Guirao

Paloma Collado is a Professor of Psychobiology at the National University of Distance Education (UNED) in Spain. She obtained her PhD in Psychobiology in 1990. Her research since has been focused on the field of physiological psychology, and for the last fifteen years, she has explored the mechanisms involved in the development of the cerebral circuits that control food intake in rodents, and particularly on the vulnerability of hypothalamic circuits that regulate energy homeostasis in under- and over-nutrition, as well as the potential modulating factors during the development of adverse effects that an inadequate nutrition produces. She has developed this research as the Principal Investigator of different grants in collaboration with researchers from the University of Cambridge, the University of Turin, and the Department of Endocrinology of Universitary Hospital Niño Jesus. Her research has been published in international peer-reviewed journals.

Preface to "Neuroendocrine Control of Energy Metabolism"

The control of energy metabolism is a central event for cell, organ, and organism survival. There are many control levels in energy metabolism, although in this Special Issue (edited by G.C. Panzica, S. Gotti, and P. Collado), we concentrated on the neuroendocrine control which is operated through specialized neural circuits controlling both food intake and energy expenditure.

A series of reviews covered different topics, in particular, the involvement of circumventricular organs in the neuroendocrine control of metabolism (Jeong et al.), the involvement of the amygdala (Pined et al.), the glial cell, and especially astrocites (De Bernardis Murat and Garcia-Caceres) through the lactate cycle (Veloz Castillo et al.), and the relationships among the brown fat tissue and the neuroendocrine circuits (Lizcano and Arroyave).

In addition, several research papers have been added to this Special Issue, involving the impact of the diet on the IGF system (Guerra-Cantera et al.), as well as on the gut microbiota (Merkley et al.). Three papers are dedicated to the action of endocrine disruptors in adult life (Marraudino et al. a), and on the effects of perinatal administered genistein (Fernandez-Garcia et al., Marraudino et al. b). Finally, the last paper discusses the effects of growth restrictions on metabolism (Taylor et al.).

The topics discussed in this Special Issue are of particular interest for those involved in the study of relationships among metabolic diseases and neuroendocrine circuits.

Giancarlo Panzica, Stefano Gotti, and Paloma Collado Guirao
Editors

Review

Sensory Circumventricular Organs, Neuroendocrine Control, and Metabolic Regulation

Jin Kwon Jeong , Samantha A. Dow and Colin N. Young *

Department of Pharmacology and Physiology, School of Medicine and Health Sciences, The George Washington University, 2300 I St NW, Washington, DC 20037, USA; jinkwon0911@gwu.edu (J.K.J.); sdow3@gwu.edu (S.A.D.)
* Correspondence: colinyoung@gwu.edu; Tel.: +1-202-994-9575; Fax: +1-202-994-287

Abstract: The central nervous system is critical in metabolic regulation, and accumulating evidence points to a distributed network of brain regions involved in energy homeostasis. This is accomplished, in part, by integrating peripheral and central metabolic information and subsequently modulating neuroendocrine outputs through the paraventricular and supraoptic nucleus of the hypothalamus. However, these hypothalamic nuclei are generally protected by a blood-brain-barrier limiting their ability to directly sense circulating metabolic signals—pointing to possible involvement of upstream brain nuclei. In this regard, sensory circumventricular organs (CVOs), brain sites traditionally recognized in thirst/fluid and cardiovascular regulation, are emerging as potential sites through which circulating metabolic substances influence neuroendocrine control. The sensory CVOs, including the subfornical organ, organum vasculosum of the lamina terminalis, and area postrema, are located outside the blood-brain-barrier, possess cellular machinery to sense the metabolic interior milieu, and establish complex neural networks to hypothalamic neuroendocrine nuclei. Here, evidence for a potential role of sensory CVO-hypothalamic neuroendocrine networks in energy homeostasis is presented.

Keywords: subfornical organ; organum vasculosum of the lamina terminalis; area postrema; hypothalamus; metabolism

Citation: Jeong, J.K.; Dow, S.A.; Young, C.N. Sensory Circumventricular Organs, Neuroendocrine Control, and Metabolic Regulation. *Metabolites* **2021**, *11*, 494. https://doi.org/10.3390/metabo11080494

Academic Editors: Giancarlo Panzica, Stefano Gotti, Paloma Collado Guirao and Peter Meikle

Received: 10 June 2021
Accepted: 27 July 2021
Published: 29 July 2021

1. Introduction

Precise and reciprocal interactions between the central nervous system (CNS) and peripheral organs plays an integral role in whole body metabolic homeostasis, and impairments in this CNS-peripheral communication are clearly implicated in the development of metabolic disorders. This encompasses a wide range of conditions including obesity, type II diabetes, hypertriglyceridemia, non-alcoholic fatty liver disease, and insulin resistance, to name a few [1–6]. Within the CNS, a network of brain regions are involved in metabolic regulation, however, it is generally accepted that metabolic information from both peripheral and central inputs will eventually be integrated into the hypothalamus [4,5,7]. Hypothalamic nuclei, in particular the paraventricular nucleus (PVN) and supraoptic nucleus (SON), possess a wide array of neuroendocrine neurons, and therefore are considered as regions central to neuroendocrine regulation. However, the majority of circulating factors (hormones, adipokines, metabolites, etc.) cannot directly access these hypothalamic nuclei as they are protected by the blood-brain barrier (BBB) and/or substances are transported in limited quantity across the BBB; specialized endothelial cells located between the bloodstream and brain as a protective barrier against circulating toxins and pathogens [8,9]. This suggests involvement of other brain region(s) upstream of the PVN/SON in neuroendocrine-dependent metabolic homeostasis. In this regard, the sensory circumventricular organs (CVOs) are a key candidate, considering that: (1) They are located outside the BBB; (2) They possess the cellular machinery to detect circulating information, and; (3) They establish direct and/or indirect synaptic networks to hypothalamic neuroendocrine nuclei (Figure 1). Here, we will discuss existing anatomical, functional, and circuit level evidence pointing to the involvement of sensory CVOs in neuroendocrine regulated control of metabolism.

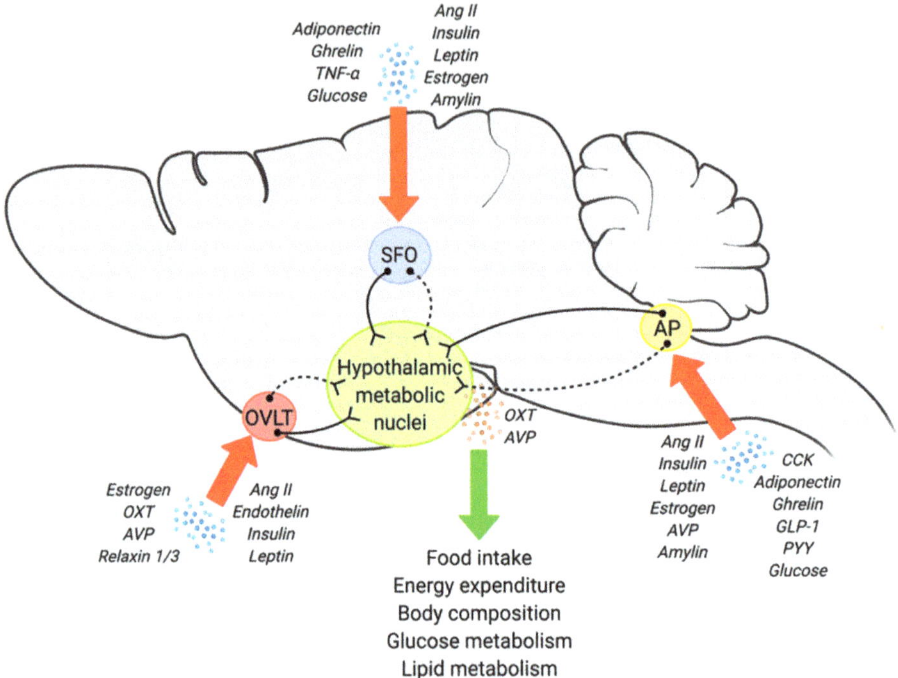

Figure 1. Schematic illustration showing potential sensory CVO-hypothalamic networks involved in metabolism regulation. Each of the sensory CVOs possesses the cellular machinery to sense multiple metabolic factors, a few of which are shown in the image. At the same time, sensory CVOs also establish direct (solid line) as well as indirect synapses (dashed line) to hypothalamic metabolic nuclei including the PVN and SON. Multiple investigations have demonstrated the involvement of the sensory CVOs in metabolism regulation, and further suggest that hypothalamic AVP and oxytocin (OXT) may play a key role. Image was created with Biorender.com.

2. Arcuate Nucleus Involvement in Metabolic Regulation

Before discussing a neuroendocrine-metabolic role of the sensory CVOs, it is important to consider what has been the predominant focus of the field. Since the identification of dense leptin receptors in hypothalamic nuclei [10], numerous investigations have focused on hypothalamic neural circuits in whole body metabolic regulation, in particular an arcuate nucleus-dependent axis [3,11]. The arcuate nucleus is a small region located in the mediobasal hypothalamus adjacent to the third ventricle (3V) and median eminence. While some studies have proposed the arcuate nucleus as a part of the CVOs [12], this region in fact possesses an intact BBB, and is therefore, fully protected from the circulation [13]. Nevertheless, the arcuate nucleus plays a key role in metabolic regulation, due to an ability of circulating factors to access the region through the median eminence and/or median eminence-3V complex [14,15].

The arcuate possesses two functionally opposing neuronal populations: neurons expressing proopiomelanocortin (POMC) and those producing agouti-related peptide (AgRP) and neuropeptide Y (NPY) [3,16,17]. Although these neuronal populations synaptically innervate multiple brain regions, hypothalamic neuroendocrine nuclei, particularly the PVN, are main targets. Conversely, POMC and AgRP/NPY receive dense inputs from regions throughout the CNS (detailed in ref [18]). When activated by satiety signals, such as leptin, estrogen, and insulin, POMC neurons produce and release alpha-melanocyte stimulating hormone (α-MSH) into other brain regions (e.g., PVN) as a neurotransmitter to decrease appetite while also increasing energy expenditure [3,19]. On the other hand, AgRP/NPY neurons release hunger factor-induced inhibitory neurotransmitters to negatively regulate POMC neuronal activity (Jeong 2014). In brief, a balance between POMC and AgRP/NPY neurons is thought to be key to the modulation of energy homeostasis [16,17].

In addition to neuronal populations within the arcuate nucleus, astrocytes and tanycytes, specialized glial cells located on the bottom of the 3V wall, also express a broad array of metabolic receptors. Multiple investigations have suggested a role for these glial cells as a means to communicate and introduce circulating metabolic cues to arcuate neurons [6,15,20]. The arcuate nucleus also receives metabolic information from the gastrointestinal tract indirectly via brainstem nuclei [19]. Collectively, while the critical role of the arcuate in metabolic regulation is well established, it is important to consider distributed CNS networks that operate in concert or independently from the arcuate in metabolic regulation. In this context, emerging evidence points to a unique role of the sensory CVOs, as detailed below.

3. Anatomy and Potential Metabolic Role of the Sensory CVOs

Most capillaries in the brain establish a BBB—a complex cellular physical barrier to protect the brain from the circulation [21]. While endothelial cells that are connected to each other through tight junctions are the basic component of the BBB, other neuronal and non-neuronal cells also form the BBB, which results in minimal fenestration and/or requires transport of select molecules [8,9]. However, the BBB in certain brain regions is "more loose" and permeable with discontinuous tight junctions, and therefore, blood-derived molecules can easily access the brain. These brain structures that lack a normal BBB are called the CVOs. The CVOs are comprised of secretory and sensory nuclei, of which the latter includes the subfornical organ (SFO), organum vasculosum of the lamina terminalis (OVLT), and area postrema (AP) [22,23]. Each of the sensory CVOs establishes neural networks, directly or indirectly, to the hypothalamus, and accumulating evidence suggests that signaling in the sensory CVOs may modulate broad metabolic parameters through hypothalamic control [23–25]. The unique characteristics and existing evidence that points to a neuroendocrine-dependent metabolic regulatory role of the SFO, OVLT, and AP is summarized below.

3.1. The SFO

The SFO is a sensory CVO located at the midline of the brain within the lateral ventricle and is comprised of two anatomically distinct subregions including the outer shell and ventromedial core [8,26–28]. Evidence suggests differential arrangement of tight-junction molecules within these SFO subregions, which impact size-dependent permeability of blood-borne molecules [8]. For example, peripheral administration of permeability indicators revealed that small molecules (<3000 kDa) accumulated primarily within the collagen IV-enriched ventromedial core, while the laminin-dominant outer shell was more selective for larger molecules (>10,000 kDa). Another example is that the hormone angiotensin-II (Ang-II) activates primarily the ventromedial core of the SFO, as represented by c-Fos expression following peripheral Ang-II administration, despite Ang-II type 1a receptors ($AT_{1a}R$) being broadly distributed throughout the SFO [28]. However, it is still unclear whether anatomically distinct SFO subregions are responsible for differential physiological outputs. Instead, multiple cell phenotypes within the entire SFO have been demonstrated to play a pivotal role in metabolism regulation [29–31].

The SFO is well recognized for its role in cardiovascular and fluid balance regulation [32–34]. However, emerging evidence from transgenic reporter mouse models and transcriptomics also suggests a role in metabolic control due to a wide distribution of cellular receptors within the SFO, including receptors for insulin, leptin, estrogen, ghrelin, and adiponectin [1,35–39]. Furthermore, dynamic regulation of SFO metabolic receptors in response to fasting and overnutrition has been demonstrated with transcriptome analysis [38,39]. In addition, multiple electrophysiological investigations have also demonstrated the responsiveness of the SFO to multiple metabolic and inflammatory factors, such as leptin, amylin, ghrelin, and tumor necrosis factor-α [40–45]. Dynamic responsiveness of SFO cells to multiple metabolic factors is also evident [35,36,41–44,46–48]. For example, while some SFO neurons were activated in response to glucose, insulin, or adiponectin, other SFO neurons were either deactivated or non-responsive to the same stimulus [35,36,43,47]. SFO neuronal responsiveness to adiponectin has also been shown to be modulated by food deprivation [35]. These results indicate metabolic status-dependent, selective, and dynamic SFO cellular plasticity in response to metabolic substances.

While the aforementioned evidence collectively points to a role for the SFO in metabolic regulation, to date, in vivo evidence is rather limited. However, acute electrical stimulation of the SFO induces feeding in satiated animals [40], and peripheral administration of a synthetic melanocortin receptor agonist has been suggested to reduce overnight food intake in rats via the SFO [49]. These limited findings suggest possible involvement of the SFO in the regulation of feeding behavior, although future studies are clearly warranted. Moreover, hormonal signaling within the SFO may modulate whole body energy homeostasis independent of food intake. For example, selective removal of SFO insulin receptors in mice results in a metabolic syndrome-like phenotype accompanied by moderate elevations in body weight, adiposity, and the development of hepatic steatosis [1]. In addition, central administration of the adipokine leptin induces weight loss and upregulates sympathetically-mediated brown adipose tissue thermogenesis; responses that are dependent on SFO Ang-II signaling [29]. In line with this, multiple investigations have suggested possible involvement of the SFO in the development of metabolic disorders including obesity [30,50] and associated conditions such as non-alcoholic fatty liver disease [51]. For instance, neuroinflammation is strongly implicated in obesity development in rodents and humans [52], and investigations in rodents suggested involvement of SFO Ang-II signaling, at least in part, in high fat diet-induced neuroinflammation and obesity development [30]. Collectively, this emerging evidence points to a key role for the SFO in metabolic regulation, including potentially complex interactions between different hormones, although further work is necessitated.

3.2. The OVLT

Located at the rostral end of the third ventricle, the OVLT is a hypothalamic sensory CVO [53] that is divided by two anatomically and functionally independent subregions including the inner capillary plexus and outer lateral zone [8]. Small molecules in the circulation access the capillary plexus and then sequentially diffuse to the lateral zone within the OVLT; this phenomenon has been associated with heterogeneous expression of capillary tight junction molecules between the two OVLT subregions [8]. However, several anatomical studies have suggested that functional regulation by the OVLT may occur primarily in the lateral zone [8,28,54,55]. For example, both mRNA and protein levels of $AT_{1a}R$ were detected throughout the entire OVLT [56,57], but peripheral administration of Ang-II results in c-Fos expression predominantly within the lateral zone [28,54]. Additionally, astrocytes, which are critical for the sensing of circulating factors in this brain region [58–60] are primarily distributed in the lateral zone [8]. Even within the lateral zone, estrogen receptor-alpha (ERα) expression, a potential area where interactions between sex hormones and metabolic/cardiovascular/fluid information occurs, is exclusively clustered at the dorsal cap area [55]. Therefore, it is plausible that the inner capillary plexus is an entrance for circulating substances, and the outer lateral zone integrates this information to drive OVLT-mediated outputs to downstream regions including hypothalamic neuroendocrine nuclei.

Multiple anatomical and biochemical investigations have demonstrated receptors for insulin, leptin, Ang-II, endothelin, estrogen, oxytocin, arginine vasopressin (AVP), and relaxin 1/3 in the OVLT [1,55,56,61–67]. Responsiveness of the OVLT to these circulating factors is also evident. For example, application of AVP into primary OVLT cell culture medium evoked increases in intracellular calcium [63,64]. Additionally, c-Fos expression in the OVLT was elevated by intracerebroventricular administration of leptin in rats on a normal chow diet [68]. These findings, along with others [55,61,62], collectively suggest that the OVLT possesses the ability to monitor and respond to overall metabolic status. Interestingly, the cellular expression profile of these metabolic receptors is rather complex. For example, oxytocin and AVP V_1 receptors are present in both neurons and glial cells [63]. Additionally, both OVLT neurons and glia are able to sense extracellular osmotic changes [59,69,70]. However, AVP V_2 receptors and ERα have been suggested to be expressed solely on neurons [63,65], while the expression of endothelin receptor-1 and toll like receptor-4 (TLR-4) are predominantly on glial cells [58,62]. It is further possible that multiple metabolic factors may interact within the same OVLT cell. For example, the majority of OVLT ERα-expressing neurons (i.e., responsive to estrogen) are also osmosensitive, and dehydration-evoked hypertonicity induces c-Fos expression within ERα-expressing cells [55]. Therefore, OVLT-mediated metabolic regulation could be determined by complex intra-OVLT cellular interactions whereby circulating substances act upon similar and/or discrete cell types.

In spite of the expression of numerous metabolic receptors in the OVLT, detailed in vivo investigations into OVLT-dependent metabolic regulation are currently lacking. This may be partially because the OVLT is a tiny structure located deep in the brain, and therefore, it is technically challenging to modify cell- and/or receptor-specific signaling pathways in this nucleus. However, several studies point to a potential role for the OVLT in energy homeostasis. For example, administration of the ovarian hormone relaxin peripherally or the neurohormone relaxin-3 directly into the brain induced OVLT neuronal activation and resulted in an increase in food intake in rats [67,71–73]. On the other hand, chemical blockade of the OVLT with acute administration of colchicine reduced food intake and blunted body weight gain [74]. In line with this, several investigations have also suggested OVLT involvement in food anticipatory behavior [75,76]. For example, in rabbit pups, increases in OVLT neuronal activity (i.e., c-Fos) was observed prior to scheduled nursing time [76]. It is also worthy to consider that the OVLT is well-recognized for its role in fluid balance. Metabolic and fluid regulation are closely related, and body fluid conditions can directly influence metabolic parameters, such as energy expenditure and food intake, both in humans and rodents [77–81]. Therefore, the OVLT may play a central role in whole body energy homeostasis by combining circulating fluid and metabolic information, although in-depth and targeted studies are clearly needed.

3.3. The AP

Similar to the SFO and OVLT, the AP possesses a specialized anatomy that allows it to monitor and regulate circulating factors, including those involved in metabolic function. Situated in the wall of the fourth ventricle, the AP is the most caudal sensory CVO and consists of three anatomically distinct areas: the perivascular, central, and lateral zones [8]. It has been suggested that the AP possesses a vascular portal system very similar to the neurohypophysis, connecting the vessels to the capillary plexus of the neuropil [82]. Sinusoidal vessels in the central zone of the AP, which is where most neurons and axon terminals reside, are much more fenestrated than the peripheral capillaries [83]. Thus, circulating molecules can directly access the central zone and then diffuse into the perivascular and lateral zones [8]. In line with this, glial cell bodies and fibers are dense in the lateral and perivascular zones, while the central zone shows very sparse glial immunoreactivity [8].

The majority of AP receptor expression is for hormones with anorexigenic effects, including amylin, CCK, GLP-1, peptide YY (PYY), adiponectin, and leptin. However, the AP is also equipped to detect orexigenic ghrelin [46,84]. Additionally, receptors for

Ang-II, AVP, estrogen, and potentially insulin have also been identified in the AP [84–88]. Further characterization of receptor expression in specific cell types has revealed that amylin, leptin, Ang-II, GLP-1, adiponectin 1/2, CCK, and ghrelin receptors are expressed in AP neurons [89–96]. Leptin, TLR-4, glial-cell derived neurotrophic factor receptor α-like (GFRAL) and complement type 3 (a receptor linked to hypoxia-induced emesis) receptors are also localized on glial cells in this brain region [97–100]. mRNA expression of AVP V_1a and PYY Y_1 receptors have been detected in the AP; however, the specific AP cell types expressing these receptors is currently unclear [101,102].

A role for the AP in metabolic regulation is further supported by histological and electrophysiological findings demonstrating responsiveness to the administration of various anorexigenic hormones. Peripheral administration of amylin, CCK, GLP-1, PYY, and adiponectin all lead to increased c-Fos expression in AP neurons [103–105]. Furthermore, amylin, CCK, PYY, insulin, and adiponectin have all been found to influence the excitability of AP neurons. For example, the use of the α-amino-3-hydroxy-5-methyl-4-isoxazolepropionic acid receptor antagonist cyanquixaline to block amylin-induced excitatory responses revealed that administration of amylin excites AP neurons by facilitating glutamate release from glutamatergic inputs to AP neurons. Similar effects have been identified for CCK [94,106]. Interestingly, heterogenous responsiveness of AP neurons to metabolic factors has also been demonstrated. For example, in culture, low concentrations of $PYY_{1\text{-}36}$ depolarize whereas high concentrations of $PYY_{3\text{-}36}$ hyperpolarize AP neurons [84]. In addition, adiponectin and leptin primarily result in depolarization of most AP neurons. However, a subpopulation hyperpolarizes in response to these adipokines [46,90]. Smith et al. also demonstrated that the same subpopulation of AP neurons was responsive to both amylin and leptin, which was further supported by the demonstration that 94% of tested AP neurons were excited by both glucose and amylin [46,107]. In addition to anorexigenic hormones, the AP also appears to respond in a potentially complex manner to orexigenic peptides. Specifically, ghrelin induces hyperpolarization in 50% of AP neurons via modulation of voltage-gated K^+ currents whereas the remaining ghrelin-sensitive neurons depolarize through a nonspecific cation current [108]. Collectively, these findings indicate that the AP is well-situated to integrate multiple circulating factors and responds to anorexigenic/orexigenic hormones, glucose, and adipokines, although the intricacies of the AP's responsiveness to metabolic factors warrant further interrogation.

In line with the aforementioned receptor expression, and the well-recognized role of the AP as a "chemoreceptor trigger zone" due to its role in emesis [109], numerous studies have demonstrated AP activation following peripheral injection of various hormones. Peripheral administration of anorexigenic hormones amylin, CCK, GLP-1, PYY, and adiponectin all lead to increased c-Fos expression in AP neurons [103–105]. Furthermore, these hormones suppress feeding behavior in rodents, and this effect requires an intact AP and receptor activation [110–112]. For example, AP-specific blockade with the amylin receptor antagonist AC187 inhibited amylin-induced-feeding suppression, as well as feeding-induced c-fos expression in fasted rats. [103,110]. A role the AP in response to "newer" anorexigenic factors is also emerging. Specifically, growth differentiation factor 15 (GDF15), a stress response cytokine that signals via GFRAL, inhibits feeding [113]. Activation of GFRAL receptors induce AP neuron activation, suggesting that GDF15-induced suppressed food intake may be mediated by the AP [114,115]. Hormones at the AP also influence other metabolic outcomes including thermogenesis and glucose homeostasis. For example, retrograde tracing from interscapular brown adipose tissue has implicated the AP in brown adipose tissue thermogenesis [116]. Additionally, mice with knockout of certain amylin receptor subunits become glucose intolerant [117]. Glucose intolerance also occurs in GFRAL knockout mice challenged by high-fat diet, which may be mediated directly by the AP or indirectly through the adjacent nucleus tractus solitarius (NTS) [118]. Together, these in vivo findings support the AP's role in integrating circulating metabolic factors to regulate various physiological outcomes, potentially through direct AP-hypothalamic pathways or indirectly through AP-brainstem-hypothalamic neural pathways [114,118] as discussed below.

4. Sensory CVOs and Hypothalamic Circuits in Metabolic Regulation

As described above, the sensory CVOs are equipped with an array of receptors and responsive to numerous stimuli, making them a key entry point for circulating metabolic factors to influence the brain. Once detected and integrated in the sensory CVOs, this information will then be transmitted via neuronal efferents to hypothalamic metabolic centers, including the PVN and SON. Evidence for sensory CVOs-hypothalamic neuroendocrine neural networks is discussed below.

SFO neurons establish direct as well as indirect synaptic connectivity with hypothalamic metabolic nuclei. For example, the SFO provides direct excitatory synaptic inputs to the PVN and SON [119–121]. In particular, cells within the dorsolateral peripheral subregion of the SFO project to the magnocellular portion of the PVN where numerous AVP and oxytocin cells are distributed [121]. The SFO also establishes excitatory and inhibitory synaptic communication with other hypothalamic nuclei, including the bed nucleus of the stria terminalis, arcuate nucleus, OVLT, and median preoptic nucleus (MnPO) [119,120,122–124]—neuronal networks that also allow the SFO to communicate indirectly with the PVN and SON. Interestingly, the cellular and synaptic architecture from the SFO to hypothalamus is very complex. For example, separate populations of SFO neurons project to the PVN and MnPO, although a weak number of SFO neurons provide collateral projections to both regions [125]. Importantly, the SFO and MnPO establish reciprocal connections, and SFO cells that receive inputs from the MnPO project to the PVN [126], suggesting a possible feedback loop between the SFO and MnPO to regulate an SFO-PVN axis. However, anatomical and synaptic projection information for specific SFO cell types, particularly in the context of "metabolic receptor" expressing neurons, is largely unavailable, and therefore needs to be addressed in the future.

Similar to SFO, OVLT-dependent metabolic regulation is most likely mediated by complex OVLT neural networks to multiple hypothalamic nuclei. However, in depth investigations are lacking, particularly as related to traditional metabolic mediators (e.g., adipokines, hepatokines, insulin, etc.). Nevertheless, insights from other areas of investigation provide insight into potential OVLT networks. In the context of thirst control and drinking behavior, the OVLT provides both excitatory and inhibitory inputs to the MnPO [124], and this information is further transmitted to the PVN [127]. Similarly, OVLT neurons expressing ERα, relaxin, $AT_{1a}R$, and cholinergic receptors are also connected to the PVN and SON, presumably through the MnPO [28,65,71,128,129]. On the other hand, OVLT neurons that respond to extracellular sodium concentrations establish monosynaptic projections to the PVN [59,130]. Although indirect, given that fluid balance and metabolic regulation are closely related in humans as well as non-human species [77–81], these findings point to possible OVLT-hypothalamic networks that may be involved in metabolism regulation.

Anatomical studies using retrograde tracers indicate that the AP sends efferent projections to the PVN and SON [131,132]. In line with this, hypertonic saline induces c-Fos expression in the PVN via the AP [133], indicating the existence of direct synaptic communications between the AP and hypothalamic neuroendocrine centers. However, more evidence is required to delineate the direct networks between the AP and PVN/SON. Nevertheless, the AP establishes strong bidirectional synaptic interactions with adjacent nuclei, including the NTS and dorsal motor nucleus (DMN). Numerous studies have suggested this AP-NTS-DMN cluster as a critical brainstem metabolic center [113,132,134–139]. Importantly, this brainstem metabolic complex is highly connected to hypothalamic neuroendocrine centers [140–143]. Thus, similar to the SFO and OVLT, the AP is anatomically situated to directly and/or indirectly influence metabolic regulation through hypothalamic neuroendocrine nuclei.

Within the hypothalamus, numerous neuroendocrine neuron subpopulations are distributed in the PVN and SON. To date, direct anatomical evidence into the precise neuroendocrine neuron type that the sensory CVOs project to remains uninvestigated. However, indirect evidence points to hypothalamic AVP and/or oxytocin neurons as a common downstream target of the sensory CVOs. For example, several hormones that are involved in metabolic regulation, including estrogen, relaxin, and Ang-II, have been shown to modulate gene expression and release of AVP and oxytocin through the sensory CVOs [65,71,144–148]. Additionally, SFO-specific electrical stimulation resulted in elevations in circulating AVP and oxytocin [149,150]. In line with this, pharmacological cholinergic stimulation of the SFO induced elevations in c-Fos expression within AVP cells in the PVN and SON [151]. Similarly, the OVLT, particularly OVLT neurons that directly project to the PVN, have been suggested to play a role in hyperosmolality-dependent AVP and oxytocin release [130,152,153]. In addition, relaxin administration in rodents also induces c-Fos expression in the PVN and SON that is paralleled by release of AVP and oxytocin; a response that is, at least in part, through OVLT mechanisms [71,146]. Peripheral administration of anorexigenic CCK induced SON oxytocin neuronal activity, and further, release of oxytocin into the bloodstream, which was blunted following AP lesioning [154]. Furthermore, central administration of GLP-1 increases plasma AVP levels, which is accompanied parallel increases in c-Fos in the AP, PVN, and SON [155].

The findings pointing to a sensory CVO influence on hypothalamic AVP and oxytocin neurons is intriguing given oxytocin and AVP's ability to modulate a variety of metabolic outcomes including feeding behavior, body composition, and glucose/lipid metabolism. Oxytocin has been shown to exhibit anorectic effects, as both central and peripheral oxytocin administration leads to decreased food intake in animal models and humans [124,156–161]. Not only does oxytocin influence feeding behavior, but it further affects body composition and energy expenditure. In multiple animal models, loss of central oxytocin signaling via oxytocin neuron ablation or oxytocin receptor deletion increases fat mass and decreases energy expenditure [162–165]. Furthermore, recent work suggests that exogenous oxytocin treatment is associated with increased brown adipose tissue thermogenesis and "browning" of white adipose tissue, which is consistent with the increased energy expenditure induced by oxytocin treatment [166–168]. Changes in body composition may be further attributed to oxytocin modulation of glucose and lipid metabolism. Oxytocin enhances glucose uptake in muscle and adipose tissue and augments lipolysis and β-oxidation in adipose tissue [169–172].

Similar to oxytocin, AVP also affects a broad spectrum of metabolic parameters [173]. For example, acute endogenous activation of PVN AVP neurons decreases food intake, and peripheral administration of AVP further decreases brown adipose tissue thermogenesis in healthy rodent models [174–176]. On the other hand, hypothalamic AVP expression in rats is also increased with the onset of diabetes mellitus [177], suggesting a normal and pathophysiological effect of AVP in metabolism regulation. Interestingly, while AVP V_1a receptor-deficient mice display enhanced hepatic glucose production accompanied by high plasma glucose levels [178,179], AVP V_{1b} receptor-deficient animals develop hypoglycemia [180], indicating AVP involvement in glucose homeostasis in a receptor-dependent manner. AVP also appears to prevent lipolysis and β-oxidation via V1a, as V1a-deficient mice display enhanced lipolysis in brown adipocytes and β-oxidation in muscle and liver [181]. In humans, the metabolic effects of AVP are unclear, however, several investigations have also suggested a link between AVP and metabolic disorders, such as obesity and diabetes [182–184].

5. Conclusions

It is well-accepted that hypothalamic neuroendocrine nuclei including the PVN and SON play a central role in regulating energy homeostasis. While the predominant and well-accepted focus has been on the role of arcuate nucleus influence to these regions, emerging results further suggest the involvement of non-hypothalamic brain regions including the sensory CVOs. Each of the sensory CVOs establishes direct as well as indirect synaptic communication with the PVN and SON. In addition, the sensory CVOs are located outside of the BBB and express a broad array of metabolic receptors. Therefore, the sensory CVOs are anatomically and biochemically situated to detect metabolic factors in the circulation and influence whole body energy homeostasis through downstream hypothalamic nuclei. While precise neuroendocrine modulation by the sensory CVOs continues to emerge, accumulating evidence points to AVP and oxytocin as potential neuroendocrine targets of the sensory CVOs in metabolic regulation. However, in-depth neuroanatomical and functional in vivo investigations are warranted to build upon existing work. Nevertheless, the sensory CVOs are likely brain sites that are involved in neural responses to circulating metabolic signals and play a key role in the central regulation of energy homeostasis through neuroendocrine mechanisms.

Author Contributions: J.K.J., S.A.D. and C.N.Y. contributed equally to generate and finalize the manuscript. All authors have read and agreed to the published version of the manuscript.

Funding: This work was supported by the American Heart Association to J.K.J. (19CDA34630010), the National Heart, Lung, and Blood Institute (NHBLI, NIH) to C.N.Y. (R01HL141393), and the National Institute of Diabetes and Digestive and Kidney Diseases (NIDDK, NIH) to C.N.Y. (R01DK117007). The figure was created with Biorender.com (accessed on 16 June 2021).

Institutional Review Board Statement: Not applicable.

Informed Consent Statement: Not applicable.

Data Availability Statement: Not applicable.

Conflicts of Interest: The authors declare that this manuscript was prepared in the absence of any commercial or financial relationships that could be construed as a potential conflict of interest.

References

1. Jeong, J.K.; Horwath, J.A.; Simonyan, H.; Blackmore, K.A.; Butler, S.D.; Young, C.N. Subfornical organ insulin receptors tonically modulate cardiovascular and metabolic function. *Physiol. Genom.* **2019**, *51*, 333–341. [CrossRef]
2. Hurr, C.; Simonyan, H.; Morgan, D.A.; Rahmouni, K.; Young, C.N. Liver sympathetic denervation reverses obesity-induced hepatic steatosis. *J. Physiol.* **2019**, *597*, 4565–4580. [CrossRef]
3. Jeong, J.K.; Kim, J.G.; Lee, B.J. Participation of the central melanocortin system in metabolic regulation and energy homeostasis. *Cell. Mol. Life Sci.* **2014**, *71*, 3799–3809. [CrossRef]
4. Silva, S.C.; Cavadas, C. Hypothalamic Dysfunction in Obesity and Metabolic Disorders. *Adv. Neurobiol.* **2017**, *19*, 73–116. [CrossRef]
5. Timper, K.; Brüning, J.C. Hypothalamic circuits regulating appetite and energy homeostasis: Pathways to obesity. *Dis. Model. Mech.* **2017**, *10*, 679–689. [CrossRef] [PubMed]
6. Kim, J.G.; Suyama, S.; Koch, M.; Jin, S.; Arizón, P.A.; Argente, J.; Liu, Z.-W.; Zimmer, M.R.; Jeong, J.K.; Szigeti-Buck, K.; et al. Leptin signaling in astrocytes regulates hypothalamic neuronal circuits and feeding. *Nat. Neurosci.* **2014**, *17*, 908–910. [CrossRef] [PubMed]
7. Myers, M.G., Jr. Metabolic sensing and regulation by the hypothalamus. *Am. J. Physiol. Endocrinol. Metab.* **2008**, *294*, E809. [CrossRef]
8. Morita, S.; Furube, E.; Mannari, T.; Okuda, H.; Tatsumi, K.; Wanaka, A.; Miyata, S. Heterogeneous vascular permeability and alternative diffusion barrier in sensory circumventricular organs of adult mouse brain. *Cell Tissue Res.* **2015**, *363*, 497–511. [CrossRef] [PubMed]
9. Engelhardt, B. Development of the blood-brain barrier. *Cell Tissue Res.* **2003**, *314*, 119–129. [CrossRef] [PubMed]
10. Fei, H.; Okano, H.J.; Li, C.; Lee, G.-H.; Zhao, C.; Darnell, R.; Friedman, J.M. Anatomic localization of alternatively spliced leptin receptors (Ob-R) in mouse brain and other tissues. *Proc. Natl. Acad. Sci. USA* **1997**, *94*, 7001–7005. [CrossRef] [PubMed]
11. Williams, G.; Bing, C.; Cai, X.J.; Harrold, J.A.; King, P.J.; Liu, X.H. The hypothalamus and the control of energy homeostasis: Different circuits, different purposes. *Physiol. Behav.* **2001**, *74*, 683–701. [CrossRef]
12. Ciofi, P. The arcuate nucleus as a circumventricular organ in the mouse. *Neurosci. Lett.* **2011**, *487*, 187–190. [CrossRef]
13. M Smith, P.; V Ferguson, A. Metabolic Signaling to the Central Nervous System: Routes Across the Blood Brain Barrier. *Curr. Pharm. Des.* **2014**, *20*, 1392–1399. [CrossRef] [PubMed]
14. Jiang, H.; Gallet, S.; Klemm, P.; Scholl, P.; Folz-Donahue, K.; Altmüller, J.; Alber, J.; Heilinger, C.; Kukat, C.; Loyens, A.; et al. MCH Neurons Regulate Permeability of the Median Eminence Barrier. *Neuron* **2020**, *107*, 306–319. [CrossRef]
15. Balland, E.; Dam, J.; Langlet, F.; Caron, E.; Steculorum, S.; Messina, A.; Rasika, S.; Falluel-Morel, A.; Anouar, Y.; Dehouck, B.; et al. Hypothalamic Tanycytes Are an ERK-Gated Conduit for Leptin into the Brain. *Cell Metab.* **2014**, *19*, 293–301. [CrossRef] [PubMed]

16. Bouret, S.G. Development of hypothalamic circuits that control food intake and energy balance. In *Appetite and Food Intake: Central Control*, 2nd ed.; Harris, R., Ed.; CRC Press: Boca Raton, FL, USA, 2017; pp. 135–154.
17. Nuzzaci, D.; Laderrière, A.; Lemoine, A.; Nédélec, E.; Pénicaud, L.; Rigault, C.; Benani, A. Plasticity of the Melanocortin System: Determinants and Possible Consequences on Food Intake. *Front. Endocrinol.* **2015**, *6*, 143. [CrossRef]
18. Wang, D.; He, X.; Zhao, Z.; Feng, Q.; Lin, R.; Sun, Y.; Ding, T.; Xu, F.; Luo, M.; Zhan, C. Whole-brain mapping of the direct inputs and axonal projections of POMC and AgRP neurons. *Front. Neuroanat.* **2015**, *9*, 40. [CrossRef] [PubMed]
19. Valassi, E.; Scacchi, M.; Cavagnini, F. Neuroendocrine control of food intake. *Nutr. Metab. Cardiovasc. Dis.* **2008**, *18*, 158–168. [CrossRef]
20. Prevot, V.; Dehouck, B.; Sharif, A.; Ciofi, P.; Giacobini, P.; Clasadonte, J. The Versatile Tanycyte: A Hypothalamic Integrator of Reproduction and Energy Metabolism. *Endocr. Rev.* **2018**, *39*, 333–368. [CrossRef]
21. Daneman, R.; Prat, A. The Blood–Brain Barrier. *Cold Spring Harb. Perspect. Biol.* **2015**, *7*, a020412. [CrossRef]
22. Kaur, C.; Ling, E.A. The circumventricular organs. *Histol. Histopathol.* **2017**, *32*, 879–892. [CrossRef]
23. Ganong, W.F. Circumventricular Organs: Definition And Role In The Regulation Of Endocrine And Autonomic Function. *Clin. Exp. Pharmacol. Physiol.* **2000**, *27*, 422–427. [CrossRef] [PubMed]
24. Johnson, A.K.; Gross, P.M. Sensory circumventricular organs and brain homeostatic pathways. *FASEB J.* **1993**, *7*, 678–686. [CrossRef] [PubMed]
25. Fry, M.; Hoyda, T.D.; Ferguson, A.V. Making sense of it: Roles of the sensory circumventricular organs in feeding and regulation of energy homeostasis. *Exp. Biol. Med.* **2007**, *232*, 14–26.
26. McKinley, M.J.; McAllen, R.; Davern, P.; Giles, M.E.; Penschow, J.; Sunn, N.; Uschakov, A.; Oldfield, B. The Sensory Circumventricular Organs of the Mammalian Brain. *Adv. Anat. Embryol. Cell Biol.* **2003**, *172*, 1–122. [CrossRef]
27. Sisó, S.; Jeffrey, M.; González, L. Sensory circumventricular organs in health and disease. *Acta Neuropathol.* **2010**, *120*, 689–705. [CrossRef]
28. Sunn, N.; McKinley, M.J.; Oldfield, B.J. Circulating angiotensin II activates neurones in circumventricular organs of the lamina terminalis that project to the bed nucleus of the stria terminalis. *J. Neuroendocr.* **2003**, *15*, 725–731. [CrossRef]
29. Young, C.N.; Morgan, D.A.; Butler, S.D.; Rahmouni, K.; Gurley, S.B.; Coffman, T.M.; Mark, A.L.; Davisson, R.L. Angiotensin type 1a receptors in the forebrain subfornical organ facilitate leptin-induced weight loss through brown adipose tissue thermogenesis. *Mol. Metab.* **2015**, *4*, 337–343. [CrossRef]
30. de Kloet, A.D.; Pioquinto, D.J.; Nguyen, D.; Wang, L.; Smith, J.A.; Hiller, H.; Sumners, C. Obesity induces neuroinflammation mediated by altered expression of the renin–angiotensin system in mouse forebrain nuclei. *Physiol. Behav.* **2014**, *136*, 31–38. [CrossRef]
31. Matsuda, T.; Hiyama, T.Y.; Kobayashi, K.; Kobayashi, K.; Noda, M. Distinct CCK-positive SFO neurons are involved in persistent or transient suppression of water intake. *Nat. Commun.* **2020**, *11*, 1–15. [CrossRef] [PubMed]
32. Young, C.N.; Li, A.; Dong, F.N.; Horwath, J.A.; Clark, C.G.; Davisson, R.L. Endoplasmic reticulum and oxidant stress mediate nuclear factor-κB activation in the subfornical organ during angiotensin II hypertension. *Am. J. Physiol. Cell Physiol.* **2015**, *308*, C803–C812. [CrossRef]
33. Zimmerman, C.; Lin, Y.C.; Leib, D.E.; Guo, L.; Huey, E.L.; Daly, G.E.; Chen, Y.; Knight, Z.A. Thirst neurons anticipate the homeostatic consequences of eating and drinking. *Nature* **2016**, *537*, 680–684. [CrossRef] [PubMed]
34. Nation, H.L.; Nicoleau, M.; Kinsman, B.J.; Browning, K.N.; Stocker, S.D. DREADD-induced activation of subfornical organ neurons stimulates thirst and salt appetite. *J. Neurophysiol.* **2016**, *115*, 3123–3129. [CrossRef]
35. Alim, I.; Fry, M.; Walsh, M.H.; Ferguson, A.V. Actions of adiponectin on the excitability of subfornical organ neurons are altered by food deprivation. *Brain Res.* **2010**, *1330*, 72–82. [CrossRef]
36. Lakhi, S.; Snow, W.; Fry, M. Insulin modulates the electrical activity of subfornical organ neurons. *NeuroReport* **2013**, *24*, 329–334. [CrossRef] [PubMed]
37. Gonzalez, A.; Wang, G.; Waters, E.; Gonzales, K.; Speth, R.; Van Kempen, T.; Marques-Lopes, J.; Young, C.; Butler, S.; Davisson, R.; et al. Distribution of angiotensin type 1a receptor-containing cells in the brains of bacterial artificial chromosome transgenic mice. *Neuroscience* **2012**, *226*, 489–509. [CrossRef]
38. Hindmarch, C.; Ferguson, A.V. Physiological roles for the subfornical organ: A dynamic transcriptome shaped by autonomic state. *J. Physiol.* **2016**, *594*, 1581–1589. [CrossRef]
39. Peterson, C.S.; Huang, S.; Lee, S.A.; Ferguson, A.; Fry, W.M. The transcriptome of the rat subfornical organ is altered in response to early postnatal overnutrition. *IBRO Rep.* **2018**, *5*, 17–23. [CrossRef]
40. Smith, P.M.; Rozanski, G.; Ferguson, A.V. Acute electrical stimulation of the subfornical organ induces feeding in satiated rats. *Physiol. Behav.* **2010**, *99*, 534–537. [CrossRef] [PubMed]
41. Riediger, T.; Rauch, M.; Schmid, H.A. Actions of amylin on subfornical organ neurons and on drinking behavior in rats. *Am. J. Physiol. Regul. Integr. Comp. Physiol.* **1999**, *276*, R514–R521. [CrossRef]
42. Pulman, K.J.; Fry, W.M.; Cottrell, G.T.; Ferguson, A.V. The Subfornical Organ: A Central Target for Circulating Feeding Signals. *J. Neurosci.* **2006**, *26*, 2022–2030. [CrossRef]
43. Paes-Leme, B.; Dos-Santos, R.C.; Mecawi, A.S.; Ferguson, A.V. Interaction between angiotensinIIand glucose sensing at the subfornical organ. *J. Neuroendocr.* **2018**, *30*, e12654. [CrossRef] [PubMed]

44. Simpson, N.J.; Ferguson, A.V. Tumor necrosis factor-α potentiates the effects of angiotensin II on subfornical organ neurons. *Am. J. Physiol. Regul. Integr. Comp. Physiol.* **2018**, *315*, R425–R433. [CrossRef] [PubMed]
45. Simpson, N.J.; Ferguson, A.V. The proinflammatory cytokine tumor necrosis factor-α excites subfornical organ neurons. *J. Neurophysiol.* **2017**, *118*, 1532–1541. [CrossRef] [PubMed]
46. Smith, P.M.; Brzezinska, P.; Hubert, F.; Mimee, A.; Maurice, D.H.; Ferguson, A.V. Leptin influences the excitability of area postrema neurons. *Am. J. Physiol. Regul. Integr. Comp. Physiol.* **2016**, *310*, R440–R448. [CrossRef]
47. Medeiros, N.; Dai, L.; Ferguson, A. Glucose-responsive neurons in the subfornical organ of the rat—A novel site for direct CNS monitoring of circulating glucose. *Neuroscience* **2012**, *201*, 157–165. [CrossRef] [PubMed]
48. Vallières, L.; Lacroix, S.; Rivest, S. Influence of interleukin-6 on neural activity and transcription of the gene encoding corticotrophin-releasing factor in the rat brain: An effect depending upon the route of administration. *Eur. J. Neurosci.* **1997**, *9*, 1461–1472. [CrossRef] [PubMed]
49. Trivedi, P.; Jiang, M.; Tamvakopoulos, C.C.; Shen, X.; Yu, H.; Mock, S.; Fenyk-Melody, J.; Van der Ploeg, L.H.; Guan, X.-M. Exploring the site of anorectic action of peripherally administered synthetic melanocortin peptide MT-II in rats. *Brain Res.* **2003**, *977*, 221–230. [CrossRef]
50. Cruz, J.C.; Flôr, A.F.L.; França-Silva, M.S.; Balarini, C.M.; Braga, V.D.A. Reactive Oxygen Species in the Paraventricular Nucleus of the Hypothalamus Alter Sympathetic Activity During Metabolic Syndrome. *Front. Physiol.* **2015**, *6*, 384. [CrossRef] [PubMed]
51. Horwath, J.A.; Hurr, C.; Butler, S.D.; Guruju, M.; Cassell, M.D.; Mark, A.L.; Davisson, R.L.; Young, C.N. Obesity-induced hepatic steatosis is mediated by endoplasmic reticulum stress in the subfornical organ of the brain. *JCI Insight* **2017**, *2*, 2. [CrossRef]
52. Thaler, J.P.; Yi, C.X.; Schur, E.A.; Guyenet, S.J.; Hwang, B.H.; Dietrich, M.; Zhao, X.; Sarruf, D.A.; Izgur, V.; Maravilla, K.R.; et al. Obesity is associated with hypothalamic injury in rodents and humans. *J. Clin. Investig.* **2012**, *122*, 153–162. [CrossRef]
53. Prager-Khoutorsky, M.; Bourque, C.W. Anatomical organization of the rat organum vasculosum laminae terminalis. *Am. J. Physiol. Regul. Integr. Comp. Physiol.* **2015**, *309*, R324–R337. [CrossRef] [PubMed]
54. Kinsman, B.J.; Simmonds, S.S.; Browning, K.N.; Wenner, M.M.; Farquhar, W.B.; Stocker, S.D. Integration of Hypernatremia and Angiotensin II by the Organum Vasculosum of the Lamina Terminalis Regulates Thirst. *J. Neurosci.* **2020**, *40*, 2069–2079. [CrossRef] [PubMed]
55. Somponpun, S.J.; Johnson, A.K.; Beltz, T.; Sladek, C.D. Estrogen receptor-α expression in osmosensitive elements of the lamina terminalis: Regulation by hypertonicity. *Am. J. Physiol. Regul. Integr. Comp. Physiol.* **2004**, *287*, R661–R669. [CrossRef]
56. Lenkei, Z.; Corvol, P.; Llorens-Cortès, C. The angiotensin receptor subtype AT1A predominates in rat forebrain areas involved in blood pressure, body fluid homeostasis and neuroendocrine control. *Mol. Brain Res.* **1995**, *30*, 53–60. [CrossRef]
57. Phillips, M.I.; Shen, L.; Richards, E.M.; Raizada, M.K. Immunohistochemical mapping of angiotensin AT1 receptors in the brain. *Regul. Pept.* **1993**, *44*, 95–107. [CrossRef]
58. Nakano, Y.; Furube, E.; Morita, S.; Wanaka, A.; Nakashima, T.; Miyata, S. Astrocytic TLR4 expression and LPS-induced nuclear translocation of STAT3 in the sensory circumventricular organs of adult mouse brain. *J. Neuroimmunol.* **2015**, *278*, 144–158. [CrossRef]
59. Nomura, K.; Hiyama, T.; Sakuta, H.; Matsuda, T.; Lin, C.-H.; Kobayashi, K.; Kobayashi, K.; Kuwaki, T.; Takahashi, K.; Matsui, S.; et al. [Na+] Increases in Body Fluids Sensed by Central Nax Induce Sympathetically Mediated Blood Pressure Elevations via H+-Dependent Activation of ASIC1a. *Neuron* **2019**, *101*, 60–75.e6. [CrossRef]
60. Mannari, T.; Morita, S.; Furube, E.; Tominaga, M.; Miyata, S. Astrocytic TRPV1 ion channels detect blood-borne signals in the sensory circumventricular organs of adult mouse brains. *Glia* **2013**, *61*, 957–971. [CrossRef]
61. Gebke, E.; Müller, A.R.; Jurzak, M.; Gerstberger, R. Angiotensin II-induced calcium signalling in neurons and astrocytes of rat circumventricular organs. *Neuroscience* **1998**, *85*, 509–520. [CrossRef]
62. Gebke, E.; Müller, A.R.; Pehl, U.; Gerstberger, R. Astrocytes in sensory circumventricular organs of the rat brain express functional binding sites for endothelin. *Neuroscience* **2000**, *97*, 371–381. [CrossRef]
63. Jurzak, M.; Müller, A.R.; Gerstberger, R. Characterization of vasopressin receptors in cultured cells derived from the region of rat brain circumventricular organs. *Neuroscience* **1995**, *65*, 1145–1159. [CrossRef]
64. Jurzak, M.; Müller, A.R.; Gerstberger, R. AVP-fragment peptides induce Ca2+ transients in cells cultured from rat circumventricular organs. *Brain Res.* **1995**, *673*, 349–355. [CrossRef]
65. Voisin, D.; Simonian, S.; Herbison, A. Identification of estrogen receptor-containing neurons projecting to the rat supraoptic nucleus. *Neuroscience* **1997**, *78*, 215–228. [CrossRef]
66. Osheroff, P.L.; Phillips, H.S. Autoradiographic localization of relaxin binding sites in rat brain. *Proc. Natl. Acad. Sci. USA* **1991**, *88*, 6413–6417. [CrossRef]
67. De Ávila, C.; Chometton, S.; Lenglos, C.; Calvez, J.; Gundlach, A.L.; Timofeeva, E. Differential effects of relaxin-3 and a selective relaxin-3 receptor agonist on food and water intake and hypothalamic neuronal activity in rats. *Behav. Brain Res.* **2018**, *336*, 135–144. [CrossRef]
68. Alsuhaymi, N.; Habeeballah, H.; Stebbing, M.J.; Badoer, E. High Fat Diet Decreases Neuronal Activation in the Brain Induced by Resistin and Leptin. *Front. Physiol.* **2017**, *8*, 867. [CrossRef] [PubMed]
69. Kinsman, B.J.; Browning, K.N.; Stocker, S.D. NaCl and osmolarity produce different responses in organum vasculosum of the lamina terminalis neurons, sympathetic nerve activity and blood pressure. *J. Physiol.* **2017**, *595*, 6187–6201. [CrossRef]

70. Kinsman, B.J.; Simmonds, S.S.; Browning, K.N.; Stocker, S.D. Organum Vasculosum of the Lamina Terminalis Detects NaCl to Elevate Sympathetic Nerve Activity and Blood Pressure. *Hypertension* **2017**, *69*, 163–170. [CrossRef]
71. Sunn, N.; McKinley, M.J.; Oldfield, B.J. Identification of Efferent Neural Pathways from the Lamina Terminalis Activated by Blood-Borne Relaxin. *J. Neuroendocr.* **2001**, *13*, 432–437. [CrossRef]
72. Sinnayah, P.; Burns, P.; Wade, J.D.; Weisinger, R.S.; McKinley, M.J. Water Drinking in Rats Resulting from Intravenous Relaxin and Its Modification by Other Dipsogenic Factors. *Endocrinology* **1999**, *140*, 5082–5086. [CrossRef] [PubMed]
73. McGowan, B.; Stanley, S.A.; Smith, K.L.; White, N.E.; Connolly, M.M.; Thompson, E.L.; Gardiner, J.; Murphy, K.; Ghatei, M.A.; Bloom, S.R. Central Relaxin-3 Administration Causes Hyperphagia in Male Wistar Rats. *Endocrinology* **2005**, *146*, 3295–3300. [CrossRef]
74. Thornton, S.; Sirinathsinghji, D.; Delaney, C. The effects of a reversible colchicine-induced lesion of the anterior ventral region of the third cerebral ventricle in rats. *Brain Res.* **1987**, *437*, 339–344. [CrossRef]
75. Zimmerman, C.; Leib, D.; Knight, Z. Neural circuits underlying thirst and fluid homeostasis. *Nat. Rev. Neurosci.* **2017**, *18*, 459–469. [CrossRef] [PubMed]
76. Moreno, M.L.; Meza, E.; Morgado, E.; Juárez, C.; Ramos-Ligonio, A.; Ortega, A.; Caba, M. Activation of Organum Vasculosum of Lamina Terminalis, Median Preoptic Nucleus, and Medial Preoptic Area in Anticipation of Nursing in Rabbit Pups. *Chronobiol. Int.* **2013**, *30*, 1272–1282. [CrossRef]
77. Boschmann, M.; Steiniger, J.; Franke, G.; Birkenfeld, A.L.; Luft, F.; Jordan, J. Water Drinking Induces Thermogenesis through Osmosensitive Mechanisms. *J. Clin. Endocrinol. Metab.* **2007**, *92*, 3334–3337. [CrossRef]
78. Stookey, J.J.D. Negative, Null and Beneficial Effects of Drinking Water on Energy Intake, Energy Expenditure, Fat Oxidation and Weight Change in Randomized Trials: A Qualitative Review. *Nutrients* **2016**, *8*, 19. [CrossRef]
79. Bankir, L.; Bichet, D.G.; Morgenthaler, N.G. Vasopressin: Physiology, assessment and osmosensation. *J. Intern. Med.* **2017**, *282*, 284–297. [CrossRef]
80. Enhörning, S.; Melander, O. The Vasopressin System in the Risk of Diabetes and Cardiorenal Disease, and Hydration as a Potential Lifestyle Intervention. *Ann. Nutr. Metab.* **2018**, *72*, 21–27. [CrossRef]
81. Chang, D.C.; Basolo, A.; Piaggi, P.; Votruba, S.B.; Krakoff, J. Hydration biomarkers and copeptin: Relationship with ad libitum energy intake, energy expenditure, and metabolic fuel selection. *Eur. J. Clin. Nutr.* **2020**, *74*, 158–166. [CrossRef]
82. Roth, G.I.; Yamamoto, W.S. The microcirculation of the area postrema in the rat. *J. Comp. Neurol.* **1968**, *133*, 329–340. [CrossRef]
83. Dempsey, E.W. Neural and vascular ultrastructure of the area postrema in the rat. *J. Comp. Neurol.* **1973**, *150*, 177–199. [CrossRef] [PubMed]
84. Price, C.J.; Hoyda, T.D.; Ferguson, A.V. The Area Postrema: A Brain Monitor and Integrator of Systemic Autonomic State. *Neuroscientist* **2007**, *14*, 182–194. [CrossRef] [PubMed]
85. Simerly, R.B.; Swanson, L.W.; Chang, C.; Muramatsu, M. Distribution of androgen and estrogen receptor mRNA-containing cells in the rat brain: An in situ hybridization study. *J. Comp. Neurol.* **1990**, *294*, 76–95. [CrossRef] [PubMed]
86. Carpenter, D.O.; Briggs, D.B. Insulin excites neurons of the area postrema and causes emesis. *Neurosci. Lett.* **1986**, *68*, 85–89. [CrossRef]
87. Van Houten, M.; Posner, B.I.; Kopriwa, B.M.; Brawer, J.R. Insulin-Binding Sites in the Rat Brain: In Vivo Localization to the Circumventricular Organs by Quantitative Radioautography. *Endocrinology* **1979**, *105*, 666–673. [CrossRef]
88. Van Houten, M.; Posner, B.I. Specific Binding and Internalization of Blood-Borne [125I]-Iodoinsulin by Neurons of the Rat Area Postrema. *Endocrinology* **1981**, *109*, 853–859. [CrossRef]
89. Liberini, C.G.; Boyle, C.N.; Cifani, C.; Venniro, M.; Hope, B.T.; Lutz, T.A. Amylin receptor components and the leptin receptor are co-expressed in single rat area postrema neurons. *Eur. J. Neurosci.* **2016**, *43*, 653–661. [CrossRef] [PubMed]
90. Fry, M.; Smith, P.M.; Hoyda, T.D.; Duncan, M.; Ahima, R.S.; Sharkey, K.A.; Ferguson, A.V. Area Postrema Neurons Are Modulated by the Adipocyte Hormone Adiponectin. *J. Neurosci.* **2006**, *26*, 9695–9702. [CrossRef] [PubMed]
91. Huang, J.; Hara, Y.; Anrather, J.; Speth, R.; Iadecola, C.; Pickel, V. Angiotensin II subtype 1A (AT1A) receptors in the rat sensory vagal complex: Subcellular localization and association with endogenous angiotensin. *Neuroscience* **2003**, *122*, 21–36. [CrossRef]
92. Cork, S.; Richards, J.E.; Holt, M.; Gribble, F.; Reimann, F.; Trapp, S. Distribution and characterisation of Glucagon-like peptide-1 receptor expressing cells in the mouse brain. *Mol. Metab.* **2015**, *4*, 718–731. [CrossRef]
93. Mercer, L.D.; Beart, P.M. Immunolocalization of CCK1R in rat brain using a new anti-peptide antibody. *Neurosci. Lett.* **2004**, *359*, 109–113. [CrossRef]
94. Sugeta, S.; Hirai, Y.; Maezawa, H.; Inoue, N.; Yamazaki, Y.; Funahashi, M. Presynaptically mediated effects of cholecystokinin-8 on the excitability of area postrema neurons in rat brain slices. *Brain Res.* **2015**, *1618*, 83–90. [CrossRef]
95. Covasa, M.; Ritter, R.C. Reduced CCK-induced Fos expression in the hindbrain, nodose ganglia, and enteric neurons of rats lacking CCK-1 receptors. *Brain Res.* **2005**, *1051*, 155–163. [CrossRef]
96. Cabral, A.; Cornejo, M.P.; Fernandez, G.; De Francesco, P.N.; Romero, G.G.; Uriarte, M.; Zigman, J.M.; Portiansky, E.; Reynaldo, M.; Perello, M. Circulating Ghrelin Acts on GABA Neurons of the Area Postrema and Mediates Gastric Emptying in Male Mice. *Endocrinology* **2017**, *158*, 1436–1449. [CrossRef]
97. Nambu, Y.; Ohira, K.; Morita, M.; Yasumoto, H.; Kurganov, E.; Miyata, S. Effects of leptin on proliferation of astrocyte- and tanycyte-like neural stem cells in the adult mouse medulla oblongata. *Neurosci. Res.* **2021**. [CrossRef]

98. Al-Saleh, S.; Kaur, C.; Ling, E. Response of neurons and microglia/macrophages in the area postrema of adult rats following exposure to hypobaric hypoxia. *Neurosci. Lett.* **2003**, *346*, 77–80. [CrossRef]
99. Wuchert, F.; Ott, D.; Murgott, J.; Rafalzik, S.; Hitzel, N.; Roth, J.; Gerstberger, R. Rat area postrema microglial cells act as sensors for the toll-like receptor-4 agonist lipopolysaccharide. *J. Neuroimmunol.* **2008**, *204*, 66–74. [CrossRef] [PubMed]
100. Dowsett, G.K.; Lam, B.Y.; Tadross, J.A.; Cimino, I.; Rimmington, D.; Coll, A.P.; Polex-Wolf, J.; Knudsen, L.B.; Pyke, C.; Yeo, G.S. A survey of the mouse hindbrain in the fed and fasted states using single-nucleus RNA sequencing. *Mol. Metab.* **2021**, *53*, 101240. [CrossRef] [PubMed]
101. Gerstberger, R.; Fahrenholz, F. Autoradiographic localization of V1 vasopressin binding sites in rat brain and kidney. *Eur. J. Pharmacol.* **1989**, *167*, 105–116. [CrossRef]
102. Kishi, T.; Aschkenasi, C.J.; Choi, B.J.; Lopez, M.E.; Lee, C.E.; Liu, H.; Hollenberg, A.N.; Friedman, J.M.; Elmquist, J.K. Neuropeptide Y Y1 receptor mRNA in rodent brain: Distribution and colocalization with melanocortin-4 receptor. *J. Comp. Neurol.* **2004**, *482*, 217–243. [CrossRef]
103. Riediger, T.; Zuend, D.; Becskei, C.; Lutz, T.A. The anorectic hormone amylin contributes to feeding-related changes of neuronal activity in key structures of the gut-brain axis. *Am. J. Physiol. Integr. Comp. Physiol.* **2004**, *286*, R114–R122. [CrossRef] [PubMed]
104. Rinaman, L.; Verbalis, J.G.; Stricker, E.M.; Hoffman, G.E. Distribution and neurochemical phenotypes of caudal medullary neurons activated to express cFos following peripheral administration of cholecystokinin. *J. Comp. Neurol.* **1993**, *338*, 475–490. [CrossRef] [PubMed]
105. Bonaz, B.; Taylor, I.; Taché, Y. Peripheral peptide YY induces c-fos-like immunoreactivity in the rat brain. *Neurosci. Lett.* **1993**, *163*, 77–80. [CrossRef]
106. Fukuda, T.; Hirai, Y.; Maezawa, H.; Kitagawa, Y.; Funahashi, M. Electrophysiologically identified presynaptic mechanisms underlying amylinergic modulation of area postrema neuronal excitability in rat brain slices. *Brain Res.* **2013**, *1494*, 9–16. [CrossRef] [PubMed]
107. Riediger, T.; Schmid, H.A.; Lutz, T.A.; Simon, E. Amylin and glucose co-activate area postrema neurons of the rat. *Neurosci. Lett.* **2002**, *328*, 121–124. [CrossRef]
108. Fry, M.; Ferguson, A.V. Ghrelin modulates electrical activity of area postrema neurons. *Am. J. Physiol. Regul. Integr. Comp. Physiol.* **2009**, *296*, R485–R492. [CrossRef]
109. Miller, A.D.; Leslie, R.A. The Area Postrema and Vomiting. *Front. Neuroendocr.* **1994**, *15*, 301–320. [CrossRef] [PubMed]
110. Mollet, A.; Gilg, S.; Riediger, T.; Lutz, T.A. Infusion of the amylin antagonist AC 187 into the area postrema increases food intake in rats. *Physiol. Behav.* **2004**, *81*, 149–155. [CrossRef]
111. Cox, J.E.; Randich, A. Enhancement of feeding suppression by PYY3-36 in rats with area postrema ablations. *Peptides* **2004**, *25*, 985–989. [CrossRef]
112. Lutz, T.; Del Prete, E.; Scharrer, E. Reduction of food intake in rats by intraperitoneal injection of low doses of amylin. *Physiol. Behav.* **1994**, *55*, 891–895. [CrossRef]
113. Mullican, S.E.; Lin-Schmidt, X.; Chin, C.N.; Chavez, J.A.; Furman, J.L.; Armstrong, A.A.; Beck, S.C.; South, V.J.; Dinh, T.Q.; Cash-Mason, T.D.; et al. GFRAL is the receptor for GDF15 and the ligand promotes weight loss in mice and nonhuman primates. *Nat. Med.* **2017**, *23*, 1150–1157. [CrossRef] [PubMed]
114. Tsai, V.W.-W.; Manandhar, R.; Jørgensen, S.B.; Lee-Ng, K.K.M.; Zhang, H.P.; Marquis, C.; Jiang, L.; Husaini, Y.; Lin, S.; Sainsbury, A.; et al. The Anorectic Actions of the TGFβ Cytokine MIC-1/GDF15 Require an Intact Brainstem Area Postrema and Nucleus of the Solitary Tract. *PLoS ONE* **2014**, *9*, e100370. [CrossRef] [PubMed]
115. Emmerson, P.J.; Wang, F.; Du, Y.; Liu, Q.; Pickard, R.T.; Gonciarz, M.D.; Coskun, T.; Hamang, M.J.; Sindelar, D.K.; Ballman, K.K.; et al. The metabolic effects of GDF15 are mediated by the orphan receptor GFRAL. *Nat. Med.* **2017**, *23*, 1215–1219. [CrossRef]
116. Vaughan, C.H.; Bartness, T.J. Anterograde transneuronal viral tract tracing reveals central sensory circuits from brown fat and sensory denervation alters its thermogenic responses. *Am. J. Physiol. Regul. Integr. Comp. Physiol.* **2012**, *302*, R1049–R1058. [CrossRef]
117. Coester, B.; Le Foll, C.; Lutz, T.A. Viral depletion of calcitonin receptors in the area postrema: A proof-of-concept study. *Physiol. Behav.* **2020**, *223*, 112992. [CrossRef] [PubMed]
118. Hsu, J.-Y.; Crawley, S.; Chen, M.; Ayupova, D.A.; Lindhout, D.A.; Higbee, J.; Kutach, A.; Joo, W.; Gao, Z.; Fu, D.; et al. Erratum: Non-homeostatic body weight regulation through a brainstem-restricted receptor for GDF15. *Nature* **2017**, *551*, 398. [CrossRef] [PubMed]
119. Oka, Y.; Ye, M.; Zuker, C.S. Thirst driving and suppressing signals encoded by distinct neural populations in the brain. *Nature* **2015**, *520*, 349–352. [CrossRef]
120. Matsuda, T.; Hiyama, T.; Niimura, F.; Matsusaka, T.; Fukamizu, A.; Kobayashi, K.; Kobayashi, K.; Noda, M. Distinct neural mechanisms for the control of thirst and salt appetite in the subfornical organ. *Nat. Neurosci.* **2017**, *20*, 230–241. [CrossRef]
121. Kawano, H.; Masuko, S. Region-specific projections from the subfornical organ to the paraventricular hypothalamic nucleus in the rat. *Neuroscience* **2010**, *169*, 1227–1234. [CrossRef]
122. Yeo, S.H.; Kyle, V.; Blouet, C.; Jones, S.; Colledge, W.H. Mapping neuronal inputs to Kiss1 neurons in the arcuate nucleus of the mouse. *PLoS ONE* **2019**, *14*, e0213927. [CrossRef]
123. Kolaj, M.; Renaud, L.P. Metabotropic Glutamate Receptors in Median Preoptic Neurons Modulate Neuronal Excitability and Glutamatergic and GABAergic Inputs From the Subfornical Organ. *J. Neurophysiol.* **2010**, *103*, 1104–1113. [CrossRef]

124. Abbott, S.B.; Machado, N.L.; Geerling, J.C.; Saper, C.B. Reciprocal Control of Drinking Behavior by Median Preoptic Neurons in Mice. *J. Neurosci.* **2016**, *36*, 8228–8237. [CrossRef] [PubMed]
125. Duan, P.G.; Kawano, H.; Masuko, S. Collateral projections from the subfornical organ to the median preoptic nucleus and paraventricular hypothalamic nucleus in the rat. *Brain Res.* **2008**, *1198*, 68–72. [CrossRef] [PubMed]
126. Kawano, H. Synaptic contact between median preoptic neurons and subfornical organ neurons projecting to the paraventricular hypothalamic nucleus. *Exp. Brain Res.* **2017**, *235*, 1053–1062. [CrossRef]
127. McKinley, M.J.; Denton, D.A.; Ryan, P.; Yao, S.; Stefanidis, A.; Oldfield, B.J. From sensory circumventricular organs to cerebral cortex: Neural pathways controlling thirst and hunger. *J. Neuroendocr.* **2019**, *31*, e12689. [CrossRef] [PubMed]
128. Xu, S.H.; Honda, E.; Ono, K.; Inenaga, K. Muscarinic modulation of GABAergic transmission to neurons in the rat subfornical organ. *Am. J. Physiol. Regul. Integr. Comp. Physiol.* **2001**, *280*, R1657–R1664. [CrossRef] [PubMed]
129. Zhu, B.; Herbert, J. Calcium channels mediate angiotensin II-induced drinking behaviour and c-fos expression in the brain. *Brain Res.* **1997**, *778*, 206–214. [CrossRef]
130. Shi, P.; Martinez, M.A.; Calderon, A.S.; Chen, Q.; Cunningham, J.T.; Toney, G.M. Intra-carotid hyperosmotic stimulation increases Fos staining in forebrain organum vasculosum laminae terminalis neurones that project to the hypothalamic paraventricular nucleus. *J. Physiol.* **2008**, *586*, 5231–5245. [CrossRef] [PubMed]
131. Van Der Kooy, D.; Koda, L.Y. Organization of the projections of a circumventricular organ: The area postrema in the rat. *J. Comp. Neurol.* **1983**, *219*, 328–338. [CrossRef]
132. Shapiro, R.E.; Miselis, R.R. The central neural connections of the area postrema of the rat. *J. Comp. Neurol.* **1985**, *234*, 344–364. [CrossRef] [PubMed]
133. Carlson, S.H.; Collister, J.P.; Osborn, J.W. The area postrema modulates hypothalamic fos responses to intragastric hypertonic saline in conscious rats. *Am. J. Physiol. Regul. Integr. Comp. Physiol.* **1998**, *275*, R1921–R1927. [CrossRef] [PubMed]
134. Alvarado, B.A.; Lemus, M.; Montero, S.; Melnikov, V.; Luquín, S.; García-Estrada, J.; De Álvarez-Buylla, E.R. Nitric oxide in the nucleus of the tractus solitarius is involved in hypoglycemic conditioned response. *Brain Res.* **2017**, *1667*, 19–27. [CrossRef] [PubMed]
135. Filippi, B.M.; Yang, C.S.; Tang, C.; Lam, T.K. Insulin Activates Erk1/2 Signaling in the Dorsal Vagal Complex to Inhibit Glucose Production. *Cell Metab.* **2012**, *16*, 500–510. [CrossRef]
136. Tsai, J.P. The association of serum leptin levels with metabolic diseases. *Tzu-Chi Med. J.* **2017**, *29*, 192–196. [CrossRef]
137. Filippi, B.M.; Bassiri, A.; Abraham, M.A.; Duca, F.A.; Yue, J.T.; Lam, T.K. Insulin Signals Through the Dorsal Vagal Complex to Regulate Energy Balance. *Diabetes* **2014**, *63*, 892–899. [CrossRef] [PubMed]
138. Wu, Q.; Lemus, M.B.; Stark, R.; Bayliss, J.A.; Reichenbach, A.; Lockie, S.; Andrews, Z.B. The Temporal Pattern of cfos Activation in Hypothalamic, Cortical, and Brainstem Nuclei in Response to Fasting and Refeeding in Male Mice. *Endocrinology* **2014**, *155*, 840–853. [CrossRef] [PubMed]
139. D'Agostino, G.; Lyons, D.; Cristiano, C.; Burke, L.K.; Madara, J.C.; Campbell, J.N.; Garcia, A.P.; Land, B.B.; Lowell, B.B.; DiLeone, R.; et al. Appetite controlled by a cholecystokinin nucleus of the solitary tract to hypothalamus neurocircuit. *eLife* **2016**, *5*, e12225. [CrossRef]
140. Herman, J.P. Regulation of Hypothalamo-Pituitary-Adrenocortical Responses to Stressors by the Nucleus of the Solitary Tract/Dorsal Vagal Complex. *Cell. Mol. Neurobiol.* **2018**, *38*, 25–35. [CrossRef]
141. Katsurada, K.; Maejima, Y.; Nakata, M.; Kodaira, M.; Suyama, S.; Iwasaki, Y.; Kario, K.; Yada, T. Endogenous GLP-1 acts on paraventricular nucleus to suppress feeding: Projection from nucleus tractus solitarius and activation of corticotropin-releasing hormone, nesfatin-1 and oxytocin neurons. *Biochem. Biophys. Res. Commun.* **2014**, *451*, 276–281. [CrossRef]
142. Maniscalco, J.W.; Rinaman, L. Overnight food deprivation markedly attenuates hindbrain noradrenergic, glucagon-like peptide-1, and hypothalamic neural responses to exogenous cholecystokinin in male rats. *Physiol. Behav.* **2013**, *121*, 35–42. [CrossRef]
143. Ito, H.; Seki, M. Ascending Projections from the Area Postrema and the Nucleus of the Solitary Tract of Suncus Murinus: Anterograde tracing study using Phaseolus vulgaris leucoagglutinin. *Okajimas Folia Anat. Jpn.* **1998**, *75*, 9–31. [CrossRef] [PubMed]
144. De Kloet, A.D.; Krause, E.; Scott, K.; Foster, M.T.; Herman, J.; Sakai, R.R.; Seeley, R.; Woods, S.C. Central angiotensin II has catabolic action at white and brown adipose tissue. *Am. J. Physiol. Endocrinol. Metab.* **2011**, *301*, E1081–E1091. [CrossRef]
145. McKinley, M.J.; Burns, P.; Colvill, L.M.; Oldfield, B.J.; Wade, J.D.; Weisinger, R.S.; Tregear, G.W. Distribution of Fos immunoreactivity in the lamina terminalis and hypothalamus induced by centrally administered relaxin in conscious rats. *J. Neuroendocr.* **1997**, *9*, 431–437. [CrossRef]
146. Wilson, B.C.; Summerlee, A.J.S. Effects of exogenous relaxin on oxytocin and vasopressin release and the intramammary pressure response to central hyperosmotic challenge. *J. Endocrinol.* **1994**, *141*, 75–80. [CrossRef]
147. Zhice, X.; Herbert, J. Regional suppression by water intake of c-fos expression induced by intraventricular infusions of angiotensin II. *Brain Res.* **1994**, *659*, 157–168. [CrossRef]
148. Ueno, H.; Yoshimura, M.; Tanaka, K.; Nishimura, K.; Sonoda, S.; Motojima, Y.; Saito, R.; Maruyama, T.; Miyamoto, T.; Serino, R.; et al. Up-regulation of hypothalamic arginine vasopressin by peripherally administered furosemide in transgenic rats expressing arginine vasopressin-enhanced green fluorescent protein. *J. Neuroendocr.* **2018**, *30*, e12603. [CrossRef] [PubMed]
149. Ferguson, A.V.; Kasting, N.W. Activation of subfornical organ efferents stimulates oxytocin secretion in the rat. *Regul. Pept.* **1987**, *18*, 93–100. [CrossRef]

150. Ferguson, A.V.; Kasting, N.W. Electrical stimulation in subfornical organ increases plasma vasopressin concentrations in the conscious rat. *Am. J. Physiol. Regul. Integr. Comp. Physiol.* **1986**, *251*, R425–R428. [CrossRef] [PubMed]
151. Xu, Z.; Pekarek, E.; Ge, J.; Yao, J. Functional relationship between subfornical organ cholinergic stimulation and cellular activation in the hypothalamus and AV3V region. *Brain Res.* **2001**, *922*, 191–200. [CrossRef]
152. Thrasher, T.N.; Keil, L.C.; Ramsay, D.J. Lesions of the organum vasculosum of the lamina terminalis (ovlt) attenuate osmotically-induced drinking and vasopressin secretion in the dog. *Endocrinology* **1982**, *110*, 1837–1839. [CrossRef]
153. Russell, J.; Blackburn, R.; Leng, G. The role of the AV3V region in the control of magnocellular oxytocin neurons. *Brain Res. Bull.* **1988**, *20*, 803–810. [CrossRef]
154. Carter, D.; Lightman, S. A role for the area postrema in mediating cholecystokinin-stimulated oxytocin secretion. *Brain Res.* **1987**, *435*, 327–330. [CrossRef]
155. Larsen, P.J.; Tang-Christensen, M.; Jessop, D.S. Central Administration of Glucagon-Like Peptide-1 Activates Hypothalamic Neuroendocrine Neurons in the Rat. *Endocrinology* **1997**, *138*, 4445–4455. [CrossRef]
156. Kublaoui, B.M.; Gemelli, T.; Tolson, K.; Wang, Y.; Zinn, A.R. Oxytocin Deficiency Mediates Hyperphagic Obesity of Sim1 Haploinsufficient Mice. *Mol. Endocrinol.* **2008**, *22*, 1723–1734. [CrossRef] [PubMed]
157. Maejima, Y.; Iwasaki, Y.; Yamahara, Y.; Kodaira, M.; Sedbazar, U.; Yada, T. Peripheral oxytocin treatment ameliorates obesity by reducing food intake and visceral fat mass. *Aging* **2011**, *3*, 1169–1177. [CrossRef]
158. Morton, G.J.; Thatcher, B.S.; Reidelberger, R.D.; Ogimoto, K.; Wolden-Hanson, T.; Baskin, D.G.; Schwartz, M.W.; Blevins, J.E. Peripheral oxytocin suppresses food intake and causes weight loss in diet-induced obese rats. *Am. J. Physiol. Endocrinol. Metab.* **2012**, *302*, E134–E144. [CrossRef] [PubMed]
159. Blevins, J.E.; Graham, J.; Morton, G.J.; Bales, K.L.; Schwartz, M.W.; Baskin, D.G.; Havel, P. Chronic oxytocin administration inhibits food intake, increases energy expenditure, and produces weight loss in fructose-fed obese rhesus monkeys. *Am. J. Physiol. Regul. Integr. Comp. Physiol.* **2015**, *308*, R431–R438. [CrossRef]
160. Ott, V.; Finlayson, G.; Lehnert, H.; Heitmann, B.L.; Heinrichs, M.; Born, J.; Hallschmid, M. Oxytocin Reduces Reward-Driven Food Intake in Humans. *Diabetes* **2013**, *62*, 3418–3425. [CrossRef]
161. Thienel, M.; Fritsche, A.; Heinrichs, M.; Peter, A.; Ewers, M.; Lehnert, H.; Born, J.; Hallschmid, M. Oxytocin's inhibitory effect on food intake is stronger in obese than normal-weight men. *Int. J. Obes.* **2016**, *40*, 1707–1714. [CrossRef]
162. Altirriba, J.; Poher, A.-L.; Caillon, A.; Arsenijevic, D.; Veyrat-Durebex, C.; Lyautey, J.; Dulloo, A.; Rohner-Jeanrenaud, F. Divergent Effects of Oxytocin Treatment of Obese Diabetic Mice on Adiposity and Diabetes. *Endocrinology* **2014**, *155*, 4189–4201. [CrossRef]
163. Takayanagi, Y.; Kasahara, Y.; Onaka, T.; Takahashi, N.; Kawada, T.; Nishimori, K. Oxytocin receptor-deficient mice developed late-onset obesity. *Neuroreport* **2008**, *19*, 951–955. [CrossRef]
164. Camerino, C. Low Sympathetic Tone and Obese Phenotype in Oxytocin-deficient Mice. *Obesity* **2009**, *17*, 980–984. [CrossRef]
165. Wu, Z.; Xu, Y.; Zhu, Y.; Sutton, A.K.; Zhao, R.; Lowell, B.B.; Olson, D.P.; Tong, Q. An Obligate Role of Oxytocin Neurons in Diet Induced Energy Expenditure. *PLoS ONE* **2012**, *7*, e45167. [CrossRef]
166. Kasahara, Y.; Sato, K.; Takayanagi, Y.; Mizukami, H.; Ozawa, K.; Hidema, S.; So, K.-H.; Kawada, T.; Inoue, N.; Ikeda, I.; et al. Oxytocin Receptor in the Hypothalamus Is Sufficient to Rescue Normal Thermoregulatory Function in Male Oxytocin Receptor Knockout Mice. *Endocrinology* **2013**, *154*, 4305–4315. [CrossRef]
167. Lawson, E.A.; Olszewski, P.K.; Weller, A.; Blevins, J.E. The role of oxytocin in regulation of appetitive behaviour, body weight and glucose homeostasis. *J. Neuroendocr.* **2020**, *32*, e12805. [CrossRef]
168. Roberts, Z.S.; Wolden-Hanson, T.; Matsen, M.E.; Ryu, V.; Vaughan, C.H.; Graham, J.L.; Havel, P.J.; Chukri, D.W.; Schwartz, M.W.; Morton, G.J.; et al. Chronic hindbrain administration of oxytocin is sufficient to elicit weight loss in diet-induced obese rats. *Am. J. Physiol. Regul. Integr. Comp. Physiol.* **2017**, *313*, R357–R371. [CrossRef]
169. Ding, C.; Leow, M.K.S.; Magkos, F. Oxytocin in metabolic homeostasis: Implications for obesity and diabetes management. *Obes. Rev.* **2018**, *20*, 22–40. [CrossRef]
170. Hanif, K.; Goren, H.J.; Hollenberg, M.D.; Lederis, K. Oxytocin action: Lipid metabolism in adipocytes from homozygous diabetes insipidus rats (Brattleboro strain). *Can. J. Physiol. Pharmacol.* **1982**, *60*, 993–997. [CrossRef]
171. Deblon, N.; Veyrat-Durebex, C.; Bourgoin, L.; Caillon, A.; Bussier, A.L.; Petrosino, S.; Piscitelli, F.; Legros, J.J.; Geenen, V.; Foti, M.; et al. Mechanisms of the Anti-Obesity Effects of Oxytocin in Diet-Induced Obese Rats. *PLoS ONE* **2011**, *6*, e25565. [CrossRef]
172. Lee, E.S.; Uhm, K.O.; Lee, Y.M.; Kwon, J.; Park, S.H.; Soo, K.H. Oxytocin stimulates glucose uptake in skeletal muscle cells through the calcium–CaMKK–AMPK pathway. *Regul. Pept.* **2008**, *151*, 71–74. [CrossRef]
173. Yoshimura, M.; Conway-Campbell, B.; Ueta, Y. Arginine vasopressin: Direct and indirect action on metabolism. *Peptides* **2021**, *142*, 170555. [CrossRef]
174. Shido, O.; Kifune, A.; Nagasaka, T. Baroreflexive suppression of heat production and fall in body temperature following peripheral administration of vasopressin in rats. *Jpn. J. Physiol.* **1984**, *34*, 397–406. [CrossRef] [PubMed]
175. Pei, H.; Sutton, A.K.; Burnett, K.H.; Fuller, P.M.; Olson, D.P. AVP neurons in the paraventricular nucleus of the hypothalamus regulate feeding. *Mol. Metab.* **2014**, *3*, 209–215. [CrossRef]
176. Yoshimura, M.; Nishimura, K.; Nishimura, H.; Sonoda, S.; Ueno, H.; Motojima, Y.; Saito, R.; Maruyama, T.; Nonaka, Y.; Ueta, Y. Activation of endogenous arginine vasopressin neurons inhibit food intake: By using a novel transgenic rat line with DREADDs system. *Sci. Rep.* **2017**, *7*, 1–10. [CrossRef]

177. Yi, S.S.; Hwang, I.K.; Na Kim, Y.; Kim, I.Y.; Pak, S.-I.; Lee, I.S.; Seong, J.K.; Yoon, Y.S. Enhanced Expressions of Arginine Vasopressin (Avp) in the Hypothalamic Paraventricular and Supraoptic Nuclei of Type 2 Diabetic Rats. *Neurochem. Res.* **2007**, *33*, 833–841. [CrossRef]
178. Aoyagi, T.; Birumachi, J.-I.; Hiroyama, M.; Fujiwara, Y.; Sanbe, A.; Yamauchi, J.; Tanoue, A. Alteration of Glucose Homeostasis in V1a Vasopressin Receptor-Deficient Mice. *Endocrinology* **2007**, *148*, 2075–2084. [CrossRef]
179. Nakamura, K.; Velho, G.; Bouby, N. Vasopressin and metabolic disorders: Translation from experimental models to clinical use. *J. Intern. Med.* **2017**, *282*, 298–309. [CrossRef]
180. Nakamura, K.; Aoyagi, T.; Hiroyama, M.; Kusakawa, S.; Mizutani, R.; Sanbe, A.; Yamauchi, J.; Kamohara, M.; Momose, K.; Tanoue, A. Both V1A and V1B vasopressin receptors deficiency result in impaired glucose tolerance. *Eur. J. Pharmacol.* **2009**, *613*, 182–188. [CrossRef]
181. Hiroyama, M.; Aoyagi, T.; Fujiwara, Y.; Birumachi, J.; Shigematsu, Y.; Kiwaki, K.; Tasaki, R.; Endo, F.; Tanoue, A. Hypermetabolism of Fat in V1a Vasopressin Receptor Knockout Mice. *Mol. Endocrinol.* **2007**, *21*, 247–258. [CrossRef] [PubMed]
182. Velho, G.; Bouby, N.; Hadjadj, S.; Matallah, N.; Mohammedi, K.; Fumeron, F.; Potier, L.; Bellili-Munoz, N.; Taveau, C.; Alhenc-Gelas, F.; et al. Plasma Copeptin and Renal Outcomes in Patients With Type 2 Diabetes and Albuminuria. *Diabetes Care* **2013**, *36*, 3639–3645. [CrossRef]
183. Velho, G.; El Boustany, R.; Lefèvre, G.; Mohammedi, K.; Fumeron, F.; Potier, L.; Bankir, L.; Bouby, N.; Hadjadj, S.; Marre, M.; et al. Plasma Copeptin, Kidney Outcomes, Ischemic Heart Disease, and All-Cause Mortality in People With Long-standing Type 1 Diabetes. *Diabetes Care* **2016**, *39*, 2288–2295. [CrossRef]
184. Roussel, R.; El Boustany, R.; Bouby, N.; Potier, L.; Fumeron, F.; Mohammedi, K.; Balkau, B.; Tichet, J.; Bankir, L.; Marre, M.; et al. Plasma Copeptin, AVP Gene Variants, and Incidence of Type 2 Diabetes in a Cohort From the Community. *J. Clin. Endocrinol. Metab.* **2016**, *101*, 2432–2439. [CrossRef]

Review

Extrahypothalamic Control of Energy Balance and Its Connection with Reproduction: Roles of the Amygdala

Rafael Pineda [1,2,3,4,*], Encarnacion Torres [1,2,3] and Manuel Tena-Sempere [1,2,3,4,5,*]

1 Instituto Maimónides de Investigación Biomédica de Cordoba (IMIBIC), 14004 Cordoba, Spain; bo2tojie@uco.es
2 Department of Cell Biology, Physiology and Immunology, University of Cordoba, 14004 Cordoba, Spain
3 Hospital Universitario Reina Sofia, 14004 Cordoba, Spain
4 CIBER Fisiopatología de la Obesidad y Nutrición, Instituto de Salud Carlos III, 14004 Cordoba, Spain
5 Institute of Biomedicine, University of Turku, FIN-20520 Turku, Finland
* Correspondence: v92pirer@uco.es (R.P.); fi1tesem@uco.es (M.T.-S.)

Abstract: Body energy and metabolic homeostasis are exquisitely controlled by multiple, often overlapping regulatory mechanisms, which permit the tight adjustment between fuel reserves, internal needs, and environmental (e.g., nutritional) conditions. As such, this function is sensitive to and closely connected with other relevant bodily systems, including reproduction and gonadal function. The aim of this mini-review article is to summarize the most salient experimental data supporting a role of the amygdala as a key brain region for emotional learning and behavior, including reward processing, in the physiological control of feeding and energy balance. In particular, a major focus will be placed on the putative interplay between reproductive signals and amygdala pathways, as it pertains to the control of metabolism, as complementary, extrahypothalamic circuit for the integral control of energy balance and gonadal function.

Keywords: metabolism; amygdala; estrogens; neuropeptides; kisspeptins; food intake; energy balance; body weight

Citation: Pineda, R.; Torres, E.; Tena-Sempere, M. Extrahypothalamic Control of Energy Balance and Its Connection with Reproduction: Roles of the Amygdala. *Metabolites* **2021**, *11*, 837. https://doi.org/10.3390/metabo11120837

Academic Editor: Amedeo Lonardo

Received: 19 November 2021
Accepted: 2 December 2021
Published: 3 December 2021

Publisher's Note: MDPI stays neutral with regard to jurisdictional claims in published maps and institutional affiliations.

1. Introduction: The Amygdala

The amygdala, also known as amygdaloid complex (AC), is a brain region that in mammals comprises several nuclei or groups of nuclei, distinguished and labelled on the basis of their cytoarchitecture, histochemistry, and inter-connections. In rodents, these different nuclei are divided into three main groups: (i) the deep or basolateral group, which includes the lateral, basal, and accessory basal nuclei; (ii) the superficial or cortical group, which includes the cortical nuclei and nucleus of the lateral olfactory tract; and (iii) the centromedial group, which includes the medial and central nuclei. In addition, other accessory nuclei, including the intercalated cell masses and the amygdalo-hippocampal area, have been also described [1–3].

The amygdala has been the focus of active research in different domains of neurosciences and neuroendocrinology since its discovery in the early 19th century. While exhaustive recapitulation of the physiological roles of AC as a key center for emotional learning and behavior, is beyond the scope of this review and can be found elsewhere [1–3]; for the interest of this work, it is important to stress that compelling evidence highlights that the amygdala is involved in the control several behaviors related to feeding, such as food intake, appetitive conditioning, gustatory neophobia, taste aversion, and food conditional-place preference. Thus, the AC seemingly contributes in a dual manner to the control of homeostatic and hedonic/reward eating [4–6]; while the former is driven mainly by energy needs, the latter is not directed primarily to satisfy energy demands, but rather responds to hedonic cues, and hence can favor overweightness [7,8]. This has drawn considerable attention and research interest, given the escalating prevalence of obesity and its severe disease burden in human health.

In fact, initial studies mapping the physiological roles of different amygdala areas, based on chemical or physical lesions of specific regions of the AC, have already highlighted the connection of amygdala with feeding control. This is epitomized by the seminal work of King and co-workers, showing that amygdaloid lesions, e.g., in the posterodorsal amygdala, lead to hyperphagia and weight gain in male and female rats [9–11]. These studies not only documented a role of AC in feeding control, but also in the regulation of other key metabolic parameters, such as glycemia, and supported the role of amygdala not only in the homeostatic control of body weight, but also in food preferences and selection of macronutrients [11]. Admittedly, however, while this work paved the way for elucidation of the role of amygdala in the control of feeding and energy homeostasis, it suffered from important technical limitations, mainly related to the lack of precise resolution and the inability to discriminate the effects derived from the lesion of nuclei located in the regions damaged or of projections passing through them. Hence, more sophisticated approaches, involving neuronal tracing, functional genomics, and viral/pharmacogenetics, have been implemented to tease apart the roles of specific amygdala pathways in the control of metabolic homeostasis. As relevant recent example, elegant neurophysiological and functional studies in mice have documented that a GABAergic neural pathway, expressing type 2a serotonin receptors, located in the central nucleus of the amygdala (CeA) plays a key role in the control of food consumption, in close connection with other brain regions involved in feeding control [12].

2. A Tight Connection: The Link between Energy Homeostasis and Reproductive Function

Reproduction is an essential function for perpetuation of the species and, hence, its control is subjected, as is also the case of feeding, to sophisticated regulatory mechanisms to ensure, whenever feasible, maximal reproductive efficiency. However, fertility is dispensable at the individual level, and, considering its high metabolic and energy demands, it makes biological sense that it can be only achieved or maintained when sufficient body energy reserves are attained to afford its metabolic drainage, especially in the female (i.e., during pregnancy and lactation). Hence, situations of body energy deficit and/or metabolic distress, ranging from anorexia to severe obesity, are often associated with pubertal perturbations and sub/infertility [13]. Nevertheless, the suppressive effect of adverse metabolic conditions on the reproductive axis, as a means to minimize energy disposal and maximize reproductive efficiency, must be controlled accurately, to ensure resuming of fertility as soon as a more favorable metabolic status is achieved. Importantly, not only metabolic cues modulate reproductive function, but, conversely, gonadal signals are also important metabolic modulators, with capacity to control key aspects of energy homeostasis [14].

While detailed recapitulation of the mechanisms whereby metabolism and reproduction are tightly connected exceeds the scope of this review and can be found elsewhere [13,15], for the objectives of this review it is important to stress that this complex physiological phenomenon relies on an array of regulatory networks, involving different hypothalamic and extra-hypothalamic circuits, as well as numerous peripheral factors, that adjust reproductive maturation and function to the endogenous metabolic conditions and environmental cues. These include, prominently, brain pathways controlling both reproduction and metabolism, as well as metabolic and gonadal hormones, which impinge upon the so-called hypothalamic–pituitary–gonadal (HPG) axis. In this axis, hypothalamic neurons that synthesize the decapeptide, gonadotropin-releasing hormone (GnRH), are a major hierarchical component, as they act as final integrators of different central and peripheral signals, and major output pathway for the brain control of the downstream elements of the HPG axis [16,17]. Of note, GnRH neurons appear to be devoid of key metabolic sensors, suggesting that afferents to GnRH neurons are responsible for sensing and transmitting the modulatory effects of different metabolic signals [13,17].

Importantly, while the metabolic gating of female reproduction is intuitively explained by the considerable metabolic demands of pregnancy and lactation, it must be stressed that key aspects of reproductive behaviors, affecting prominently males, such as aggression, mating, territoriality, and dominance, also depend on sufficient energy reserves [15]. Thus, metabolic control of reproduction, in an ample sense, does not only depend on mechanisms controlling the hormonal reproductive axis, but also on pathways controlling key behaviors. In this context, the amygdala has emerged as relevant brain area for the regulation of key metabolic and reproductive neuroendocrine functions and behaviors.

In this mini-review, we will focus our attention on three elements that may participate in such integral regulatory mechanisms, also engaging the amygdala, which include: (i) metabolic neuro-peptide pathways; (ii) the kisspeptin system; and (iii) gonadal hormones. These will be reviewed in the following sections. As a search method, we have implemented a comprehensive MEDLINE search, using PubMed as main interface, of research articles and reviews, published mainly between 2005 and 2021, using previously published guidelines [18]. In detail, search was implemented using multiple keywords, including amygdala, metabolism, estrogens, androgens, reproduction, neuropeptides, kisspeptin, food intake, energy balance, and body weight, with the Boolean operators AND/OR, focusing mainly on preclinical data connecting amygdala, energy balance, and reproductive function. In addition, studies addressing the role of the amygdala in the physiological control of feeding and energy balance were also considered, and, when relevant, the web application Connected Papers (https://www.connectedpapers.com/, accessed on 15 November 2021) was used to comprehensively cover all key references in specific topics of the review.

3. Metabolic Neuropeptide Pathways and the Amygdala: Roles of NPY/AgRP and POMC

A major circuit for the homeostatic control of body weight and energy balance is placed in the hypothalamic arcuate nucleus (ARC), and involves the reciprocal interplay of two populations of neurons, with opposite roles in the control of feeding, namely neurons expressing proopiomelanocortin (aka, POMC neurons, which conduct anorexigenic actions) and neurons expressing neuropeptide Y/agouti-related peptide (aka, NPY/AgRP neurons, with dominant orexigenic actions) [19–21]. In a broader perspective, POMC and NPY/AgRP neurons have been also shown to participate in the modulation of the reproductive axis, and may contribute to the integral control of reproduction and metabolism [13]. In this section, we will briefly summarize available evidence supporting a role of these neuropeptide pathways in the control of amygdala and related functions.

Anorexigenic POMC neurons in the ARC have been defined as critical in the control of body weight and energy homeostasis [22]. Of note, a second population of POMC neurons has been found in the nucleus of solitary tract (NTS), but its role controlling energy homeostasis seems less relevant than that of the ARC population [23]. ARC POMC neurons express a panoply of neuropeptides, such as melanocortins, β-endorphin, and cocaine and amphetamine-regulated transcript (CART), as well as γ-aminobutyric acid (GABA) and glutamate neurotransmitters [22,24]. In addition to adreno-corticotropic hormone (ACTH) produced in the anterior pituitary, melanocortin peptides include α-, β-, and γ-melanocyte-stimulating hormones (MSH), derived from post-translational processing of POMC [22], which are all involved in the control of energy homeostasis [25]. Nevertheless, the major ARC POMC neuronal product is α-MSH, which operates via two of the five melanocortin receptor (MCR) subtypes, namely MC3R and MC4R. These MCR are expressed in the hypothalamus, as well as in other brain regions [26].

Admittedly, the connection of POMC signaling pathways and the amygdala remains unfolded, but fragmentary evidence suggests a potential bidirectional interplay between POMC neurons and amygdala circuits. Thus, MC4R has been shown to be highly expressed in the amygdala [27], particularly in the MeA [28]. Moreover, central infusion of the MC3R/MC4R agonist, melanotan II, into the CeA caused a marked and long-lasting decrease of food intake in rats, in a dose dependent-manner; responses that were higher in animals fed a high-fat diet (HFD). Conversely, injection of the MCR antagonist, SHU-9119, caused hyperphagia [29]. These pharmacological data argue in favor of a role of direct α-MSH effects in the amygdala to modulate feeding. This contention has been recently documented by elegant studies from Kwon and Jo, showing by a combination of molecular tracing and optogenetic experiments, that a circuit originating from ARC POMC neurons projects to neurons in the medial amygdala (MeA), which express not only MC4R, but also estrogen receptors, whose activation reduces food intake, in a MC4R-dependent manner [30]. In the same vein, we present herein our previously unpublished evidence in the rat for the presence of α-MSH immunoreactive fibers surrounding/in close contact with another neuronal population in the amygdala, namely Kiss1 neurons (Figure 1). While this Kiss1 neuronal population will be described in detail in Section 4 of this review, it is interesting to note that other subpopulations of Kiss1 neurons, in the rostral hypothalamus, have been shown to express MC4R [31]; whether the same applies to amygdala Kiss1 cells awaits future investigation.

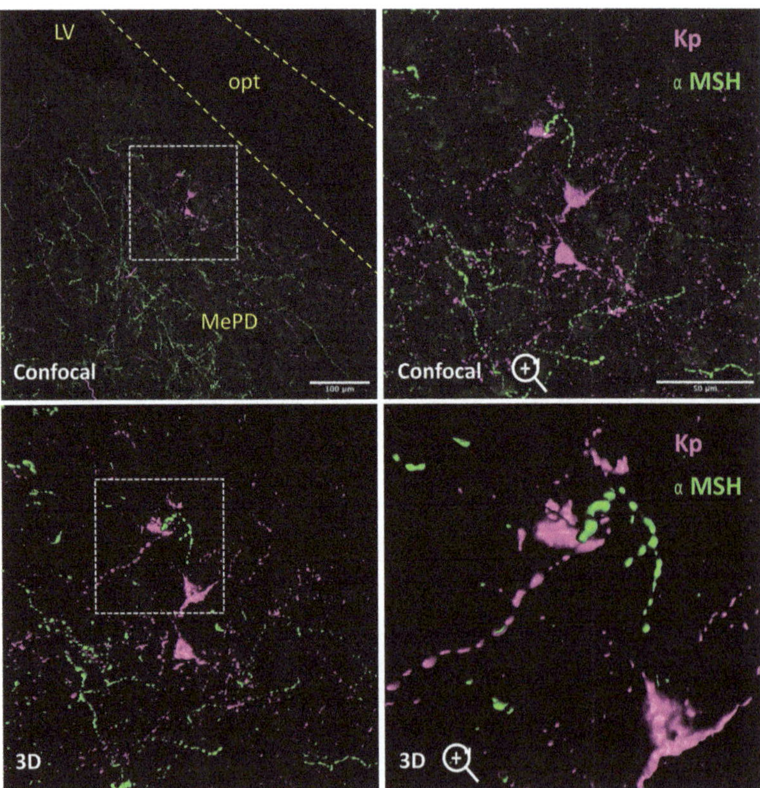

Figure 1. Rat amygdala Kiss1 neurons receive melanocortin inputs. Confocal images and 3D reconstructions of amygdala Kiss1 neurons (magenta) receiving α-MSH (green) appositions. Coronal section of an adult male rat at bregma level—3.60 mm. Materials (antibodies) and methods are described in detail elsewhere [32]. LV = Lateral ventricle; opt = Optic track; MePD = Postero-dorsal area of the medial amygdala (MeA).

In addition to the projections of ARC POMC neurons to the amygdala, it has been demonstrated that POMC neurons in the NTS receive projections from the amygdala [33]; this population of NTS POMC neurons participates in the short-term inhibitory control of feeding [34], in contrast to the long-term anorexigenic actions of ARC POMC neurons. In fact, our unpublished data show that acute fasting in rats suppresses *POMC* expression more potently in the NTS than in the ARC (Pineda and Torres, unpublished), supporting a major role of POMC neurons in the brain stem in the acute control of feeding. The fact that this population receives projections from the amygdala strongly suggests that this pathway might contribute to the role of amygdala in the modulation of feeding responses in conditions of acute metabolic stress [35].

As counterbalance to the anorectic actions of melanocortins, NPY is a highly conserved, widely distributed neuropeptide, one of the most abundantly expressed in the mammalian brain, which conducts strong orexigenic actions, acting mainly via two subtypes of the Y receptors, namely Y1 and Y5 [36,37]. Neurons expressing NPY are found in several brain areas, including prominently the ARC, and also the amygdala [38]. Of note, ARC NPY cells co-express AgRP, which acts as functional antagonist of MC3R and MC4R [39], thereby also driving a potent orexigenic effect. Accordingly, NPY/AgRP expression markedly increases in the hypothalamus under conditions of food deprivation [40]. In the rat, NPY neurons are also found in the AC, especially in the medial and lateral amygdaloid nuclei [41]; expression of NPY receptors has been also reported in the centromedial amygdala [38]. Of note, amygdala NPY neurons are found in roughly the same coordinates/region as amygdaloid Kiss1 neurons [42]; for further details see Section 4. However, co-expression of these two neuropeptides in the same amygdala cells and/or their interactions have not been documented yet. Similarly, whether amygdala NPY expression is modulated under conditions of nutritional stress (e.g., fasting) is yet to be clarified. Interestingly, intra-amygdala injection of NPY (in the CeA) has been shown to alter food preference and macronutrient selection in fed and overnight fasted rats, decreasing preference for a high-fat content diet, but without changing total calorie intake [43]. These data suggest that amygdaloid NPY signaling may play specific roles in feeding control, beyond the orexigenic, energy homeostatic actions of ARC NPY. In this context, very recent studies have highlighted that the amygdala population of NPY neurons originating in the CeA may play a relevant role in promoting feeding, specifically under conditions of chronic stress. Thus, selective over-expression of NPY in the CeA led to increased feeding and decreased energy expenditure, thereby promoting an obesity phenotype, when chronic stress and high fat diet were combined [44]. Again, these findings would argue for specific roles of NPY signaling originating from the amygdala in the control of feeding and energy balance.

Less is known about the putative roles of AgRP signaling in the control of AC, although ARC AgRP neurons have been shown to project to several extrahypothalamic areas, including the CeA. Of note, however, the physiological roles of these amygdala projections are yet to be elucidated, as specific activation of this pathway was insufficient, per se, to evoke feeding [45].

Notably, while there is ample consensus that peripheral metabolic hormones operate primarily on hypothalamic circuits to convey their regulatory actions on feeding and energy homeostasis, fragmentary evidence suggests that amygdala circuits might be also modulated (directly or indirectly) by key metabolic hormones to conduct at least part of their regulatory actions. This is the case of insulin, a key metabolic hormone with potent anorectic effects acting at central levels. Very recent evidence has documented that suppression of insulin signaling on CeA NPY neurons is a major mechanism for the development of hyperphagia and obesity in a mouse model of HFD and chronic stress [44]. In addition, loss of insulin signaling in the CeA has been shown to decrease core body temperature through direct regulation of brown adipose tissue activity [46], thereby modulating cold-induced thermogenesis and energy balance.

As final note in this section, it must be stressed that both POMC and NPY/ AgRP neurons in the hypothalamus have been reported to participate in the control of the reproductive axis [13,32,47]. However, whether the amygdala circuits involving these neuropeptides actually contribute to such regulatory function is yet to be fully clarified and warrants future investigation. For additional comments on this issue, see Section 4.

4. The Kisspeptin System and the Amygdala

While different neuropeptide pathways other than NPY/AgRP and POMC have been found in the amygdala, with potential connections with the control of metabolism and/or reproduction, the recent discovery of the expression and putative functions of the Kiss1 system in the amygdala has drawn considerable interest, as this might illuminate the underlying circuits connecting different essential behaviors closely related with reproduction and, eventually, metabolic homeostasis.

Kisspeptins are a family of structurally related peptides with a distinctive RF-amide motif at the C-terminus, encoded by the Kiss1 gene, that operate via the G protein-coupled receptor Kiss1R (also known as Gpr54). Discovery of the reproductive dimension of kisspeptins in late 2003 is now considered a major breakthrough in reproductive endocrinology, as kisspeptins play central roles in virtually all major aspects of reproductive maturation and function, from puberty onset to adult fertility. While the physiology of kisspeptins has been extensively reviewed elsewhere [18,48,49], for the purpose of this review, it is important to stress that the Kiss1 system has been shown to play crucial roles in the metabolic control of the reproductive axis [50], and may participate in the direct modulation of different aspects of metabolism, from body weight to glucose homeostasis and thermogenesis [51,52], whose physiological relevance is yet to be fully defined.

In rodents, two major populations of Kiss1 neurons are found in the ARC and the anteroventral periventricular nuclei (AVPV) of the hypothalamus. These are sensitive to sex steroid hormones, and play fundamental roles in the control of the pulsatile and surge modes of secretion of GnRH and gonadotropins by mediating the negative and positive (this is in females only) feedback actions of gonadal steroids [53]. Notably, a third population of Kiss1-expressing neurons has been identified in the amygdala, particularly in the MeA [31,42,54–59]. As it is the case for hypothalamic Kiss1 neurons, this amygdala Kiss1 neuronal population is also sexually dimorphic, but contrary to the AVPV, amygdala Kiss1 expression is higher in males [57], with null or negligible expression in early post-natal periods in rats and mice [58,60]. These data suggest that, in contrast to the hypothalamus, the Kiss1 neuronal population in the amygdala arises during puberty, possibly driven by the escalating levels of circulating sex steroids coming from maturing gonads.

The neuroanatomy of the Kiss1 neuronal population in the amygdala has begun to be elucidated in recent years. Thus, in situ hybridization and immunohistochemical studies have documented that Kiss1 neurons at this site are located at the postero-dorsal domain of MeA [42,57]. By a combination of tracing techniques, it has been shown that this population of Kiss1 neurons receive appositions from vasopressin and dopaminergic neurons, and display reciprocal connectivity with the accessory olfactory bulb [42]. Moreover, Kiss1 neurons in the MeA also project to GnRH neurons, a pathway that may contribute to the modulation of the gonadotropic axis by environmental cues, such as odor stimuli [42]. In addition, we have found close contacts of MeA Kiss1 neurons and α-MSH fibers in rats (see Figure 1), whose physiological role is yet to be defined.

Accumulating evidence from functional studies has pointed out that the amygdala population of Kiss1 neurons does play a role in the control of the reproductive axis. Thus, while peripheral administration of kisspeptin decreased neuronal activity in the amygdala, intra-amygdala injection of kisspeptin, at the MeA, induced LH secretion [61], whereas intra-MeA infusion of a kisspeptin antagonist reduced LH secretion and pulse frequency in rodents [61]. In good agreement, optogenetic activation of Kiss1 neurons in the MeA resulted in increased LH pulsatility in female mice [62]. Similarly, chemogenetic activation of Kiss1 neurons in the postero-dorsal medial amygdala has been also shown to induce LH secretion [63]. In any event, the roles of amygdala kisspeptin signaling are not restricted to the control of gonadotropin secretion in adulthood, and may involve the modulation of pubertal timing [64], and, more importantly, relevant sex behaviors [65], including partner preference and pheromonal responses [66,67]. While fragmentary evidence has suggested that kisspeptins (or Kiss1 neurons) may modulate feeding in rodents [51], whether this particular amygdala Kiss1 pathway participates in the control of feeding behavior has not been addressed to date.

Interestingly, amygdala Kiss1 neurons are sensitive to the sex steroid milieu, and both estradiol and testosterone (possibly via aromatization to estrogen) upregulates Kiss1 mRNA expression in the MeA [57], an effect that is conducted via estrogen receptor-α (ERα), but not ERβ [68]. In fact, a recent report has documented that a large proportion of Kiss1 neurons in the MeA co-express ERα [69], therefore providing the basis for the estrogenic regulation of this neuronal circuit. Of note, as will be described in detail in the following section, targeted ablation of ERα in the amygdala using the SIM1-Cre mouse revealed a role of estrogen signaling at this site in the modulation of body weight, so that mice with selective ablation of ERα in SIM1 neurons displayed an obesity phenotype [70]. Considering that SIM1- and Kiss1 neurons in the amygdala share a similar neuroanatomical location, and both populations abundantly express ERα, it remains plausible that the above genetic approach might have also caused ablation of ERα in amygdala Kiss1 neurons, which may, thereby, contribute to mediate at least part of the effects of estrogen, acting at this site, on body weight homeostasis. However, this intriguing possibility is yet to be experimentally tested. Overall, while the evidence suggesting a role of amygdala Kiss1 neurons in the control of reproductive hormones and behaviors is solid, whether this pathway participates also in the bidirectional connection between metabolism and reproduction remains largely unexplored, and requires further investigation.

5. Sex Steroids and the Amygdala: Roles of Estrogens and Androgens

Sex steroids, as major products of the gonads, are not only essential players in the control of the neuroendocrine axis governing reproduction, but are also relevant modulators of key aspects of metabolic homeostasis, from food intake to thermogenesis [14]. While a substantial fraction of these metabolic actions are conducted at the level of the hypothalamus [14], we will review the evidence supporting a role of sex steroids in the amygdala, as putative mechanism for the control of body weight and energy balance.

Estrogens exist in three major forms: estrone (E1), 17β-estradiol (E2), and estriol (E3). These exert their effects through three main receptors: the classical ERα, ERβ, and the G protein-coupled receptor, GPER/GPR30 [71]. Besides their well-known effects in the control of reproductive function, estrogens are relevant elements in metabolic regulation, with prominent roles in the control of feeding (with anorexigenic effects) and energy expenditure (increasing thermogenesis and energy consumption) [14,72]. The effects of estrogens on energy homeostasis are mainly conveyed via ERα, as genetic deletion of this receptor subtype blunts the effects of estradiol on feeding and body weight [73]. In good agreement, women with polymorphisms in the ERα gene [74], or men with genetic inactivation of ERα [75,76], suffer increased adiposity. A substantial component of such metabolic effects of ERα signaling are conducted in the brain, as demonstrated by studies involving selective deletion of this receptor in populations of hypothalamic neurons [77].

The first evidence suggesting the contribution of estrogen signaling in the amygdala to energy homeostasis came from classical experiments showing that placement of implants of estradiol benzoate into the amygdala suppressed food intake in female rats [78]. In the same vein, it has been shown that estradiol modulates the neuronal activation induced by feeding, as measured by c-Fos, in different brain areas, including the CeA [79]. Functional genomic approaches have further refined these observations. As mentioned in the previous section, seminal studies by Xu and co-workers demonstrated that selective congenital removal of ERα from SIM1 neurons, which are mainly found in the MeA and co-express ERα (>40% of them), increased body weight gain [70]. Moreover, when ERα was selectively deleted in the MeA of adult male mice, using an adeno-associated virus approach, the same phenotype was produced [70]. However, the obesity-inducing impact of ablation of ERα signaling from the amygdala was not apparently related with changes in feeding, but rather with a decrease in physical activity, which enhanced the susceptibility to develop obesity in both sexes, and further increased body weight gain after exposure to HFD [70].

In good agreement, over-expression of ERα in the MeA reduced the obesity phenotype, while pharmacogenetic activation of MeA SIM1 neurons increased physical activity [70]. Future work involving chemo- or optogenetic activation/inactivation of SIM1 vs. ERα-expressing neurons in the amygdala will help to further delineate the physiological role of this pathway in the control of body weight and energy balance.

In addition to estrogens, androgens are also known to modulate energy balance. Yet, in contrast to estrogens, androgens have been shown to stimulate feeding in several species [80–84]. Of note, brain effects of testosterone, as the main male sex steroid, are largely conducted via conversion to estradiol, in a region and cell-specific manner, by the action of the enzyme, aromatase, encoded by the CYP19 gene [85]. While the major actions of androgens in the central control of metabolism are thought to be conducted at the hypothalamus, high levels of androgen receptor (AR) expression have been reported in the MeA, with higher levels in males than in females [86,87]. In addition, expression of aromatase is also found in the MeA [88,89], therefore providing the basis for local conversion of androgens into estrogens, which, in turn, might further influence body energy balance. However, whether androgen signaling in the amygdala may contribute to regulation of energy homeostasis remains unexplored.

6. Summary and Conclusions

Compelling evidence has demonstrated that the amygdala, as a key brain area involved in emotional learning and behavioral control, including reward processing, plays a salient role in the regulation of various aspects of feeding behavior [19], a phenomenon also observed in humans, especially as it pertains to food choices and hedonic decisions [90]. As direct consequence, the amygdala contributes also to maintaining body energy balance, and possibly physical activity, and participates in responses to different forms of metabolic stress, ranging from starvation to obesity. This function is seemingly conducted, at least partially, by the interplay with metabolic neuropeptide systems, with key roles in energy homeostasis, such as POMC and NPY/AgRP, and is modulated by sex steroids, prominently estrogens. Given the proven roles of these signals in the control of reproductive function, and the known interplay between gonadal function and metabolism, it is tenable to propose that these amygdala circuits, as well as other related pathways, such as possibly amygdala Kiss1 neurons, might also contribute to the integral control of reproduction and metabolism. Admittedly, however, most of the data suggesting such a role are indirect or circumstantial, and hence, further research is needed to fully characterize the actual roles of amygdala circuits in defining these behaviors (i.e., feeding, reproductive) which are essential for survival, and the interplay between them.

Funding: The work from the authors' laboratory reviewed herein was supported by grant BFU2017-83934-P and PID2020-118660GB-I00 (Agencia Estatal de Investigación, Spain; co-funded with EU funds from FEDER Program); Projects P18-RT-4093 (to M.T.-S.) and grant P18-RTJ-4163 (to R.P.; (Junta de Andalucía, Spain), and Project 1254821 (University of Cordoba-FEDER). CIBER is an initiative of Instituto de Salud Carlos III (Ministerio de Sanidad, Spain).

Acknowledgments: The authors are indebted with the members of their research team at the Physiology Department of the University of Cordoba and IMIBIC, who actively participated in the generation of some experimental data discussed herein.

Conflicts of Interest: The authors declare no conflict of interest.

References

1. Pabba, M. Evolutionary development of the amygdaloid complex. *Front. Neuroanat* **2013**, *7*, 27. [CrossRef]
2. LeDoux, J. The amygdala. *Curr. Biol.* **2007**, *17*, R868–R874. [CrossRef]
3. Sah, P.; Faber, E.S.; Lopez De Armentia, M.; Power, J. The amygdaloid complex: Anatomy and physiology. *Physiol. Rev.* **2003**, *83*, 803–834. [CrossRef]
4. Berthoud, H.R. Metabolic and hedonic drives in the neural control of appetite: Who is the boss? *Curr. Opin. Neurobiol.* **2011**, *21*, 888–896. [CrossRef] [PubMed]
5. Kenny, P.J. Reward mechanisms in obesity: New insights and future directions. *Neuron* **2011**, *69*, 664–679. [CrossRef] [PubMed]

6. Hebebrand, J.; Albayrak, O.; Adan, R.; Antel, J.; Dieguez, C.; de Jong, J.; Leng, G.; Menzies, J.; Mercer, J.G.; Murphy, M.; et al. "Eating addiction", rather than "food addiction", better captures addictive-like eating behavior. *Neurosci. Biobehav. Rev.* **2014**, *47*, 295–306. [CrossRef] [PubMed]
7. Stice, E.; Figlewicz, D.P.; Gosnell, B.A.; Levine, A.S.; Pratt, W.E. The contribution of brain reward circuits to the obesity epidemic. *Neurosci. Biobehav. Rev.* **2013**, *37*, 2047–2058. [CrossRef] [PubMed]
8. Ziauddeen, H.; Alonso-Alonso, M.; Hill, J.O.; Kelley, M.; Khan, N.A. Obesity and the neurocognitive basis of food reward and the control of intake. *Adv. Nutr.* **2015**, *6*, 474–486. [CrossRef]
9. King, B.M.; Kass, J.M.; Cadieux, N.L.; Sam, H.; Neville, K.L.; Arceneaux, E.R. Hyperphagia and obesity in female rats with temporal lobe lesions. *Physiol. Behav.* **1993**, *54*, 759–765. [CrossRef]
10. King, B.M.; Rollins, B.L.; Stines, S.G.; Cassis, S.A.; McGuire, H.B.; Lagarde, M.L. Sex differences in body weight gains following amygdaloid lesions in rats. *Am. J. Physiol.* **1999**, *277*, R975–R980. [CrossRef]
11. King, B.M.; Rossiter, K.N.; Stines, S.G.; Zaharan, G.M.; Cook, J.T.; Humphries, M.D.; York, D.A. Amygdaloid-lesion hyperphagia: Impaired response to caloric challenges and altered macronutrient selection. *Am. J. Physiol.* **1998**, *275*, R485–R493. [CrossRef] [PubMed]
12. Douglass, A.M.; Kucukdereli, H.; Ponserre, M.; Markovic, M.; Grundemann, J.; Strobel, C.; Alcala Morales, P.L.; Conzelmann, K.K.; Luthi, A.; Klein, R. Central amygdala circuits modulate food consumption through a positive-valence mechanism. *Nat. Neurosci.* **2017**, *20*, 1384–1394. [CrossRef] [PubMed]
13. Manfredi-Lozano, M.; Roa, J.; Tena-Sempere, M. Connecting metabolism and gonadal function: Novel central neuropeptide pathways involved in the metabolic control of puberty and fertility. *Front. Neuroendocrinol.* **2018**, *48*, 37–49. [CrossRef] [PubMed]
14. Lopez, M.; Tena-Sempere, M. Estrogens and the control of energy homeostasis: A brain perspective. *Trends Endocrinol. Metab.* **2015**, *26*, 411–421. [CrossRef] [PubMed]
15. Hill, J.W.; Elias, C.F. Neuroanatomical Framework of the Metabolic Control of Reproduction. *Physiol. Rev.* **2018**, *98*, 2349–2380. [CrossRef] [PubMed]
16. Herbison, A.E. Control of puberty onset and fertility by gonadotropin-releasing hormone neurons. *Nat. Rev. Endocrinol.* **2016**, *12*, 452–466. [CrossRef] [PubMed]
17. Pinilla, L.; Aguilar, E.; Dieguez, C.; Millar, R.P.; Tena-Sempere, M. Kisspeptins and reproduction: Physiological roles and regulatory mechanisms. *Physiol. Rev.* **2012**, *92*, 1235–1316. [CrossRef]
18. Motschall, E.; Falck-Ytter, Y. Searching the MEDLINE literature database through PubMed: A short guide. *Onkologie* **2005**, *28*, 517–522. [CrossRef]
19. Schwartz, G.J.; Zeltser, L.M. Functional organization of neuronal and humoral signals regulating feeding behavior. *Annu. Rev. Nutr.* **2013**, *33*, 1–21. [CrossRef]
20. Gao, Q.; Horvath, T.L. Neurobiology of feeding and energy expenditure. *Annu. Rev. Neurosci.* **2007**, *30*, 367–398. [CrossRef] [PubMed]
21. Joly-Amado, A.; Cansell, C.; Denis, R.G.; Delbes, A.S.; Castel, J.; Martinez, S.; Luquet, S. The hypothalamic arcuate nucleus and the control of peripheral substrates. *Best Pract. Res. Clin. Endocrinol. Metab.* **2014**, *28*, 725–737. [CrossRef]
22. Anderson, E.J.P.; Çakir, I.; Carrington, S.J.; Cone, R.D.; Ghamari-Langroudi, M.; Gillyard, T.; Gimenez, L.E.; Litt, M.J. 60 YEARS OF POMC: Regulation of feeding and energy homeostasis by α-MSH. *J. Mol. Endocrinol.* **2016**, *56*, T157–T174. [CrossRef]
23. Huo, L.; Grill, H.J.; Bjørbaek, C. Divergent regulation of proopiomelanocortin neurons by leptin in the nucleus of the solitary tract and in the arcuate hypothalamic nucleus. *Diabetes* **2006**, *55*, 567–573. [CrossRef] [PubMed]
24. Hentges, S.T.; Otero-Corchon, V.; Pennock, R.L.; King, C.M.; Low, M.J. Proopiomelanocortin expression in both GABA and glutamate neurons. *J. Neurosci.* **2009**, *29*, 13684–13690. [CrossRef] [PubMed]
25. Irani, B.G.; Haskell-Luevano, C. Feeding effects of melanocortin ligands—A historical perspective. *Peptides* **2005**, *26*, 1788–1799. [CrossRef] [PubMed]
26. Dores, R.M.; Londraville, R.L.; Prokop, J.; Davis, P.; Dewey, N.; Lesinski, N. Molecular evolution of GPCRs: Melanocortin/melanocortin receptors. *J. Mol. Endocrinol.* **2014**, *52*, T29–T42. [CrossRef] [PubMed]
27. Gantz, I.; Miwa, H.; Konda, Y.; Shimoto, Y.; Tashiro, T.; Watson, S.J.; DelValle, J.; Yamada, T. Molecular cloning, expression, and gene localization of a fourth melanocortin receptor. *J. Biol. Chem.* **1993**, *268*, 15174–15179. [CrossRef]
28. Mountjoy, K.G.; Mortrud, M.T.; Low, M.J.; Simerly, R.B.; Cone, R.D. Localization of the melanocortin-4 receptor (MC4-R) in neuroendocrine and autonomic control circuits in the brain. *Mol. Endocrinol. (Baltim. Md.)* **1994**, *8*, 1298–1308. [CrossRef] [PubMed]
29. Boghossian, S.; Park, M.; York, D.A. Melanocortin activity in the amygdala controls appetite for dietary fat. *AJP: Regul. Integr. Comp. Physiol.* **2010**, *298*, R385–R393. [CrossRef] [PubMed]
30. Kwon, E.; Jo, Y.H. Activation of the ARC(POMC)->MeA Projection Reduces Food Intake. *Front. Neural. Circuits* **2020**, *14*, 595783. [CrossRef] [PubMed]
31. Cravo, R.M.; Margatho, L.O.; Osborne-Lawrence, S.; Donato, J.; Atkin, S.; Bookout, A.L.; Rovinsky, S.; Frazão, R.; Lee, C.E.; Gautron, L.; et al. Characterization of Kiss1 neurons using transgenic mouse models. *Neuroscience* **2011**, *173*, 37–56. [CrossRef]
32. Manfredi-Lozano, M.; Roa, J.; Ruiz-Pino, F.; Piet, R.; García-Galiano, D.; Pineda, R.; Zamora, A.; Leon, S.; Sánchez-Garrido, M.A.; Romero-Ruiz, A.; et al. Defining a novel leptin-melanocortin-kisspeptin pathway involved in the metabolic control of puberty. *Mol. Metab.* **2016**, *5*, 844–857. [CrossRef] [PubMed]

33. Wang, D.; He, X.; Zhao, Z.; Feng, Q.; Lin, R.; Sun, Y.; Ding, T.; Xu, F.; Luo, M.; Zhan, C. Whole-brain mapping of the direct inputs and axonal projections of POMC and AgRP neurons. *Front. Neuroanat.* **2015**, *9*, 40. [CrossRef] [PubMed]
34. Zhan, C.; Zhou, J.; Feng, Q.; Zhang, J.-E.; Lin, S.; Bao, J.; Wu, P.; Luo, M. Acute and long-term suppression of feeding behavior by POMC neurons in the brainstem and hypothalamus, respectively. *J. Neurosci.* **2013**, *33*, 3624–3632. [CrossRef] [PubMed]
35. Roozendaal, B.; McEwen, B.S.; Chattarji, S. Stress, memory and the amygdala. *Nat. Reviews. Neurosci.* **2009**, *10*, 423–433. [CrossRef] [PubMed]
36. Lin, S.; Boey, D.; Herzog, H. NPY and Y receptors: Lessons from transgenic and knockout models. *Neuropeptides* **2004**, *38*, 189–200. [CrossRef] [PubMed]
37. Pedragosa Badia, X.; Stichel, J.; Beck-Sickinger, A.G. Neuropeptide Y receptors: How to get subtype selectivity. *Front. Endocrinol.* **2013**, *4*, 5. [CrossRef]
38. Wood, J.; Verma, D.; Lach, G.; Bonaventure, P.; Herzog, H.; Sperk, G.; Tasan, R.O. Structure and function of the amygdaloid NPY system: NPY Y2 receptors regulate excitatory and inhibitory synaptic transmission in the centromedial amygdala. *Brain Struct. Funct.* **2016**, *221*, 3373–3391. [CrossRef] [PubMed]
39. Ollmann, M.M.; Wilson, B.D.; Yang, Y.K.; Kerns, J.A.; Chen, Y.; Gantz, I.; Barsh, G.S. Antagonism of central melanocortin receptors in vitro and in vivo by agouti-related protein. *Science* **1997**, *278*, 135–138. [CrossRef]
40. Hahn, T.M.; Breininger, J.F.; Baskin, D.G.; Schwartz, M.W. Coexpression of Agrp and NPY in fasting-activated hypothalamic neurons. *Nat. Neurosci.* **1998**, *1*, 271–272. [CrossRef] [PubMed]
41. Chronwall, B.M.; DiMaggio, D.A.; Massari, V.J.; Pickel, V.M.; Ruggiero, D.A.; O'Donohue, T.L. The anatomy of neuropeptide-Y-containing neurons in rat brain. *NSC* **1985**, *15*, 1159–1181. [CrossRef]
42. Pineda, R.; Plaisier, F.; Millar, R.P.; Ludwig, M. Amygdala Kisspeptin Neurons: Putative Mediators of Olfactory Control of the Gonadotropic Axis. *Neuroendocrinology* **2017**, *104*, 223–238. [CrossRef]
43. Primeaux, S.D.; York, D.A.; Bray, G.A. Neuropeptide Y administration into the amygdala alters high fat food intake. *Peptides* **2006**, *27*, 1644–1651. [CrossRef] [PubMed]
44. Ip, C.K.; Zhang, L.; Farzi, A.; Qi, Y.; Clarke, I.; Reed, F.; Shi, Y.-C.; Enriquez, R.; Dayas, C.; Graham, B.; et al. Amygdala NPY Circuits Promote the Development of Accelerated Obesity under Chronic Stress Conditions. *Cell Metab.* **2019**, *30*, 111–128.e116. [CrossRef]
45. Betley, J.N.; Cao, Z.F.H.; Ritola, K.D.; Sternson, S.M. Parallel, redundant circuit organization for homeostatic control of feeding behavior. *Cell* **2013**, *155*, 1337–1350. [CrossRef] [PubMed]
46. Soto, M.; Cai, W.; Konishi, M.; Kahn, C.R. Insulin signaling in the hippocampus and amygdala regulates metabolism and neurobehavior. *Proc. Natl. Acad. Sci. USA* **2019**, *116*, 6379–6384. [CrossRef] [PubMed]
47. Padilla, S.L.; Qiu, J.; Nestor, C.C.; Zhang, C.; Smith, A.W.; Whiddon, B.B.; Ronnekleiv, O.K.; Kelly, M.J.; Palmiter, R.D. AgRP to Kiss1 neuron signaling links nutritional state and fertility. *Proc. Natl. Acad. Sci. USA* **2017**, *114*, 2413–2418. [CrossRef]
48. Pineda, R.; Aguilar, E.; Pinilla, L.; Tena-Sempere, M. Physiological roles of the kisspeptin/GPR54 system in the neuroendocrine control of reproduction. *Prog. Brain. Res.* **2010**, *181*, 55–77. [CrossRef]
49. Wolfe, A.; Hussain, M.A. The Emerging Role(s) for Kisspeptin in Metabolism in Mammals. *Front. Endocrinol.* **2018**, *9*, 184. [CrossRef] [PubMed]
50. Navarro, V.M. Metabolic regulation of kisspeptin—The link between energy balance and reproduction. *Nat. Rev. Endocrinol.* **2020**, *16*, 407–420. [CrossRef]
51. Hudson, A.D.; Kauffman, A.S. Metabolic actions of kisspeptin signaling: Effects on body weight, energy expenditure, and feeding. *Pharmacol. Ther.* **2021**, 107974. [CrossRef]
52. Velasco, I.; Leon, S.; Barroso, A.; Ruiz-Pino, F.; Heras, V.; Torres, E.; Leon, M.; Ruohonen, S.T.; Garcia-Galiano, D.; Romero-Ruiz, A.; et al. Gonadal hormone-dependent vs. -independent effects of kisspeptin signaling in the control of body weight and metabolic homeostasis. *Metabolism* **2019**, *98*, 84–94. [CrossRef] [PubMed]
53. Garcia-Galiano, D.; Pinilla, L.; Tena-Sempere, M. Sex steroids and the control of the Kiss1 system: Developmental roles and major regulatory actions. *J. Neuroendocrinol.* **2012**, *24*, 22–33. [CrossRef]
54. Smith, J.T.; Cunningham, M.J.; Rissman, E.F.; Clifton, D.K.; Steiner, R.A. Regulation of Kiss1 gene expression in the brain of the female mouse. *Endocrinology* **2005**, *146*, 3686–3692. [CrossRef]
55. Smith, J.T.; Dungan, H.M.; Stoll, E.A.; Gottsch, M.L.; Braun, R.E.; Eacker, S.M.; Clifton, D.K.; Steiner, R.A. Differential regulation of KiSS-1 mRNA expression by sex steroids in the brain of the male mouse. *Endocrinology* **2005**, *146*, 2976–2984. [CrossRef]
56. Yeo, S.H.; Kyle, V.; Morris, P.G.; Jackman, S.; Sinnett-Smith, L.C.; Schacker, M.; Chen, C.; Colledge, W.H. Visualisation of Kiss1 Neurone Distribution Using a Kiss1-CRE Transgenic Mouse. *J. Neuroendocrinol.* **2016**, *28*, 713. [CrossRef]
57. Kim, J.; Semaan, S.J.; Clifton, D.K.; Steiner, R.A.; Dhamija, S.; Kauffman, A.S. Regulation of Kiss1 Expression by Sex Steroids in the Amygdala of the Rat and Mouse. *Endocrinology* **2011**, *152*, 2020–2030. [CrossRef]
58. Di Giorgio, N.P.; Semaan, S.J.; Kim, J.; López, P.V.; Bettler, B.; Libertun, C.; Lux-Lantos, V.A.; Kauffman, A.S. Impaired GABAB receptor signaling dramatically up-regulates Kiss1 expression selectively in nonhypothalamic brain regions of adult but not prepubertal mice. *Endocrinology* **2014**, *155*, 1033–1044. [CrossRef]
59. Clarkson, J.; d'Anglemont de Tassigny, X.; Colledge, W.H.; Caraty, A.; Herbison, A.E. Distribution of kisspeptin neurones in the adult female mouse brain. *J. Neuroendocrinol.* **2009**, *21*, 673–682. [CrossRef]

60. Cao, J.; Patisaul, H.B. Sex-specific expression of estrogen receptors α and β and Kiss1 in the postnatal rat amygdala. *J. Comp. Neurol.* **2013**, *521*, 465–478. [CrossRef]
61. Comninos, A.N.; Anastasovska, J.; Sahuri-Arisoylu, M.; Li, X.; Li, S.; Hu, M.; Jayasena, C.N.; Ghatei, M.A.; Bloom, S.R.; Matthews, P.M.; et al. Kisspeptin signaling in the amygdala modulates reproductive hormone secretion. *Brain Struct. Funct.* **2016**, *221*, 2035–2047. [CrossRef] [PubMed]
62. Lass, G.; Li, X.F.; de Burgh, R.A.; He, W.; Kang, Y.; Hwa-Yeo, S.; Sinnett-Smith, L.C.; Manchishi, S.M.; Colledge, W.H.; Lightman, S.L.; et al. Optogenetic stimulation of kisspeptin neurones within the posterodorsal medial amygdala increases luteinising hormone pulse frequency in female mice. *J. Neuroendocrinol.* **2020**, *32*, e12823. [CrossRef] [PubMed]
63. Fergani, C.; Leon, S.; Padilla, S.L.; Verstegen, A.M.J.; Palmiter, R.D.; Navarro, V.M. NKB signaling in the posterodorsal medial amygdala stimulates gonadotropin release in a kisspeptin-independent manner in female mice. *eLife* **2018**, *7*, e40476. [CrossRef]
64. Adekunbi, D.A.; Li, X.F.; Li, S.; Adegoke, O.A.; Iranloye, B.O.; Morakinyo, A.O.; Lightman, S.L.; Taylor, P.D.; Poston, L.; O'Byrne, K.T. Role of amygdala kisspeptin in pubertal timing in female rats. *PLoS ONE* **2017**, *12*, e0183596. [CrossRef]
65. Gresham, R.; Li, S.; Adekunbi, D.A.; Hu, M.; Li, X.F.; O'Byrne, K.T. Kisspeptin in the medial amygdala and sexual behavior in male rats. *Neurosci. Lett.* **2016**, *627*, 13–17. [CrossRef]
66. Adekunbi, D.A.; Li, X.F.; Lass, G.; Shetty, K.; Adegoke, O.A.; Yeo, S.H.; Colledge, W.H.; Lightman, S.L.; O'Byrne, K.T. Kisspeptin neurones in the posterodorsal medial amygdala modulate sexual partner preference and anxiety in male mice. *J. Neuroendocrinol.* **2018**, *30*, e12572. [CrossRef]
67. Aggarwal, S.; Tang, C.; Sing, K.; Kim, H.W.; Millar, R.P.; Tello, J.A. Medial Amygdala Kiss1 Neurons Mediate Female Pheromone Stimulation of Luteinizing Hormone in Male Mice. *Neuroendocrinology* **2019**, *108*, 172–189. [CrossRef] [PubMed]
68. Stephens, S.B.Z.; Chahal, N.; Munaganuru, N.; Parra, R.A.; Kauffman, A.S. Estrogen Stimulation of Kiss1 Expression in the Medial Amygdala Involves Estrogen Receptor-α But Not Estrogen Receptor-β. *Endocrinology* **2016**, *157*, 4021–4031. [CrossRef] [PubMed]
69. Lima, L.B.; Haubenthal, F.T.; Silveira, M.A.; Bohlen, T.M.; Metzger, M.; Donato, J.; Frazao, R. Conspecific odor exposure predominantly activates non-kisspeptin cells in the medial nucleus of the amygdala. *Neurosci. Lett.* **2018**, *681*, 12–16. [CrossRef]
70. Xu, P.; Cao, X.; He, Y.; Zhu, L.; Yang, Y.; Saito, K.; Wang, C.; Yan, X.; Hinton Jr, A.O.; Zou, F.; et al. Estrogen receptor–α in medial amygdala neurons regulates body weight. *J. Clin. Investig.* **2015**, *125*, 2861–2876. [CrossRef]
71. Hewitt, S.C.; Korach, K.S. Estrogen Receptors: New Directions in the New Millennium. *Endocr. Rev.* **2018**, *39*, 664–675. [CrossRef] [PubMed]
72. Mauvais-Jarvis, F.; Clegg, D.J.; Hevener, A.L. The role of estrogens in control of energy balance and glucose homeostasis. *Endocr. Rev.* **2013**, *34*, 309–338. [CrossRef]
73. Geary, N.; Asarian, L.; Korach, K.S.; Pfaff, D.W.; Ogawa, S. Deficits in E2-dependent control of feeding, weight gain, and cholecystokinin satiation in ER-alpha null mice. *Endocrinology* **2001**, *142*, 4751–4757. [CrossRef]
74. Okura, T.; Koda, M.; Ando, F.; Niino, N.; Ohta, S.; Shimokata, H. Association of polymorphisms in the estrogen receptor alpha gene with body fat distribution. *Int. J. Obes. Relat. Metab. Disord.* **2003**, *27*, 1020–1027. [CrossRef]
75. Smith, E.P.; Boyd, J.; Frank, G.R.; Takahashi, H.; Cohen, R.M.; Specker, B.; Williams, T.C.; Lubahn, D.B.; Korach, K.S. Estrogen resistance caused by a mutation in the estrogen-receptor gene in a man. *N. Engl. J. Med.* **1994**, *331*, 1056–1061. [CrossRef]
76. Grumbach, M.M.; Auchus, R.J. Estrogen: Consequences and implications of human mutations in synthesis and action. *J. Clin. Endocrinol. Metab.* **1999**, *84*, 4677–4694. [CrossRef] [PubMed]
77. Xu, Y.; Nedungadi, T.P.; Zhu, L.; Sobhani, N.; Irani, B.G.; Davis, K.E.; Zhang, X.; Zou, F.; Gent, L.M.; Hahner, L.D.; et al. Distinct hypothalamic neurons mediate estrogenic effects on energy homeostasis and reproduction. *Cell Metab.* **2011**, *14*, 453–465. [CrossRef]
78. Donohoe, T.P.; Stevens, R. Modulation of food intake by amygdaloid estradiol benzoate implants in female rats. *Physiol. Behav.* **1981**, *27*, 105–114. [CrossRef]
79. Eckel, L.A.; Geary, N. Estradiol treatment increases feeding-induced c-Fos expression in the brains of ovariectomized rats. *Am. J. Physiol. Regul. Integr. Comp. Physiol.* **2001**, *281*, R738–R746. [CrossRef]
80. Rowland, D.L.; Perrings, T.S.; Thommes, J.A. Comparison of androgenic effects on food intake and body weight in adult rats. *Physiol. Behav.* **1980**, *24*, 205–209. [CrossRef]
81. Chai, J.K.; Blaha, V.; Meguid, M.M.; Laviano, A.; Yang, Z.J.; Varma, M. Use of orchiectomy and testosterone replacement to explore meal number-to-meal size relationship in male rats. *Am. J. Physiol.* **1999**, *276*, R1366–R1373. [CrossRef] [PubMed]
82. Anukulkitch, C.; Rao, A.; Dunshea, F.R.; Blache, D.; Lincoln, G.A.; Clarke, I.J. Influence of photoperiod and gonadal status on food intake, adiposity, and gene expression of hypothalamic appetite regulators in a seasonal mammal. *Am. J. Physiol. Regul. Integr. Comp. Physiol.* **2007**, *292*, R242–R252. [CrossRef] [PubMed]
83. Rana, K.; Fam, B.C.; Clarke, M.V.; Pang, T.P.S.; Zajac, J.D.; MacLean, H.E. Increased adiposity in DNA binding-dependent androgen receptor knockout male mice associated with decreased voluntary activity and not insulin resistance. *AJP Endocrinol. Metab.* **2011**, *301*, E767–E778. [CrossRef]
84. Borgquist, A.; Meza, C.; Wagner, E.J. The role of AMP-activated protein kinase in the androgenic potentiation of cannabinoid-induced changes in energy homeostasis. *AJP: Endocrinol. Metab.* **2015**, *308*, E482–E495. [CrossRef]
85. Kamat, A.; Hinshelwood, M.M.; Murry, B.A.; Mendelson, C.R. Mechanisms in tissue-specific regulation of estrogen biosynthesis in humans. *Trends Endocrinol. Metab. TEM* **2002**, *13*, 122–128. [CrossRef]

86. Simerly, R.B.; Chang, C.; Muramatsu, M.; Swanson, L.W. Distribution of androgen and estrogen receptor mRNA-containing cells in the rat brain: An in situ hybridization study. *J. Comp. Neurol.* **1990**, *294*, 76–95. [CrossRef] [PubMed]
87. McAbee, M.D.; DonCarlos, L.L. Ontogeny of region-specific sex differences in androgen receptor messenger ribonucleic acid expression in the rat forebrain. *Endocrinology* **1998**, *139*, 1738–1745. [CrossRef]
88. Roselli, C.E.; Abdelgadir, S.E.; Ronnekleiv, O.K.; Klosterman, S.A. Anatomic distribution and regulation of aromatase gene expression in the rat brain. *Biol. Reprod.* **1998**, *58*, 79–87. [CrossRef] [PubMed]
89. Wagner, C.K.; Morrell, J.I. Distribution and steroid hormone regulation of aromatase mRNA expression in the forebrain of adult male and female rats: A cellular-level analysis using in situ hybridization. *J. Comp. Neurol.* **1996**, *370*, 71–84. [CrossRef]
90. Tiedemann, L.J.; Alink, A.; Beck, J.; Büchel, C.; Brassen, S. Valence Encoding Signals in the Human Amygdala and the Willingness to Eat. *J. Neurosci.* **2020**, *40*, 5264–5272. [CrossRef] [PubMed]

MDPI

Review

Astrocyte Gliotransmission in the Regulation of Systemic Metabolism

Cahuê De Bernardis Murat [1,2] and Cristina García-Cáceres [1,2,3,*]

[1] Helmholtz Diabetes Center, Helmholtz Zentrum München, German Research Center for Environmental Health (GmbH), Institute for Diabetes and Obesity, 85764 Neuherberg, Germany; cahue.bernardis@helmholtz-muenchen.de
[2] German Center for Diabetes Research (DZD), 85764 Neuherberg, Germany
[3] Medizinische Klinik and Poliklinik IV, Klinikum der Universität, Ludwig-Maximilians-Universität München, 80336 Munich, Germany
* Correspondence: garcia-caceres@helmholtz-muenchen.de

Abstract: Normal brain function highly relies on the appropriate functioning of astrocytes. These glial cells are strategically situated between blood vessels and neurons, provide significant substrate support to neuronal demand, and are sensitive to neuronal activity and energy-related molecules. Astrocytes respond to many metabolic conditions and regulate a wide array of physiological processes, including cerebral vascular remodeling, glucose sensing, feeding, and circadian rhythms for the control of systemic metabolism and behavior-related responses. This regulation ultimately elicits counterregulatory mechanisms in order to couple whole-body energy availability with brain function. Therefore, understanding the role of astrocyte crosstalk with neighboring cells via the release of molecules, e.g., gliotransmitters, into the parenchyma in response to metabolic and neuronal cues is of fundamental relevance to elucidate the distinct roles of these glial cells in the neuroendocrine control of metabolism. Here, we review the mechanisms underlying astrocyte-released gliotransmitters that have been reported to be crucial for maintaining homeostatic regulation of systemic metabolism.

Keywords: astrocytes; calcium signaling; energy balance; gliotransmission; systemic metabolism

Citation: Murat, C.D.B.; García-Cáceres, C. Astrocyte Gliotransmission in the Regulation of Systemic Metabolism. *Metabolites* **2021**, *11*, 732. https://doi.org/10.3390/metabo11110732

Academic Editors: Giancarlo Panzica, Stefano Gotti and Paloma Collado Guirao

Received: 8 October 2021
Accepted: 26 October 2021
Published: 26 October 2021

Publisher's Note: MDPI stays neutral with regard to jurisdictional claims in published maps and institutional affiliations.

1. Introduction

The field of neuroscience has experienced significant advancement in knowledge on how the brain processes information as a result of the growing evidence supporting that glial cells, as with astrocytes, are fully integrated into neuronal networks, thus forming one functional regulatory circuit required for brain function [1]. In addition to serving as a support system, active functions have been assigned to astrocytes, including the control of cerebral vascular remodeling and blood flow [2], and the regulation of all aspects of neuronal function, such as neurogenesis [3], neuronal transmission [4], and synapse formation/elimination and homeostasis [5], among others. By providing energy substrates and neurotransmitter precursor molecules via the astrocyte-neuron lactate shuttle [6] and the glutamate/γ-aminobutyric acid (GABA)-glutamine cycle [7], astrocytes ensure adequate neuronal metabolism, connectivity, and brain functioning. The characteristic star-like shape of astrocytes possesses specific non-overlapping territorial domains and hence fills the local environment, interacting with a large number of synapses that can dynamically change depending on the surrounding microenvironment in response to neuronal activity and/or metabolic status for the regulation of physiological responses [1,8]. Remarkably, astrocytes play a key role in neurotransmitter clearance [1] and spatial K^+ buffering [9], which support neurotransmission homeostasis. Astrocytes, as with neurons, sense and respond to metabolic [8] and synaptic cues [10] through specific metabolic and neurotransmitter receptors/transporters expressed along their membranes, thus influencing the active state of synapses to which they are often intimately associated [11]. Seminal findings demonstrated that astrocytes display changes in their intracellular Ca^{2+} concentration [12], a signal that appears to be the relevant signal for astrocytic responses [4,13]. As secretory cells, astrocytes possess the molecular machinery to send molecules and ions back and forth, which are essential in regulating all physiological processes (e.g., synaptic connectivity) required for a normal brain function [1,14].

Astrocyte Gliotransmission: The Hallmark of Astrocyte Communication

Despite the absence of membrane electrical excitability, astrocytes exhibit a marked ionic handling in response to diverse stimuli, which is crucial for proper regulation of physiological processes controlled by the brain [1]. For instance, internal K^+, Na^+, Ca^{2+}, and H^+ fluctuations in astrocytes are associated with increased synaptic activity whereas Cl^- permeability is involved in astrocyte volume changes. Since compelling evidence suggests the existence of Ca^{2+}-dependent astrocyte-neuron communication [4,13], intracellular Ca^{2+} signaling has been extensively studied in astrocytes. Several works have reported that an enhancement in synaptic activity may result in astrocyte Ca^{2+} rises following the activation of specific metabotropic G-protein coupled receptors (GPCRs) by synaptic neurotransmitter spillover, such as glutamate [15,16], GABA [17], ATP [18], acetylcholine [19,20], and dopamine [21]. Intracellular Ca^{2+} events at the soma, primary branches, and branchlets of astrocytes are greatly mediated by the inositol 1,4,5-trisphosphate receptor type 2 (IP_3R2) signaling pathway that mobilizes Ca^{2+} from the endoplasmic reticulum to the cytosol. However, astrocyte Ca^{2+} responses may occur in an IP_3R2-independent manner, especially at their fine processes, i.e., astrocyte leaflets, in which Ca^{2+}-permeating channels [22] and Na^+/Ca^{2+} exchangers [23] underlie the mechanism of Ca^{2+} entry into the cytosol. Ca^{2+} transients in astrocyte processes may also occur via Ca^{2+} efflux from mitochondrial membrane permeability transition pores (mPTPs) [24]. Therefore, there are multiple sources of Ca^{2+} that contribute to the increased cytosolic Ca^{2+} content in response to synaptic activity (Figure 1), highlighting the complex Ca^{2+} dynamics within astrocyte cellular compartments, ranging from slow, global Ca^{2+} events to rapid, local Ca^{2+} transients [25–28]. In turn, astrocyte Ca^{2+} elevation promotes the release of signaling molecules, such as glutamate, ATP, D-serine, and GABA, which can influence the activity of neighboring neurons and other cells to ultimately modulate local metabolism and the information processing within neuronal networks, a process known as gliotransmission [4]. Such signaling molecules are released from astrocytes through several intracellular pathways including vesicle-mediated exocytosis and diffusion through channels (Box 1). The complex intracellular Ca^{2+} dynamics in astrocytes and the great variety of mechanisms releasing their gliotransmitters suggest distinct, specific roles of astrocyte gliotransmission depending on the spatial location, quality, and intensity of the stimulus as well as the gliotransmitter releasing site and astrocyte interactions with the surrounding microenvironment (Figure 1).

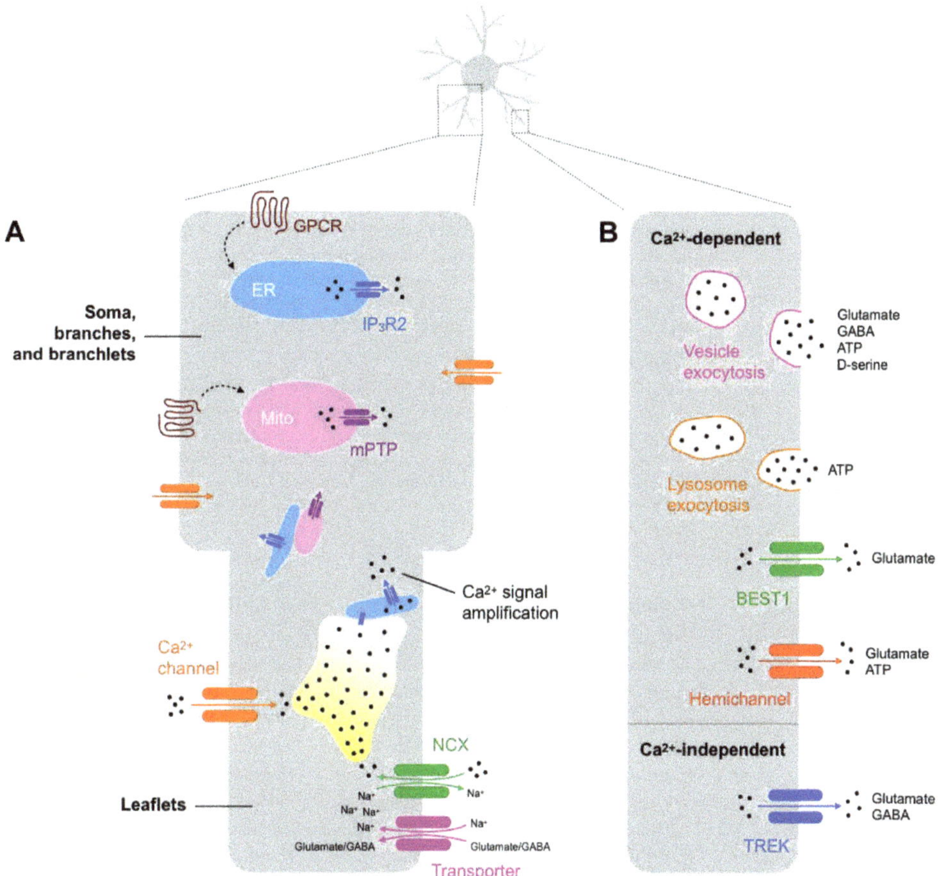

Figure 1. Mechanisms underlying intracellular Ca^{2+} rises and gliotransmitter release from astrocytes. (**A**) At the soma, branches and branchlets of astrocytes, the activation of G-protein-coupled receptors (GPCRs) following increased synaptic activity triggers cytosolic Ca^{2+} rises by several mechanisms, such as via the 1,4,5-trisphosphate receptor type 2 receptor (IP_3R2)-dependent mobilization Ca^{2+} from the endoplasmic reticulum (ER) or the efflux of Ca^{2+} through mitochondrial membrane permeability transition pores (mPTPs). At fine processes of astrocytes, i.e., leaflets, increased synaptic activity may promote the co-transport of neurotransmitters and Na^+ into the cytosol, the latter increasing the activity of Na^+/Ca^{2+} exchangers (NCX) that results in cytosolic Ca^{2+} elevations. Additionally, Ca^{2+}-permeating channels contribute to the influx of Ca^{2+} in astrocytic leaflets. The Ca^{2+} influx into leaflets may trigger local Ca^{2+} transients and propagate the Ca^{2+} signaling to distant domains via its signal amplification mediated by a Ca^{2+}-dependent Ca^{2+} release from ERs via the IP_3R2 pathway; (**B**) Several mechanisms account for the release of gliotransmitters from astrocytes. Mostly, these processes occur in a Ca^{2+}-dependent manner via the exocytosis of vesicles. Lysosome exocytosis, bestrophin1 (BEST1) channels and hemichannels have also been described to participate in Ca^{2+}-dependent gliotransmitter release mechanisms. Moreover, the involvement of a Ca^{2+}-independent release of gliotransmitters via two-pore domain K^+ (TREK) channels is reported.

Box 1. Astrocytes release gliotransmitters via several pathways.

Vesicle-mediated exocytosis

The soluble N-ethylmaleimide-sensitive factor attachment protein receptor (SNARE)-mediated vesicular exocytosis is likely the major mechanism for the Ca^{2+}-sensitive release of gliotransmitters from astrocytes. Using ex vivo brain slices from mice and human, it was observed that Ca^{2+}-dependent astrocyte-released glutamate induces the activation of N-methyl-D-aspartate receptors (NMDARs) in neurons triggering slow inward currents [29–32], an effect greatly attenuated by disrupting the SNARE complex [33–35]. These currents have also been shown to be associated with changes in neuronal excitability and neurotransmission. Accordingly, vesicular glutamate transporters and SNARE proteins are localized in astrocyte processes adjacent to neurons [35]. The blockade of vesicular exocytosis also impairs the release of ATP from astrocytes, which may influence synaptic transmission and behavioral responses [36–39]. Likewise, the exocytosis of lysosomes is also thought to participate in ATP release from astrocytes [40,41].

Diffusion through channels

In addition to exocytotic mechanisms, the release of astrocyte gliotransmitters may occur through ion channels. For instance, glutamate can be released via Ca^{2+}-activated bestrophin 1 (BEST1) channels localized at astrocyte microdomains [42] to modulate synaptic plasticity [43,44]. BEST1 channels are also permeable to GABA, which may tonically inhibit neighboring neurons [45–47] and drive pathological mechanisms following its impaired release [48,49]. Moreover, astrocytes are able to release gliotransmitters via hemichannels [50–54] and Ca^{2+}-independent pathways, such as two-pore domain K^+ channels [42,55].

2. Physiological Processes by Which Astrocytes Regulate Systemic Metabolism

Although diverse studies have reported the mechanisms underlying Ca^{2+} responses and gliotransmitter release from astrocytes in the regulation of local metabolism and synapse physiology [1,2,4], their influence on the control of systemic metabolism has recently begun to be explored. Understanding the communication between astrocytes and neighboring cells involved in whole-body counterregulatory responses to metabolic challenges may add relevant insights on how physiological processes are controlled by the brain. In this section, we aim to describe the contribution of astrocyte gliotransmission to the modulation of the surrounding microenvironment and synaptic transmission that are related to the homeostatic regulation of systemic metabolism and behavior.

2.1. Cerebral Vascular Integrity and Remodeling

The brain stores a low amount of energy [56] and largely depends on oxidative metabolism for supporting its energy requirements [57]. Therefore, a constant, adequate supply of glucose and oxygen from the brain vasculature is needed in order to match the high metabolic demand of neurotransmission and brain function [58–60]. In this regard, astrocytes are situated in a strategic position to control continuous fuel supply to the brain by enveloping virtually all brain blood vessels with their endfeet [61] and making close contact with synapses by their processes [11], thus regulating the cerebral vascular tone to accomplish neuronal function in both resting and active states. In the last decade, multiple studies have highlighted the active role of vascular endothelial growth factors (VEGFs) in angiogenesis and vascular architecture in the brain [62] by modulating tight-junction proteins in blood vessels for controlling blood-brain barrier (BBB) permeability [63–66]. Astrocytes have been shown to be the predominant source of VEGF within the brain [67,68], as the blockade of astrocytic VEGF-dependent releasing mechanisms attenuates BBB leakage in animal models [68–70]. Other studies have pointed out that angiogenesis correlates with higher astrocyte density and elevated VEGF expression levels in the brain of mice and humans [71]. Notably, additional studies have reported that a hypercaloric diet rapidly increases the number of astrocytes in the hypothalamus [72] and promotes angiogenesis and endothelial dysfunction in both rodents and humans [73,74]. Recently, it has been revealed that VEGF-derived hypothalamic astrocytes are directly involved in obesity-induced hypothalamic microvasculature remodeling and elevated systemic blood pressure via sympathetic outflow, an effect dependent on leptin signaling and concomitant with the onset of obesity [75] (Figure 2A). Further, the selective disruption of the hypoxia-inducible factor 1α-VEGF signaling cascade in astrocytes protected mice against obesity-induced hypothalamic angiopathy, increased sympathetic drive, and arterial hypertension [75]. These findings reveal the astrocyte-released gliotransmitter VEGF as a relevant molecule involved in the tuning of sympathetic outflow controlling cardiovascular function and challenge the traditional view that microvascular complications in the brain are derived from arterial hypertension [76].

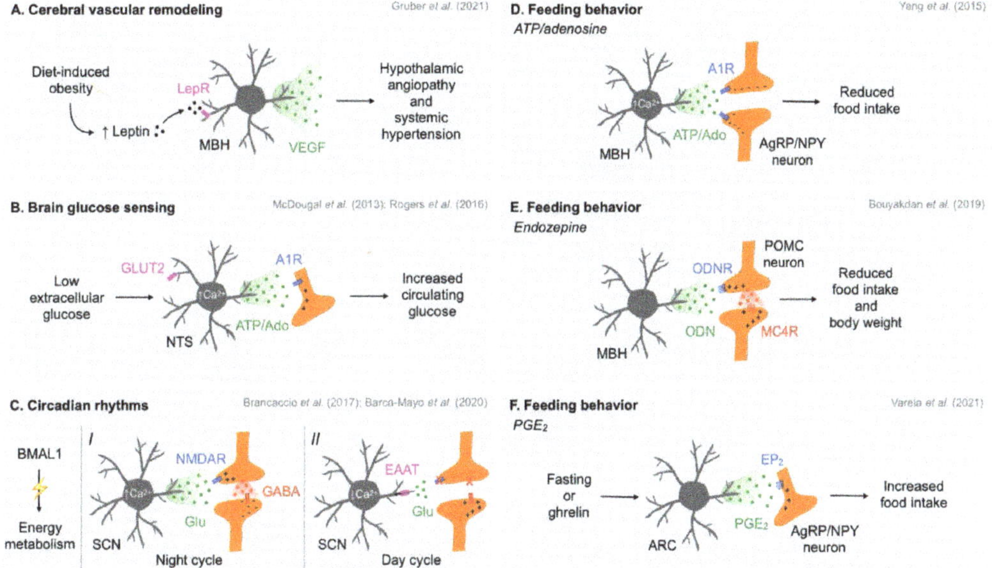

Figure 2. The action of astrocyte-released gliotransmitters in the control of systemic metabolism. (**A**) Diet-induced obesity promotes hyperleptinemia, which hyperactivates leptin receptors (LepRs) in astrocytes from the mediobasal hypothalamus (MBH) and leads to the release of vascular endothelial growth factors (VEGFs), promoting hypothalamic angiopathy and systemic hypertension; (**B**) Astrocytes from the brainstem nucleus tractus solitarius (NTS) sense extracellular glucose concentration drops via glucose transporter type 2 (GLUT2) and respond with ATP/adenosine (Ado) release, leading to the activation of adenosine A1 receptors (A1Rs) in neighboring neurons to restore normoglycemia; (**C**) The disruption of the brain and muscle ARNT-like protein-1 (BMAL1) signaling in astrocytes impairs energy metabolism. (I) During the night cycle, astrocytes from the suprachiasmatic nucleus (SCN) show increased Ca^{2+} transients, which induce the release of glutamate that binds to N-methyl-D-aspartate receptors (NMDARs) subtype 2C in presynaptic neurons resulting in increased γ-aminobutyric acid (GABA)-mediated neurotransmission; (II) In the day cycle, astrocytes are silent and the glutamate near the synaptic cleft is taken up by astrocytic excitatory amino acid transporters (EAATs), therefore reducing the GABAergic tone onto SNC neurons; (**D**) Ca^{2+} rises in astrocytes from the MBH promote the release of ATP/Ado that acts in presynaptic neurons and/or postsynaptic agouti-related protein/neuropeptide Y (AgRP/NPY) neurons to reduce food consumption; (**E**) Astrocytes from the MBH release the endozepine octadecaneuropeptide (ODN), which acts on its receptor in proopiomelanocortin (POMC) neurons, leading to the activation of the upstream melanocortin-4 receptor (MC4R) signaling to reduce food intake and body weight; (**F**) Fasting/ghrelin may activate astrocytes from the arcuate nucleus of the hypothalamus (ARC), promoting the release of prostaglandin E_2 (PGE_2) to increase the activity of AgRP/NPY neurons, ultimately inducing food intake.

2.2. Brain Glucose Sensing

Astrocytes are highly glycolytic cells [6] and exhibit higher glucose transport and utilization in comparison to neurons [77]. Using a fast-responsive machinery, astrocytes do not only sense extracellular glucose drops but also monitor interstitial glucose presumably to elicit autonomic responses to restore normoglycemia. Among several glucose transporters (GLUTs) expressed in astrocytes [78], GLUT1 is the predominant active isoform at the cell membrane and plays a marked role in basal glucose uptake [79]. Astrocytes also express GLUT2, which has a low affinity for glucose [80–82], providing a wide range of sensitivity to changes in glucose availability. Notably, GLUT2 expression in astrocytes, but not in neurons, has been reported to be necessary and sufficient to increase plasma glucagon levels in response to hypoglycemic conditions in mice [83]. The hypothalamus and the hindbrain are well-known glucose-sensing central areas [84], particularly due to their close location to brain ventricles. Here, we report evidence from the literature that describes how hypothalamic and hindbrain astrocytes may modulate local circuits and systemic metabolism in response to glucose concentration fluctuations.

2.2.1. Hypothalamus

Hypothalamic neurons are capable of directly responding to changes in systemic glucose levels [85,86]. Application of glucose induces Ca^{2+} rises in tanycytes—specialized glial cells lining the floor of the third ventricle located exclusively in the mediobasal hypothalamus—which promote the release of ATP via connexin 43 (Cx43) hemichannels acting on neighboring tanycytes through purinergic P2Y1 receptor to result in cellular activation by an IP_3R-mediated Ca^{2+} signaling [87,88]. Although astrocytic Ca^{2+} rises in response to glucose fluctuations have not been demonstrated in the hypothalamus yet, hypothalamic astrocytes are markedly involved in the regulation of glucose homeostasis [8]. Particularly, insulin signaling in hypothalamic astrocytes is essential for adequate glucose transport into the brain and systemic glucose handling [89]. Other findings have also pointed out that elevated glucose levels lead to reductions in astrocyte coverage on proopiomelanocortin (POMC) neurons—an effect associated with increased excitatory synaptic input onto these neurons [90]. Moreover, hypothalamic astrocytes induce insulin secretion in response to acute intracarotid injection of glucose [91], presumably via Cx43-containing gap-junction functioning [92].

2.2.2. Hindbrain

Similar to the hypothalamus, the hindbrain is strongly involved in counterregulatory responses to hypoglycemia [84]. The nucleus tractus solitarius (NTS) is the primary central site receiving afferent glycolytic inputs from peripheral domains [93]. The NTS also contains astrocytes sensitive to extracellular glucose fluctuations [94], as is the case in neurons [95,96]. Intriguingly, glucose deprivation triggers Ca^{2+} rises in astrocytes via the phospholipase C-IP_3 signaling pathway [97], an effect preceding the Ca^{2+} responses in neighboring neurons [94]. Recent studies have also reported that astrocyte purinergic signaling underlies counterregulatory responses to limited glucose availability via an NTS-arcuate nucleus of the hypothalamus (ARC) circuit. In particular, infusion of 2-deoxyglucose (2-DG), a non-metabolizable glucose analog that mimics hypoglycemic conditions, into the fourth ventricle induces blood glucose elevation in rats, an effect dependent on astrocyte integrity and adenosine A1 receptor (A1R) signaling [98] (Figure 2B). Moreover, functional astrocytes are required for purinergic P2 receptor-dependent activation of tyrosine hydroxylase (TH)-expressing NTS neurons in response to glucose deprivation [99]. Importantly, NTS^{TH} neurons can bidirectionally modulate the electrical activity of orexigenic agouti-related protein/neuropeptide Y (AgRP/NPY) and anorexigenic POMC-expressing neurons in the ARC to promote food intake in response to glucoprivic conditions [100]. Notwithstanding, the ability of ATP-mediated astrocyte signaling in tuning an NTS-ARC neuronal circuitry to ultimately modulate feeding behavior remains to be shown.

2.3. Feeding Circuits

Feeding is driven by an intricate neuronal network that encompasses homeostatic energy balance and hedonic responses [101]. External sensory information, vagal inputs, and circulating nutritional signals converge and are processed in the brain to then adjust feeding behavior according to whole-body energy demands [102]. Remarkably, the melanocortin system has been greatly studied as being the main integrator and control center of hunger circuits [103], and its dysfunction is directly linked with the development of metabolic diseases [104]. Two melanocortin neuron populations in the ARC with opposite functions play essential roles in the control of energy intake and expenditure: activation of AgRP/NPY-expressing neurons induces rapid and marked food seeking and consumption [105–108] whereas activation of POMC-expressing neurons promotes satiety and energy expenditure [105,109,110]. Notably, the postnatal genetic ablation of AgRP/NPY [111,112] or POMC neurons [112–114] results in starvation-induced death or obesity, respectively. A great deal of evidence supports that astrocytes are active players as regulators of these feeding responses by interacting with melanocortin neurons [89,115–117]. Specifically, several studies have reported that astrocytes within the mediobasal hypothalamus (MBH) are capable of responding to energy-related signals, such as hormones and nutrients, in order to modulate neuronal and behavioral responses required for maintaining whole-body energy homeostasis [8]. Indeed, the postnatal ablation of leptin receptors (LepRs) in astrocytes reduces hypothalamic astrogenesis [118] and leads to a retraction in primary processes coverage on melanocortin neurons in the ARC—the latter of which is associated with changes in neuronal excitability and alterations in feeding behavior [116]. Accordingly, astrocyte-specific LepR knockout induces astrogliosis in the hypothalamus of mice, blunts hypothalamic pSTAT3 signaling, and contributes to diet-induced obesity [119]. As with leptin, the disruption of insulin signaling in hypothalamic astrocytes also promotes metabolic alterations mainly due to a defect in brain glucose sensing, resulting in an aberrant systemic glucose handling [89]. Furthermore, the same line of studies has observed that the ingestion of high caloric meals triggers rapid astrocyte-neuron rearrangements, including astrocyte reactivity and alterations in the synaptology of melanocortin neurons [115]; most of these cellular events were observed prior to body weight gain [72], suggesting their potential role in promoting obesity.

2.3.1. Identified Gliotransmitters by Which Astrocytes Regulate Feeding Behavior

ATP/Adenosine

Astrocytes have been reported to mediate feeding control via purinergic gliotransmission. Specifically, it was reported that mice reduce food consumption in response to chemogenetic Ca^{2+}-dependent activation of MBH astrocytes, an effect associated with decreased firing activity of AgRP neurons following adenosine A1R activation [120] (Figure 2D). Accordingly, optogenetic stimulation of MBH astrocytes leads to an increase in extracellular adenosine content, preventing long-term fasting-induced food intake, which is abolished by A1R antagonist injection [121]. These results indicate that MBH astrocytes can release ATP—being converted to adenosine in the extracellular compartment—or adenosine itself [122] to promote anorexigenic effects by decreasing the activity of AgRP/NPY neurons. Nevertheless, it is not clear whether adenosine directly reduces the excitability of AgRP/NPY neurons, presumably by the opening of G-protein-coupled inwardly rectifying K^+ channels associated with A1Rs [123–125], or inhibits presynaptic glutamatergic neurons via A1R activation, as observed in other brain regions [21,126,127]. On the contrary, other studies have shown opposing results using a similar approach with chemogenetic activation of astrocytes, but only those exclusively located in the ARC. In this case, the authors have observed that astrocyte activation promotes food consumption by increasing the orexigenic drive of AgRP/NPY neurons [128], although no potential gliotransmitter involved in this mechanism was reported. The divergent findings when exploring the role of astrocytes in the control of feeding behavior might reside in the intricate nature of neuronal circuits confined to the MBH requiring hypothalamic nuclei with opposing roles in the control of metabolism. Therefore, millimetric stereotaxic variations in the affected area may target distinct astrocytic-neuronal circuits involved in the diverse effects of feeding responses.

Other hypothalamic centered lines of investigation have shown that astrocytes located in the dorsomedial nucleus of the hypothalamus (DMH) are involved in the satiety effect of cholecystokinin (CCK), a well-known anorexigenic gut-derived peptide hormone, via purinergic gliotransmission [129]. Astrocytes respond to CCK through their CCK receptors (CCKRs) expressed along the membrane [130,131] via a Ca^{2+}-dependent mechanism [129,131]. Specifically, CCKR type 2-dependent astrocyte activation triggers the release of ATP that in turn activates P2X receptors in inhibitory neurons, culminating in increased GABA release at the synapse level. Additionally, astrocyte mGluR5 was shown to be necessary for the CCK-mediated effects on GABAergic neurotransmission [129]. Indeed, astrocyte mGluR5 acts as a sensor of synaptic transmission and is markedly involved in astrocyte-neuron gliotransmission [15,16,31]. Overall, these findings suggest that the detection of glutamatergic activity by astrocytes at nearby synaptic clefts may modulate the release of ATP from astrocytes to fine-tune the information processing triggered by CCK signaling in the DMH.

Additional studies have shown the involvement of extra-hypothalamic astrocytes in feeding regulation. In this regard, the selective activation of astrocytes within the brainstem dorsal vagal complex (DVC) induces morphological changes in NTS astrocytes and reduces food-seeking behavior and food consumption, even following overnight fasting [132]. The latter effect was associated with increased c-Fos immunoreactivity, as subrogate marker for neuronal activation, in neurons from the DVC and lateral parabranchial nucleus but not in the paraventricular nucleus of the hypothalamus [132], suggesting that the astrocyte-mediated anorexigenic drive from the brainstem DVC may activate alternative circuitries to the melanocortin system.

Endozepines

The acyl-CoA-binding protein (ACBP) is a ubiquitously expressed cytosolic molecule that acts: (i) in intracellular pathways controlling lipid metabolism [133] or (ii) to generate and release regulatory peptides namely endozepines, such as ACBP itself, octadecaneuropeptide (ODN), and C-terminal octapeptide (OP) [134]. Remarkably, ACBP and ODN expression levels are enriched in the hypothalamus [135,136], particularly in glial cells [137–140]. Indeed, multiple evidence support that astroglial-released endozepines play a key role in the regulation of energy homeostasis. Particularly, it was shown that central administration of ODN or OP decreases food consumption in rodents and fish [140–143] by reducing NPY and enhancing POMC mRNA expression levels in the ARC [144]. Moreover, the hyperphagic response to central infusion of 2-DG is attenuated by co-infusion of OP [137]. In vitro studies from rodents also support that astrocytes are able to release endozepines upon stimulation [145,146]. Amongst several brain areas of action, astrocytes from the MBH were demonstrated to be required for triggering an anorexigenic effect via endozepine release [138]. A selective genetic manipulation of ACBP in astrocytes from the ARC is sufficient to modulate feeding behavior and body weight control. Interestingly, ACBP-expressing astrocytes are in close opposition with POMC neurons in the ARC [138], and ODN or OP application activates hypothalamic POMC neurons, as observed in ex vivo brain slices [138,140]. Given that ODN-induced food intake reduction is abolished in melanocortin-4 receptor (MC4R) knockout mice [138], astrocyte-released endozepines appear to drive an anorexigenic effect via the melanocortin system by modulating POMC neuron excitability and MC4R-dependent signaling transmission (Figure 2E). Likewise, it is thought that ODN binds to melanocortin neurons via an uncharacterized GPCR [142,147]. Central infusion of ODN-GPCR agonists attenuates food intake in mice and fish [138,140,142,143], which is associated with increased excitation of POMC neurons in the ARC of mice [138]. Accordingly, the central administration of an ODN-GPCR antagonist suppresses ODN-induced anorexigenic effects [138,142,143]. Emerging findings also suggest that leptin signaling in tanycytes is required for ODN-induced anti-obesogenic effects in mice [140], indicating the importance of the crosstalk between astrocytes and other glial cells for satiety control. Nevertheless, astrocyte-derived endozepine actions in feeding behavior appear not to be restricted to hypothalamic areas. Astrocytes from the brainstem area postrema and NTS within the DVC have been found to be enriched with ACBP and ODN protein levels [140,148]. Consistent with the hypothalamic centered studies, central administration of ODN or OP induces marked c-Fos immunoreactivity of NTS neurons accompanied by food intake inhibition [140], while blunting the swallowing reflex in mice [148]. Given that ACBP has also been shown to have CNS-independent effects on the promotion of appetite, energy storage, and obesity in mice [149], further investigations should be performed to disentangle the peripheral and central contributions of endozepines in whole-body energy balance.

Prostaglandin E_2

A recent study has shown that fasting, ghrelin administration, or GABA-mediated AgRP neuron signaling increases astrocyte coverage and lowers the number of inhibitory inputs onto AgRP neurons in the ARC, an effect accompanied by depolarization of the membrane potential of neighboring astrocytes [150]. Additionally, the authors observed that the application of astrocyte-derived gliotransmitter prostaglandin E_2 (PGE_2) increases the firing activity of AgRP/NPY neurons from ex vivo brain slices whereas the blockade of PGE_2 receptor EP_2 abolishes ghrelin-induced food consumption [150] (Figure 2F). These findings indicate that rearrangements between surrounding astrocytes and AgRP-dependent circuits in a pre-feeding condition could facilitate the actions of the PGE_2 in the activity of those neurons to promote feeding.

2.4. Circadian Rhythms

The circadian rhythm is present in virtually all cells of almost all living organisms. The cellular clock relies on oscillatory patterns of transcription factors based on a transcription-translation negative feedback loop (TTFL) mechanism. This process ensures the synchronization of biological mechanisms in an adequate time scale according to the active and resting phases [151]. The active phase is markedly characterized by high energy expenditure and nutrient consumption whereas the resting phase is associated with tissue repair, waste clearance, and memory consolidation [151,152]. Notably, the suprachiasmatic nucleus of the hypothalamus (SCN) is one of the major centers in coordinating the whole-body circadian rhythm [151], which influences feeding/fasting patterns and thus metabolic control [153]. In fact, lesions in the SCN elicit alterations in the daily pattern of circulating glucose, fatty acids, and insulin [154]. Besides the marked role of SCN neurons in the control of circadian behavior [151], astrocytes have recently emerged as important players in the regulation of neuronal circuits involved in the circadian rhythms, and in consequence, in whole-body energy metabolism. Specifically, the lack of the clock gene brain and muscle ARNT-like protein-1 (BMAL1) in astrocytes leads to increased food intake, body weight gain, impaired glucose handling, and shorter lifespan in mice [155]. Such changes are associated with alterations in the expression pattern of clock genes in SCN neurons and also affect circadian locomotor activity in mice [156–158]. These effects seem to be driven by the inability of astrocytes to control extracellular GABA content [155,157,159]. Considering that the vast majority of neurons in the SCN are GABAergic [160] and the synchronization of clock neurons in the SCN highly depends on GABAergic transmission [161,162], astrocytes may exert relevant modulation on the inhibitory circuitry dictating circadian oscillations via GABA homeostasis regulation. Indeed, the cooperative orchestration of the activity fluctuations of neurons and astrocytes in the SCN governing the circadian rhythm has gained new insights since the observation that neurons are active during the active phase of the circadian rhythm whereas astrocytes are active during the resting phase, as evidenced by Ca^{2+} measurements [158]. In this study, the authors also showed that Ca^{2+} variations in astrocytes match the release of glutamate, which binds to NMDAR subtype 2C in pre-synaptic GABAergic neurons and enhances the inhibitory drive onto SCN neurons to control behavioral rhythms (Figure 2C). On the other hand, GABAergic tone is reduced during the resting cycle by decreased release of glutamate and elevated glutamate clearance via excitatory amino acid transporters by astrocytes, thereby facilitating SCN neuron activity [158]. Strikingly, astrocytes can sustain their circadian molecular oscillations for many days even in culture [163]. Such oscillations in astrocytes endow autonomous cell-specific molecular patterns in vivo, which are sufficient to control circadian behavior via glutamate-mediated astrocyte gliotransmission within the SCN, regardless of the TTFL functioning in surrounding neurons [164]. Therefore, the circadian rhythm function highly relies on the tuning of GABA-mediated signaling by glutamatergic astrocyte-neuron communication in the SCN.

3. Concluding Remarks

Unlike neurons, showing long and static projections for delivering long-distance messages, astrocytes occupy small domains defined by their finger-like thin processes to influence local circuitries. Therefore, it is not surprising that astrocytes are very plastic cells with multiple functional roles and a high capacity to adapt their cytoarchitecture, gene profile, and activity in response to local neuronal demands. Despite occupying small territories, an astrocyte can physically interact with multiple synapses (estimated number > 100 synapses)—a fact that highlights the vast amount of neuronal information that a single astrocyte can process in a short amount of time. In recent years, notable progress has been made to elucidate many aspects of astrocyte physiology and gliotransmission by using the most advanced neurophysiological techniques. However, the individual distinctions of each astrocyte together with its intricate interactions with neuronal circuitries and the complex Ca^{2+} dynamics at different levels of its compartments have challenged researchers in the field to further understand how communication occurs between astrocytes and neighboring cells. Therefore, studies focused on how astrocytes decode external signals into spatial and temporal gliotransmitter release depending on the microdomain environment and its interactions would also be fundamental to shed more light on these paradigms.

Author Contributions: Writing—original draft preparation, C.D.B.M. and C.G.-C.; writing—review and editing, C.D.B.M. and C.G.-C. All authors have read and agreed to the published version of the manuscript.

Funding: This research was funded by the European Research Council ERC (CGC: STG grant AstroNeuroCrosstalk # 757393).

Acknowledgments: We thank Cassie Holleman (Helmholtz Zentrum München, Germany) for her comments.

Conflicts of Interest: The authors declare no conflict of interest.

References

1. Verkhratsky, A.; Nedergaard, M. Physiology of Astroglia. *Physiol. Rev.* **2018**, *98*, 239–389. [CrossRef]
2. Schaeffer, S.; Iadecola, C. Revisiting the neurovascular unit. *Nat. Neurosci.* **2021**, *24*, 1198–1209. [CrossRef] [PubMed]
3. Falk, S.; Gotz, M. Glial control of neurogenesis. *Curr. Opin. Neurobiol.* **2017**, *47*, 188–195. [CrossRef] [PubMed]
4. Araque, A.; Carmignoto, G.; Haydon, P.G.; Oliet, S.H.; Robitaille, R.; Volterra, A. Gliotransmitters travel in time and space. *Neuron* **2014**, *81*, 728–739. [CrossRef] [PubMed]
5. Chung, W.S.; Allen, N.J.; Eroglu, C. Astrocytes Control Synapse Formation, Function, and Elimination. *Cold Spring Harb. Perspect. Biol.* **2015**, *7*, a020370. [CrossRef] [PubMed]
6. Belanger, M.; Allaman, I.; Magistretti, P.J. Brain energy metabolism: Focus on astrocyte-neuron metabolic cooperation. *Cell Metab.* **2011**, *14*, 724–738. [CrossRef]
7. Bak, L.K.; Schousboe, A.; Waagepetersen, H.S. The glutamate/GABA-glutamine cycle: Aspects of transport, neurotransmitter homeostasis and ammonia transfer. *J. Neurochem.* **2006**, *98*, 641–653. [CrossRef]
8. Garcia-Caceres, C.; Balland, E.; Prevot, V.; Luquet, S.; Woods, S.C.; Koch, M.; Horvath, T.L.; Yi, C.X.; Chowen, J.A.; Verkhratsky, A.; et al. Role of astrocytes, microglia, and tanycytes in brain control of systemic metabolism. *Nat. Neurosci.* **2019**, *22*, 7–14. [CrossRef]
9. Kofuji, P.; Newman, E.A. Potassium buffering in the central nervous system. *Neuroscience* **2004**, *129*, 1045–1056. [CrossRef]

10. Perea, G.; Navarrete, M.; Araque, A. Tripartite synapses: Astrocytes process and control synaptic information. *Trends Neurosci.* **2009**, *32*, 421–431. [CrossRef]
11. Allen, N.J.; Eroglu, C. Cell Biology of Astrocyte-Synapse Interactions. *Neuron* **2017**, *96*, 697–708. [CrossRef]
12. Parpura, V.; Basarsky, T.A.; Liu, F.; Jeftinija, K.; Jeftinija, S.; Haydon, P.G. Glutamate-mediated astrocyte-neuron signalling. *Nature* **1994**, *369*, 744–747. [CrossRef]
13. Savtchouk, I.; Volterra, A. Gliotransmission: Beyond Black-and-White. *J. Neurosci.* **2018**, *38*, 14–25. [CrossRef]
14. Verkhratsky, A.; Matteoli, M.; Parpura, V.; Mothet, J.P.; Zorec, R. Astrocytes as secretory cells of the central nervous system: Idiosyncrasies of vesicular secretion. *EMBO J.* **2016**, *35*, 239–257. [CrossRef]
15. Panatier, A.; Vallee, J.; Haber, M.; Murai, K.K.; Lacaille, J.C.; Robitaille, R. Astrocytes are endogenous regulators of basal transmission at central synapses. *Cell* **2011**, *146*, 785–798. [CrossRef]
16. Wang, X.; Lou, N.; Xu, Q.; Tian, G.F.; Peng, W.G.; Han, X.; Kang, J.; Takano, T.; Nedergaard, M. Astrocytic Ca^{2+} signaling evoked by sensory stimulation in vivo. *Nat. Neurosci.* **2006**, *9*, 816–823. [CrossRef]
17. Perea, G.; Gomez, R.; Mederos, S.; Covelo, A.; Ballesteros, J.J.; Schlosser, L.; Hernandez-Vivanco, A.; Martin-Fernandez, M.; Quintana, R.; Rayan, A.; et al. Activity-dependent switch of GABAergic inhibition into glutamatergic excitation in astrocyte-neuron networks. *eLife* **2016**, *5*, e20362. [CrossRef] [PubMed]
18. Darabid, H.; St-Pierre-See, A.; Robitaille, R. Purinergic-Dependent Glial Regulation of Synaptic Plasticity of Competing Terminals and Synapse Elimination at the Neuromuscular Junction. *Cell Rep.* **2018**, *25*, 2070–2082.e2076. [CrossRef]
19. Papouin, T.; Dunphy, J.M.; Tolman, M.; Dineley, K.T.; Haydon, P.G. Septal Cholinergic Neuromodulation Tunes the Astrocyte-Dependent Gating of Hippocampal NMDA Receptors to Wakefulness. *Neuron* **2017**, *94*, 840–854.e847. [CrossRef]
20. Navarrete, M.; Perea, G.; Fernandez de Sevilla, D.; Gomez-Gonzalo, M.; Nunez, A.; Martin, E.D.; Araque, A. Astrocytes mediate in vivo cholinergic-induced synaptic plasticity. *PLoS Biol.* **2012**, *10*, e1001259. [CrossRef] [PubMed]
21. Corkrum, M.; Covelo, A.; Lines, J.; Bellocchio, L.; Pisansky, M.; Loke, K.; Quintana, R.; Rothwell, P.E.; Lujan, R.; Marsicano, G.; et al. Dopamine-Evoked Synaptic Regulation in the Nucleus Accumbens Requires Astrocyte Activity. *Neuron* **2020**, *105*, 1036–1047.e1035. [CrossRef]
22. Rungta, R.L.; Bernier, L.P.; Dissing-Olesen, L.; Groten, C.J.; LeDue, J.M.; Ko, R.; Drissler, S.; MacVicar, B.A. Ca(2^{+}) transients in astrocyte fine processes occur via Ca(2^{+}) influx in the adult mouse hippocampus. *Glia* **2016**, *64*, 2093–2103. [CrossRef]
23. Brazhe, A.R.; Verisokin, A.Y.; Verveyko, D.V.; Postnov, D.E. Sodium-Calcium Exchanger Can Account for Regenerative Ca(2^{+}) Entry in Thin Astrocyte Processes. *Front. Cell Neurosci.* **2018**, *12*, 250. [CrossRef]
24. Agarwal, A.; Wu, P.H.; Hughes, E.G.; Fukaya, M.; Tischfield, M.A.; Langseth, A.J.; Wirtz, D.; Bergles, D.E. Transient Opening of the Mitochondrial Permeability Transition Pore Induces Microdomain Calcium Transients in Astrocyte Processes. *Neuron* **2017**, *93*, 587–605.e587. [CrossRef]
25. Bindocci, E.; Savtchouk, I.; Liaudet, N.; Becker, D.; Carriero, G.; Volterra, A. Three-dimensional Ca(2^{+}) imaging advances understanding of astrocyte biology. *Science* **2017**, *356*, eaai8185. [CrossRef]
26. Volterra, A.; Liaudet, N.; Savtchouk, I. Astrocyte Ca(2)(+) signalling: An unexpected complexity. *Nat. Rev. Neurosci.* **2014**, *15*, 327–335. [CrossRef]
27. Rusakov, D.A. Disentangling calcium-driven astrocyte physiology. *Nat. Rev. Neurosci.* **2015**, *16*, 226–233. [CrossRef] [PubMed]
28. Semyanov, A.; Henneberger, C.; Agarwal, A. Making sense of astrocytic calcium signals—From acquisition to interpretation. *Nat. Rev. Neurosci.* **2020**, *21*, 551–564. [CrossRef]
29. Araque, A.; Parpura, V.; Sanzgiri, R.P.; Haydon, P.G. Glutamate-dependent astrocyte modulation of synaptic transmission between cultured hippocampal neurons. *Eur. J. Neurosci.* **1998**, *10*, 2129–2142. [CrossRef]
30. Kovacs, A.; Pal, B. Astrocyte-Dependent Slow Inward Currents (SICs) Participate in Neuromodulatory Mechanisms in the Pedunculopontine Nucleus (PPN). *Front. Cell Neurosci.* **2017**, *11*, 16. [CrossRef]
31. D'Ascenzo, M.; Fellin, T.; Terunuma, M.; Revilla-Sanchez, R.; Meaney, D.F.; Auberson, Y.P.; Moss, S.J.; Haydon, P.G. mGluR5 stimulates gliotransmission in the nucleus accumbens. *Proc. Natl. Acad. Sci. USA* **2007**, *104*, 1995–2000. [CrossRef]
32. Navarrete, M.; Perea, G.; Maglio, L.; Pastor, J.; Garcia de Sola, R.; Araque, A. Astrocyte calcium signal and gliotransmission in human brain tissue. *Cereb. Cortex* **2013**, *23*, 1240–1246. [CrossRef]
33. Araque, A.; Li, N.; Doyle, R.T.; Haydon, P.G. SNARE protein-dependent glutamate release from astrocytes. *J. Neurosci.* **2000**, *20*, 666–673. [CrossRef]
34. Fellin, T.; Pascual, O.; Gobbo, S.; Pozzan, T.; Haydon, P.G.; Carmignoto, G. Neuronal synchrony mediated by astrocytic glutamate through activation of extrasynaptic NMDA receptors. *Neuron* **2004**, *43*, 729–743. [CrossRef] [PubMed]
35. Bezzi, P.; Gundersen, V.; Galbete, J.L.; Seifert, G.; Steinhauser, C.; Pilati, E.; Volterra, A. Astrocytes contain a vesicular compartment that is competent for regulated exocytosis of glutamate. *Nat. Neurosci.* **2004**, *7*, 613–620. [CrossRef]
36. Lalo, U.; Palygin, O.; Rasooli-Nejad, S.; Andrew, J.; Haydon, P.G.; Pankratov, Y. Exocytosis of ATP from astrocytes modulates phasic and tonic inhibition in the neocortex. *PLoS Biol.* **2014**, *12*, e1001747. [CrossRef]
37. Angelova, P.R.; Kasymov, V.; Christie, I.; Sheikhbahaei, S.; Turovsky, E.; Marina, N.; Korsak, A.; Zwicker, J.; Teschemacher, A.G.; Ackland, G.L.; et al. Functional Oxygen Sensitivity of Astrocytes. *J. Neurosci.* **2015**, *35*, 10460–10473. [CrossRef] [PubMed]
38. Gourine, A.V.; Kasymov, V.; Marina, N.; Tang, F.; Figueiredo, M.F.; Lane, S.; Teschemacher, A.G.; Spyer, K.M.; Deisseroth, K.; Kasparov, S. Astrocytes control breathing through pH-dependent release of ATP. *Science* **2010**, *329*, 571–575. [CrossRef] [PubMed]

39. Rajani, V.; Zhang, Y.; Jalubula, V.; Rancic, V.; SheikhBahaei, S.; Zwicker, J.D.; Pagliardini, S.; Dickson, C.T.; Ballanyi, K.; Kasparov, S.; et al. Release of ATP by pre-Botzinger complex astrocytes contributes to the hypoxic ventilatory response via a Ca(2$^+$)-dependent P2Y1 receptor mechanism. *J. Physiol.* **2018**, *596*, 3245–3269. [CrossRef] [PubMed]
40. Li, D.; Ropert, N.; Koulakoff, A.; Giaume, C.; Oheim, M. Lysosomes are the major vesicular compartment undergoing Ca2$^+$-regulated exocytosis from cortical astrocytes. *J. Neurosci.* **2008**, *28*, 7648–7658. [CrossRef]
41. Zhang, Z.; Chen, G.; Zhou, W.; Song, A.; Xu, T.; Luo, Q.; Wang, W.; Gu, X.S.; Duan, S. Regulated ATP release from astrocytes through lysosome exocytosis. *Nat. Cell Biol.* **2007**, *9*, 945–953. [CrossRef]
42. Woo, D.H.; Han, K.S.; Shim, J.W.; Yoon, B.E.; Kim, E.; Bae, J.Y.; Oh, S.J.; Hwang, E.M.; Marmorstein, A.D.; Bae, Y.C.; et al. TREK-1 and Best1 channels mediate fast and slow glutamate release in astrocytes upon GPCR activation. *Cell* **2012**, *151*, 25–40. [CrossRef]
43. Park, H.; Han, K.S.; Seo, J.; Lee, J.; Dravid, S.M.; Woo, J.; Chun, H.; Cho, S.; Bae, J.Y.; An, H.; et al. Channel-mediated astrocytic glutamate modulates hippocampal synaptic plasticity by activating postsynaptic NMDA receptors. *Mol. Brain* **2015**, *8*, 7. [CrossRef] [PubMed]
44. Han, K.S.; Woo, J.; Park, H.; Yoon, B.J.; Choi, S.; Lee, C.J. Channel-mediated astrocytic glutamate release via Bestrophin-1 targets synaptic NMDARs. *Mol. Brain* **2013**, *6*, 4. [CrossRef]
45. Lee, S.; Yoon, B.E.; Berglund, K.; Oh, S.J.; Park, H.; Shin, H.S.; Augustine, G.J.; Lee, C.J. Channel-mediated tonic GABA release from glia. *Science* **2010**, *330*, 790–796. [CrossRef]
46. Woo, J.; Min, J.O.; Kang, D.S.; Kim, Y.S.; Jung, G.H.; Park, H.J.; Kim, S.; An, H.; Kwon, J.; Kim, J.; et al. Control of motor coordination by astrocytic tonic GABA release through modulation of excitation/inhibition balance in cerebellum. *Proc. Natl. Acad. Sci. USA* **2018**, *115*, 5004–5009. [CrossRef] [PubMed]
47. Yoon, B.E.; Woo, J.; Chun, Y.E.; Chun, H.; Jo, S.; Bae, J.Y.; An, H.; Min, J.O.; Oh, S.J.; Han, K.S.; et al. Glial GABA, synthesized by monoamine oxidase B, mediates tonic inhibition. *J. Physiol.* **2014**, *592*, 4951–4968. [CrossRef]
48. Jo, S.; Yarishkin, O.; Hwang, Y.J.; Chun, Y.E.; Park, M.; Woo, D.H.; Bae, J.Y.; Kim, T.; Lee, J.; Chun, H.; et al. GABA from reactive astrocytes impairs memory in mouse models of Alzheimer's disease. *Nat. Med.* **2014**, *20*, 886–896. [CrossRef]
49. Pandit, S.; Neupane, C.; Woo, J.; Sharma, R.; Nam, M.H.; Lee, G.S.; Yi, M.H.; Shin, N.; Kim, D.W.; Cho, H.; et al. Bestrophin1-mediated tonic GABA release from reactive astrocytes prevents the development of seizure-prone network in kainate-injected hippocampi. *Glia* **2020**, *68*, 1065–1080. [CrossRef]
50. Chever, O.; Lee, C.Y.; Rouach, N. Astroglial connexin43 hemichannels tune basal excitatory synaptic transmission. *J. Neurosci.* **2014**, *34*, 11228–11232. [CrossRef] [PubMed]
51. Iglesias, R.; Dahl, G.; Qiu, F.; Spray, D.C.; Scemes, E. Pannexin 1: The molecular substrate of astrocyte "hemichannels". *J. Neurosci.* **2009**, *29*, 7092–7097. [CrossRef]
52. Orellana, J.A.; Froger, N.; Ezan, P.; Jiang, J.X.; Bennett, M.V.; Naus, C.C.; Giaume, C.; Saez, J.C. ATP and glutamate released via astroglial connexin 43 hemichannels mediate neuronal death through activation of pannexin 1 hemichannels. *J. Neurochem.* **2011**, *118*, 826–840. [CrossRef]
53. Ye, Z.C.; Wyeth, M.S.; Baltan-Tekkok, S.; Ransom, B.R. Functional hemichannels in astrocytes: A novel mechanism of glutamate release. *J. Neurosci.* **2003**, *23*, 3588–3596. [CrossRef]
54. Torres, A.; Wang, F.; Xu, Q.; Fujita, T.; Dobrowolski, R.; Willecke, K.; Takano, T.; Nedergaard, M. Extracellular Ca(2)(+) acts as a mediator of communication from neurons to glia. *Sci Signal* **2012**, *5*, ra8. [CrossRef]
55. Woo, D.H.; Bae, J.Y.; Nam, M.H.; An, H.; Ju, Y.H.; Won, J.; Choi, J.H.; Hwang, E.M.; Han, K.S.; Bae, Y.C.; et al. Activation of Astrocytic mu-opioid Receptor Elicits Fast Glutamate Release Through TREK-1-Containing K2P Channel in Hippocampal Astrocytes. *Front. Cell Neurosci.* **2018**, *12*, 319. [CrossRef]
56. Obel, L.F.; Muller, M.S.; Walls, A.B.; Sickmann, H.M.; Bak, L.K.; Waagepetersen, H.S.; Schousboe, A. Brain glycogen-new perspectives on its metabolic function and regulation at the subcellular level. *Front. Neuroenerget.* **2012**, *4*, 3. [CrossRef]
57. Magistretti, P.J.; Allaman, I. A cellular perspective on brain energy metabolism and functional imaging. *Neuron* **2015**, *86*, 883–901. [CrossRef]
58. Allen, N.J.; Karadottir, R.; Attwell, D. A preferential role for glycolysis in preventing the anoxic depolarization of rat hippocampal area CA1 pyramidal cells. *J. Neurosci.* **2005**, *25*, 848–859. [CrossRef]
59. Dennis, S.H.; Jaafari, N.; Cimarosti, H.; Hanley, J.G.; Henley, J.M.; Mellor, J.R. Oxygen/glucose deprivation induces a reduction in synaptic AMPA receptors on hippocampal CA3 neurons mediated by mGluR1 and adenosine A3 receptors. *J. Neurosci.* **2011**, *31*, 11941–11952. [CrossRef]
60. Maraula, G.; Traini, C.; Mello, T.; Coppi, E.; Galli, A.; Pedata, F.; Pugliese, A.M. Effects of oxygen and glucose deprivation on synaptic transmission in rat dentate gyrus: Role of A2A adenosine receptors. *Neuropharmacology* **2013**, *67*, 511–520. [CrossRef]
61. Mathiisen, T.M.; Lehre, K.P.; Danbolt, N.C.; Ottersen, O.P. The perivascular astroglial sheath provides a complete covering of the brain microvessels: An electron microscopic 3D reconstruction. *Glia* **2010**, *58*, 1094–1103. [CrossRef]
62. Apte, R.S.; Chen, D.S.; Ferrara, N. VEGF in Signaling and Disease: Beyond Discovery and Development. *Cell* **2019**, *176*, 1248–1264. [CrossRef] [PubMed]
63. Zhang, Z.G.; Zhang, L.; Jiang, Q.; Zhang, R.; Davies, K.; Powers, C.; Bruggen, N.; Chopp, M. VEGF enhances angiogenesis and promotes blood-brain barrier leakage in the ischemic brain. *J. Clin. Invest.* **2000**, *106*, 829–838. [CrossRef]
64. Argaw, A.T.; Gurfein, B.T.; Zhang, Y.; Zameer, A.; John, G.R. VEGF-mediated disruption of endothelial CLN-5 promotes blood-brain barrier breakdown. *Proc. Natl. Acad. Sci. USA* **2009**, *106*, 1977–1982. [CrossRef] [PubMed]

65. Valable, S.; Montaner, J.; Bellail, A.; Berezowski, V.; Brillault, J.; Cecchelli, R.; Divoux, D.; Mackenzie, E.T.; Bernaudin, M.; Roussel, S.; et al. VEGF-induced BBB permeability is associated with an MMP-9 activity increase in cerebral ischemia: Both effects decreased by Ang-1. *J. Cereb. Blood Flow Metab.* **2005**, *25*, 1491–1504. [CrossRef] [PubMed]
66. Fischer, S.; Wobben, M.; Marti, H.H.; Renz, D.; Schaper, W. Hypoxia-induced hyperpermeability in brain microvessel endothelial cells involves VEGF-mediated changes in the expression of zonula occludens-1. *Microvasc. Res.* **2002**, *63*, 70–80. [CrossRef] [PubMed]
67. Boer, K.; Troost, D.; Spliet, W.G.; van Rijen, P.C.; Gorter, J.A.; Aronica, E. Cellular distribution of vascular endothelial growth factor A (VEGFA) and B (VEGFB) and VEGF receptors 1 and 2 in focal cortical dysplasia type IIB. *Acta Neuropathol.* **2008**, *115*, 683–696. [CrossRef]
68. Argaw, A.T.; Asp, L.; Zhang, J.; Navrazhina, K.; Pham, T.; Mariani, J.N.; Mahase, S.; Dutta, D.J.; Seto, J.; Kramer, E.G.; et al. Astrocyte-derived VEGF-A drives blood-brain barrier disruption in CNS inflammatory disease. *J. Clin. Invest.* **2012**, *122*, 2454–2468. [CrossRef] [PubMed]
69. Chapouly, C.; Tadesse Argaw, A.; Horng, S.; Castro, K.; Zhang, J.; Asp, L.; Loo, H.; Laitman, B.M.; Mariani, J.N.; Straus Farber, R.; et al. Astrocytic TYMP and VEGFA drive blood-brain barrier opening in inflammatory central nervous system lesions. *Brain* **2015**, *138*, 1548–1567. [CrossRef]
70. Li, Y.N.; Pan, R.; Qin, X.J.; Yang, W.L.; Qi, Z.; Liu, W.; Liu, K.J. Ischemic neurons activate astrocytes to disrupt endothelial barrier via increasing VEGF expression. *J. Neurochem.* **2014**, *129*, 120–129. [CrossRef] [PubMed]
71. Salhia, B.; Angelov, L.; Roncari, L.; Wu, X.; Shannon, P.; Guha, A. Expression of vascular endothelial growth factor by reactive astrocytes and associated neoangiogenesis. *Brain Res.* **2000**, *883*, 87–97. [CrossRef]
72. Thaler, J.P.; Yi, C.X.; Schur, E.A.; Guyenet, S.J.; Hwang, B.H.; Dietrich, M.O.; Zhao, X.; Sarruf, D.A.; Izgur, V.; Maravilla, K.R.; et al. Obesity is associated with hypothalamic injury in rodents and humans. *J. Clin. Invest.* **2012**, *122*, 153–162. [CrossRef]
73. Yi, C.X.; Gericke, M.; Kruger, M.; Alkemade, A.; Kabra, D.G.; Hanske, S.; Filosa, J.; Pfluger, P.; Bingham, N.; Woods, S.C.; et al. High calorie diet triggers hypothalamic angiopathy. *Mol. Metab.* **2012**, *1*, 95–100. [CrossRef]
74. Salameh, T.S.; Mortell, W.G.; Logsdon, A.F.; Butterfield, D.A.; Banks, W.A. Disruption of the hippocampal and hypothalamic blood-brain barrier in a diet-induced obese model of type II diabetes: Prevention and treatment by the mitochondrial carbonic anhydrase inhibitor, topiramate. *Fluids Barriers CNS* **2019**, *16*, 1. [CrossRef]
75. Gruber, T.; Pan, C.; Contreras, R.E.; Wiedemann, T.; Morgan, D.A.; Skowronski, A.A.; Lefort, S.; De Bernardis Murat, C.; Le Thuc, O.; Legutko, B.; et al. Obesity-associated hyperleptinemia alters the gliovascular interface of the hypothalamus to promote hypertension. *Cell Metab.* **2021**, *33*, 1155–1170 e1110. [CrossRef]
76. Folkow, B.; Grimby, G.; Thulesius, O. Adaptive structural changes of the vascular walls in hypertension and their relation to the control of the peripheral resistance. *Acta Physiol. Scand.* **1958**, *44*, 255–272. [CrossRef]
77. Jakoby, P.; Schmidt, E.; Ruminot, I.; Gutierrez, R.; Barros, L.F.; Deitmer, J.W. Higher transport and metabolism of glucose in astrocytes compared with neurons: A multiphoton study of hippocampal and cerebellar tissue slices. *Cereb. Cortex* **2014**, *24*, 222–231. [CrossRef]
78. Koepsell, H. Glucose transporters in brain in health and disease. *Pflugers Arch.* **2020**, *472*, 1299–1343. [CrossRef] [PubMed]
79. Simpson, I.A.; Carruthers, A.; Vannucci, S.J. Supply and demand in cerebral energy metabolism: The role of nutrient transporters. *J. Cereb. Blood Flow Metab.* **2007**, *27*, 1766–1791. [CrossRef] [PubMed]
80. Leloup, C.; Arluison, M.; Lepetit, N.; Cartier, N.; Marfaing-Jallat, P.; Ferre, P.; Penicaud, L. Glucose transporter 2 (GLUT 2): Expression in specific brain nuclei. *Brain Res.* **1994**, *638*, 221–226. [CrossRef]
81. Thorens, B. GLUT2, glucose sensing and glucose homeostasis. *Diabetologia* **2015**, *58*, 221–232. [CrossRef] [PubMed]
82. Uldry, M.; Ibberson, M.; Hosokawa, M.; Thorens, B. GLUT2 is a high affinity glucosamine transporter. *FEBS Lett.* **2002**, *524*, 199–203. [CrossRef]
83. Marty, N.; Dallaporta, M.; Foretz, M.; Emery, M.; Tarussio, D.; Bady, I.; Binnert, C.; Beermann, F.; Thorens, B. Regulation of glucagon secretion by glucose transporter type 2 (glut2) and astrocyte-dependent glucose sensors. *J. Clin. Invest.* **2005**, *115*, 3545–3553. [CrossRef]
84. Donovan, C.M.; Watts, A.G. Peripheral and central glucose sensing in hypoglycemic detection. *Physiology* **2014**, *29*, 314–324. [CrossRef] [PubMed]
85. Burdakov, D.; Luckman, S.M.; Verkhratsky, A. Glucose-sensing neurons of the hypothalamus. *Philos. Trans. R Soc. Lond. B Biol. Sci.* **2005**, *360*, 2227–2235. [CrossRef]
86. Fioramonti, X.; Contie, S.; Song, Z.; Routh, V.H.; Lorsignol, A.; Penicaud, L. Characterization of glucosensing neuron subpopulations in the arcuate nucleus: Integration in neuropeptide Y and pro-opio melanocortin networks? *Diabetes* **2007**, *56*, 1219–1227. [CrossRef]
87. Orellana, J.A.; Saez, P.J.; Cortes-Campos, C.; Elizondo, R.J.; Shoji, K.F.; Contreras-Duarte, S.; Figueroa, V.; Velarde, V.; Jiang, J.X.; Nualart, F.; et al. Glucose increases intracellular free Ca(2^+) in tanycytes via ATP released through connexin 43 hemichannels. *Glia* **2012**, *60*, 53–68. [CrossRef] [PubMed]
88. Frayling, C.; Britton, R.; Dale, N. ATP-mediated glucosensing by hypothalamic tanycytes. *J. Physiol.* **2011**, *589*, 2275–2286. [CrossRef]
89. Garcia-Caceres, C.; Quarta, C.; Varela, L.; Gao, Y.; Gruber, T.; Legutko, B.; Jastroch, M.; Johansson, P.; Ninkovic, J.; Yi, C.X.; et al. Astrocytic Insulin Signaling Couples Brain Glucose Uptake with Nutrient Availability. *Cell* **2016**, *166*, 867–880. [CrossRef]

90. Nuzzaci, D.; Cansell, C.; Lienard, F.; Nedelec, E.; Ben Fradj, S.; Castel, J.; Foppen, E.; Denis, R.; Grouselle, D.; Laderriere, A.; et al. Postprandial Hyperglycemia Stimulates Neuroglial Plasticity in Hypothalamic POMC Neurons after a Balanced Meal. *Cell Rep.* **2020**, *30*, 3067–3078.e3065. [CrossRef] [PubMed]
91. Guillod-Maximin, E.; Lorsignol, A.; Alquier, T.; Penicaud, L. Acute intracarotid glucose injection towards the brain induces specific c-fos activation in hypothalamic nuclei: Involvement of astrocytes in cerebral glucose-sensing in rats. *J. Neuroendocrinol.* **2004**, *16*, 464–471. [CrossRef]
92. Allard, C.; Carneiro, L.; Grall, S.; Cline, B.H.; Fioramonti, X.; Chretien, C.; Baba-Aissa, F.; Giaume, C.; Penicaud, L.; Leloup, C. Hypothalamic astroglial connexins are required for brain glucose sensing-induced insulin secretion. *J. Cereb. Blood Flow Metab.* **2014**, *34*, 339–346. [CrossRef] [PubMed]
93. Grill, H.J.; Hayes, M.R. The nucleus tractus solitarius: A portal for visceral afferent signal processing, energy status assessment and integration of their combined effects on food intake. *Int. J. Obes.* **2009**, *33*, S11–S15. [CrossRef]
94. McDougal, D.H.; Hermann, G.E.; Rogers, R.C. Astrocytes in the nucleus of the solitary tract are activated by low glucose or glucoprivation: Evidence for glial involvement in glucose homeostasis. *Front. Neurosci.* **2013**, *7*, 249. [CrossRef]
95. De Bernardis Murat, C.; Leao, R.M. A voltage-dependent depolarization induced by low external glucose in neurons of the nucleus of the tractus solitarius: Interaction with KATP channels. *J. Physiol.* **2019**, *597*, 2515–2532. [CrossRef]
96. Lamy, C.M.; Sanno, H.; Labouebe, G.; Picard, A.; Magnan, C.; Chatton, J.Y.; Thorens, B. Hypoglycemia-activated GLUT2 neurons of the nucleus tractus solitarius stimulate vagal activity and glucagon secretion. *Cell Metab.* **2014**, *19*, 527–538. [CrossRef] [PubMed]
97. Rogers, R.C.; Burke, S.J.; Collier, J.J.; Ritter, S.; Hermann, G.E. Evidence that hindbrain astrocytes in the rat detect low glucose with a glucose transporter 2-phospholipase C-calcium release mechanism. *Am J. Physiol. Regul. Integr. Comp. Physiol.* **2020**, *318*, R38–R48. [CrossRef] [PubMed]
98. Rogers, R.C.; Ritter, S.; Hermann, G.E. Hindbrain cytoglucopenia-induced increases in systemic blood glucose levels by 2-deoxyglucose depend on intact astrocytes and adenosine release. *Am J. Physiol. Regul. Integr. Comp. Physiol.* **2016**, *310*, R1102–R1108. [CrossRef]
99. Rogers, R.C.; McDougal, D.H.; Ritter, S.; Qualls-Creekmore, E.; Hermann, G.E. Response of catecholaminergic neurons in the mouse hindbrain to glucoprivic stimuli is astrocyte dependent. *Am J. Physiol. Regul. Integr. Comp. Physiol.* **2018**, *315*, R153–R164. [CrossRef]
100. Aklan, I.; Sayar Atasoy, N.; Yavuz, Y.; Ates, T.; Coban, I.; Koksalar, F.; Filiz, G.; Topcu, I.C.; Oncul, M.; Dilsiz, P.; et al. NTS Catecholamine Neurons Mediate Hypoglycemic Hunger via Medial Hypothalamic Feeding Pathways. *Cell Metab.* **2020**, *31*, 313–326.e315. [CrossRef]
101. Rossi, M.A.; Stuber, G.D. Overlapping Brain Circuits for Homeostatic and Hedonic Feeding. *Cell Metab.* **2018**, *27*, 42–56. [CrossRef]
102. Zeltser, L.M. Feeding circuit development and early-life influences on future feeding behaviour. *Nat. Rev. Neurosci.* **2018**, *19*, 302–316. [CrossRef] [PubMed]
103. Dietrich, M.O.; Horvath, T.L. Hypothalamic control of energy balance: Insights into the role of synaptic plasticity. *Trends Neurosci.* **2013**, *36*, 65–73. [CrossRef]
104. Timper, K.; Bruning, J.C. Hypothalamic circuits regulating appetite and energy homeostasis: Pathways to obesity. *Dis. Model. Mech.* **2017**, *10*, 679–689. [CrossRef] [PubMed]
105. Aponte, Y.; Atasoy, D.; Sternson, S.M. AGRP neurons are sufficient to orchestrate feeding behavior rapidly and without training. *Nat. Neurosci.* **2011**, *14*, 351–355. [CrossRef] [PubMed]
106. Dietrich, M.O.; Zimmer, M.R.; Bober, J.; Horvath, T.L. Hypothalamic Agrp neurons drive stereotypic behaviors beyond feeding. *Cell* **2015**, *160*, 1222–1232. [CrossRef]
107. Krashes, M.J.; Koda, S.; Ye, C.; Rogan, S.C.; Adams, A.C.; Cusher, D.S.; Maratos-Flier, E.; Roth, B.L.; Lowell, B.B. Rapid, reversible activation of AgRP neurons drives feeding behavior in mice. *J. Clin. Investig.* **2011**, *121*, 1424–1428. [CrossRef]
108. Atasoy, D.; Betley, J.N.; Su, H.H.; Sternson, S.M. Deconstruction of a neural circuit for hunger. *Nature* **2012**, *488*, 172–177. [CrossRef] [PubMed]
109. Biglari, N.; Gaziano, I.; Schumacher, J.; Radermacher, J.; Paeger, L.; Klemm, P.; Chen, W.; Corneliussen, S.; Wunderlich, C.M.; Sue, M.; et al. Functionally distinct POMC-expressing neuron subpopulations in hypothalamus revealed by intersectional targeting. *Nat. Neurosci.* **2021**, *24*, 913–929. [CrossRef]
110. Zhan, C.; Zhou, J.; Feng, Q.; Zhang, J.E.; Lin, S.; Bao, J.; Wu, P.; Luo, M. Acute and long-term suppression of feeding behavior by POMC neurons in the brainstem and hypothalamus, respectively. *J. Neurosci.* **2013**, *33*, 3624–3632. [CrossRef]
111. Luquet, S.; Perez, F.A.; Hnasko, T.S.; Palmiter, R.D. NPY/AgRP neurons are essential for feeding in adult mice but can be ablated in neonates. *Science* **2005**, *310*, 683–685. [CrossRef] [PubMed]
112. Xu, A.W.; Kaelin, C.B.; Morton, G.J.; Ogimoto, K.; Stanhope, K.; Graham, J.; Baskin, D.G.; Havel, P.; Schwartz, M.W.; Barsh, G.S. Effects of hypothalamic neurodegeneration on energy balance. *PLoS Biol.* **2005**, *3*, e415. [CrossRef] [PubMed]
113. Challis, B.G.; Coll, A.P.; Yeo, G.S.; Pinnock, S.B.; Dickson, S.L.; Thresher, R.R.; Dixon, J.; Zahn, D.; Rochford, J.J.; White, A.; et al. Mice lacking pro-opiomelanocortin are sensitive to high-fat feeding but respond normally to the acute anorectic effects of peptide-YY(3-36). *Proc. Natl. Acad. Sci. USA* **2004**, *101*, 4695–4700. [CrossRef]
114. Bumaschny, V.F.; Yamashita, M.; Casas-Cordero, R.; Otero-Corchon, V.; de Souza, F.S.; Rubinstein, M.; Low, M.J. Obesity-programmed mice are rescued by early genetic intervention. *J. Clin. Investig.* **2012**, *122*, 4203–4212. [CrossRef]

115. Horvath, T.L.; Sarman, B.; Garcia-Caceres, C.; Enriori, P.J.; Sotonyi, P.; Shanabrough, M.; Borok, E.; Argente, J.; Chowen, J.A.; Perez-Tilve, D.; et al. Synaptic input organization of the melanocortin system predicts diet-induced hypothalamic reactive gliosis and obesity. *Proc. Natl. Acad. Sci. USA* **2010**, *107*, 14875–14880. [CrossRef]
116. Kim, J.G.; Suyama, S.; Koch, M.; Jin, S.; Argente-Arizon, P.; Argente, J.; Liu, Z.W.; Zimmer, M.R.; Jeong, J.K.; Szigeti-Buck, K.; et al. Leptin signaling in astrocytes regulates hypothalamic neuronal circuits and feeding. *Nat. Neurosci.* **2014**, *17*, 908–910. [CrossRef] [PubMed]
117. Gao, Y.; Layritz, C.; Legutko, B.; Eichmann, T.O.; Laperrousaz, E.; Moulle, V.S.; Cruciani-Guglielmacci, C.; Magnan, C.; Luquet, S.; Woods, S.C.; et al. Disruption of Lipid Uptake in Astroglia Exacerbates Diet-Induced Obesity. *Diabetes* **2017**, *66*, 2555–2563. [CrossRef]
118. Rottkamp, D.M.; Rudenko, I.A.; Maier, M.T.; Roshanbin, S.; Yulyaningsih, E.; Perez, L.; Valdearcos, M.; Chua, S.; Koliwad, S.K.; Xu, A.W. Leptin potentiates astrogenesis in the developing hypothalamus. *Mol. Metab.* **2015**, *4*, 881–889. [CrossRef]
119. Wang, Y.; Hsuchou, H.; He, Y.; Kastin, A.J.; Pan, W. Role of Astrocytes in Leptin Signaling. *J. Mol. Neurosci.* **2015**, *56*, 829–839. [CrossRef]
120. Yang, L.; Qi, Y.; Yang, Y. Astrocytes control food intake by inhibiting AGRP neuron activity via adenosine A1 receptors. *Cell Rep.* **2015**, *11*, 798–807. [CrossRef]
121. Sweeney, P.; Qi, Y.; Xu, Z.; Yang, Y. Activation of hypothalamic astrocytes suppresses feeding without altering emotional states. *Glia* **2016**, *64*, 2263–2273. [CrossRef] [PubMed]
122. Chen, J.F.; Eltzschig, H.K.; Fredholm, B.B. Adenosine receptors as drug targets-what are the challenges? *Nat. Rev. Drug Discov.* **2013**, *12*, 265–286. [CrossRef]
123. Ulrich, D.; Huguenard, J.R. Purinergic inhibition of GABA and glutamate release in the thalamus: Implications for thalamic network activity. *Neuron* **1995**, *15*, 909–918. [CrossRef]
124. Thompson, S.M.; Haas, H.L.; Gahwiler, B.H. Comparison of the actions of adenosine at pre- and postsynaptic receptors in the rat hippocampus in vitro. *J. Physiol.* **1992**, *451*, 347–363. [CrossRef]
125. Luscher, C.; Jan, L.Y.; Stoffel, M.; Malenka, R.C.; Nicoll, R.A. G protein-coupled inwardly rectifying K+ channels (GIRKs) mediate postsynaptic but not presynaptic transmitter actions in hippocampal neurons. *Neuron* **1997**, *19*, 687–695. [CrossRef]
126. Brambilla, D.; Chapman, D.; Greene, R. Adenosine mediation of presynaptic feedback inhibition of glutamate release. *Neuron* **2005**, *46*, 275–283. [CrossRef]
127. Covelo, A.; Araque, A. Neuronal activity determines distinct gliotransmitter release from a single astrocyte. *eLife* **2018**, *7*, e32237. [CrossRef] [PubMed]
128. Chen, N.; Sugihara, H.; Kim, J.; Fu, Z.; Barak, B.; Sur, M.; Feng, G.; Han, W. Direct modulation of GFAP-expressing glia in the arcuate nucleus bi-directionally regulates feeding. *eLife* **2016**, *5*, e18716. [CrossRef]
129. Crosby, K.M.; Murphy-Royal, C.; Wilson, S.A.; Gordon, G.R.; Bains, J.S.; Pittman, Q.J. Cholecystokinin Switches the Plasticity of GABA Synapses in the Dorsomedial Hypothalamus via Astrocytic ATP Release. *J. Neurosci.* **2018**, *38*, 8515–8525. [CrossRef] [PubMed]
130. Hosli, L.; Hosli, E.; Winter, T.; Kaser, H. Electrophysiological evidence for the presence of receptors for cholecystokinin and bombesin on cultured astrocytes of rat central nervous system. *Neurosci. Lett.* **1993**, *163*, 145–147. [CrossRef]
131. Muller, W.; Heinemann, U.; Berlin, K. Cholecystokinin activates CCKB-receptor-mediated Ca-signaling in hippocampal astrocytes. *J. Neurophysiol.* **1997**, *78*, 1997–2001. [CrossRef] [PubMed]
132. MacDonald, A.J.; Holmes, F.E.; Beall, C.; Pickering, A.E.; Ellacott, K.L.J. Regulation of food intake by astrocytes in the brainstem dorsal vagal complex. *Glia* **2020**, *68*, 1241–1254. [CrossRef] [PubMed]
133. Neess, D.; Bek, S.; Engelsby, H.; Gallego, S.F.; Faergeman, N.J. Long-chain acyl-CoA esters in metabolism and signaling: Role of acyl-CoA binding proteins. *Prog. Lipid Res.* **2015**, *59*, 1–25. [CrossRef]
134. Tonon, M.C.; Vaudry, H.; Chuquet, J.; Guillebaud, F.; Fan, J.; Masmoudi-Kouki, O.; Vaudry, D.; Lanfray, D.; Morin, F.; Prevot, V.; et al. Endozepines and their receptors: Structure, functions and pathophysiological significance. *Pharmacol. Ther.* **2020**, *208*, 107386. [CrossRef]
135. Alho, H.; Costa, E.; Ferrero, P.; Fujimoto, M.; Cosenza-Murphy, D.; Guidotti, A. Diazepam-binding inhibitor: A neuropeptide located in selected neuronal populations of rat brain. *Science* **1985**, *229*, 179–182. [CrossRef]
136. Malagon, M.; Vaudry, H.; Van Strien, F.; Pelletier, G.; Gracia-Navarro, F.; Tonon, M.C. Ontogeny of diazepam-binding inhibitor-related peptides (endozepines) in the rat brain. *Neuroscience* **1993**, *57*, 777–786. [CrossRef]
137. Lanfray, D.; Arthaud, S.; Ouellet, J.; Compere, V.; Do Rego, J.L.; Leprince, J.; Lefranc, B.; Castel, H.; Bouchard, C.; Monge-Roffarello, B.; et al. Gliotransmission and brain glucose sensing: Critical role of endozepines. *Diabetes* **2013**, *62*, 801–810. [CrossRef] [PubMed]
138. Bouyakdan, K.; Martin, H.; Lienard, F.; Budry, L.; Taib, B.; Rodaros, D.; Chretien, C.; Biron, E.; Husson, Z.; Cota, D.; et al. The gliotransmitter ACBP controls feeding and energy homeostasis via the melanocortin system. *J. Clin. Invest.* **2019**, *129*, 2417–2430. [CrossRef]
139. Tonon, M.C.; Desy, L.; Nicolas, P.; Vaudry, H.; Pelletier, G. Immunocytochemical localization of the endogenous benzodiazepine ligand octadecaneuropeptide (ODN) in the rat brain. *Neuropeptides* **1990**, *15*, 17–24. [CrossRef]
140. Guillebaud, F.; Duquenne, M.; Djelloul, M.; Pierre, C.; Poirot, K.; Roussel, G.; Riad, S.; Lanfray, D.; Morin, F.; Jean, A.; et al. Glial Endozepines Reverse High-Fat Diet-Induced Obesity by Enhancing Hypothalamic Response to Peripheral Leptin. *Mol. Neurobiol.* **2020**, *57*, 3307–3333. [CrossRef]

141. De Mateos-Verchere, J.G.; Leprince, J.; Tonon, M.C.; Vaudry, H.; Costentin, J. The octadecaneuropeptide [diazepam-binding inhibitor (33–50)] exerts potent anorexigenic effects in rodents. *Eur. J. Pharmacol.* **2001**, *414*, 225–231. [CrossRef]
142. Do Rego, J.C.; Orta, M.H.; Leprince, J.; Tonon, M.C.; Vaudry, H.; Costentin, J. Pharmacological characterization of the receptor mediating the anorexigenic action of the octadecaneuropeptide: Evidence for an endozepinergic tone regulating food intake. *Neuropsychopharmacology* **2007**, *32*, 1641–1648. [CrossRef]
143. Matsuda, K.; Wada, K.; Miura, T.; Maruyama, K.; Shimakura, S.I.; Uchiyama, M.; Leprince, J.; Tonon, M.C.; Vaudry, H. Effect of the diazepam-binding inhibitor-derived peptide, octadecaneuropeptide, on food intake in goldfish. *Neuroscience* **2007**, *150*, 425–432. [CrossRef] [PubMed]
144. Compere, V.; Li, S.; Leprince, J.; Tonon, M.C.; Vaudry, H.; Pelletier, G. Effect of intracerebroventricular administration of the octadecaneuropeptide on the expression of pro-opiomelanocortin, neuropeptide Y and corticotropin-releasing hormone mRNAs in rat hypothalamus. *J. Neuroendocrinol.* **2003**, *15*, 197–203. [CrossRef] [PubMed]
145. Loomis, W.F.; Behrens, M.M.; Williams, M.E.; Anjard, C. Pregnenolone sulfate and cortisol induce secretion of acyl-CoA-binding protein and its conversion into endozepines from astrocytes. *J. Biol. Chem.* **2010**, *285*, 21359–21365. [CrossRef] [PubMed]
146. Tokay, T.; Hachem, R.; Masmoudi-Kouki, O.; Gandolfo, P.; Desrues, L.; Leprince, J.; Castel, H.; Diallo, M.; Amri, M.; Vaudry, H.; et al. Beta-amyloid peptide stimulates endozepine release in cultured rat astrocytes through activation of N-formyl peptide receptors. *Glia* **2008**, *56*, 1380–1389. [CrossRef]
147. Leprince, J.; Oulyadi, H.; Vaudry, D.; Masmoudi, O.; Gandolfo, P.; Patte, C.; Costentin, J.; Fauchere, J.L.; Davoust, D.; Vaudry, H.; et al. Synthesis, conformational analysis and biological activity of cyclic analogs of the octadecaneuropeptide ODN. Design of a potent endozepine antagonist. *Eur. J. Biochem.* **2001**, *268*, 6045–6057. [CrossRef] [PubMed]
148. Guillebaud, F.; Girardet, C.; Abysique, A.; Gaige, S.; Barbouche, R.; Verneuil, J.; Jean, A.; Leprince, J.; Tonon, M.C.; Dallaporta, M.; et al. Glial Endozepines Inhibit Feeding-Related Autonomic Functions by Acting at the Brainstem Level. *Front. Neurosci.* **2017**, *11*, 308. [CrossRef]
149. Bravo-San Pedro, J.M.; Sica, V.; Martins, I.; Pol, J.; Loos, F.; Maiuri, M.C.; Durand, S.; Bossut, N.; Aprahamian, F.; Anagnostopoulos, G.; et al. Acyl-CoA-Binding Protein Is a Lipogenic Factor that Triggers Food Intake and Obesity. *Cell Metab.* **2019**, *30*, 754–767.e759. [CrossRef]
150. Varela, L.; Stutz, B.; Song, J.E.; Kim, J.G.; Liu, Z.W.; Gao, X.B.; Horvath, T.L. Hunger-promoting AgRP neurons trigger an astrocyte-mediated feed-forward autoactivation loop in mice. *J. Clin. Invest.* **2021**, *131*. [CrossRef]
151. Hastings, M.H.; Maywood, E.S.; Brancaccio, M. Generation of circadian rhythms in the suprachiasmatic nucleus. *Nat. Rev. Neurosci.* **2018**, *19*, 453–469. [CrossRef] [PubMed]
152. Challet, E. The circadian regulation of food intake. *Nat. Rev. Endocrinol.* **2019**, *15*, 393–405. [CrossRef]
153. Reinke, H.; Asher, G. Crosstalk between metabolism and circadian clocks. *Nat. Rev. Mol. Cell. Biol.* **2019**, *20*, 227–241. [CrossRef]
154. Yamamoto, H.; Nagai, K.; Nakagawa, H. Role of SCN in daily rhythms of plasma glucose, FFA, insulin and glucagon. *Chronobiol. Int.* **1987**, *4*, 483–491. [CrossRef]
155. Barca-Mayo, O.; Boender, A.J.; Armirotti, A.; De Pietri Tonelli, D. Deletion of astrocytic BMAL1 results in metabolic imbalance and shorter lifespan in mice. *Glia* **2020**, *68*, 1131–1147. [CrossRef]
156. Tso, C.F.; Simon, T.; Greenlaw, A.C.; Puri, T.; Mieda, M.; Herzog, E.D. Astrocytes Regulate Daily Rhythms in the Suprachiasmatic Nucleus and Behavior. *Curr. Biol.* **2017**, *27*, 1055–1061. [CrossRef] [PubMed]
157. Barca-Mayo, O.; Pons-Espinal, M.; Follert, P.; Armirotti, A.; Berdondini, L.; De Pietri Tonelli, D. Astrocyte deletion of Bmal1 alters daily locomotor activity and cognitive functions via GABA signalling. *Nat. Commun.* **2017**, *8*, 14336. [CrossRef] [PubMed]
158. Brancaccio, M.; Patton, A.P.; Chesham, J.E.; Maywood, E.S.; Hastings, M.H. Astrocytes Control Circadian Timekeeping in the Suprachiasmatic Nucleus via Glutamatergic Signaling. *Neuron* **2017**, *93*, 1420–1435.e1425. [CrossRef]
159. Moldavan, M.; Cravetchi, O.; Allen, C.N. GABA transporters regulate tonic and synaptic GABAA receptor-mediated currents in the suprachiasmatic nucleus neurons. *J. Neurophysiol.* **2017**, *118*, 3092–3106. [CrossRef]
160. Abrahamson, E.E.; Moore, R.Y. Suprachiasmatic nucleus in the mouse: Retinal innervation, intrinsic organization and efferent projections. *Brain Res.* **2001**, *916*, 172–191. [CrossRef]
161. Liu, C.; Reppert, S.M. GABA synchronizes clock cells within the suprachiasmatic circadian clock. *Neuron* **2000**, *25*, 123–128. [CrossRef]
162. Maejima, T.; Tsuno, Y.; Miyazaki, S.; Tsuneoka, Y.; Hasegawa, E.; Islam, M.T.; Enoki, R.; Nakamura, T.J.; Mieda, M. GABA from vasopressin neurons regulates the time at which suprachiasmatic nucleus molecular clocks enable circadian behavior. *Proc. Natl. Acad. Sci. USA* **2021**, *118*, e2010168118. [CrossRef] [PubMed]
163. Prolo, L.M.; Takahashi, J.S.; Herzog, E.D. Circadian rhythm generation and entrainment in astrocytes. *J. Neurosci.* **2005**, *25*, 404–408. [CrossRef] [PubMed]
164. Brancaccio, M.; Edwards, M.D.; Patton, A.P.; Smyllie, N.J.; Chesham, J.E.; Maywood, E.S.; Hastings, M.H. Cell-autonomous clock of astrocytes drives circadian behavior in mammals. *Science* **2019**, *363*, 187–192. [CrossRef]

MDPI

Review

L-Lactate: Food for Thoughts, Memory and Behavior

María Fernanda Veloz Castillo [1], Pierre J. Magistretti [1,*] and Corrado Calì [2,3,*]

1 Biological and Environmental Science and Engineering Division, King Abdullah University of Science and Technology, Thuwal 23955-6900, Saudi Arabia; maria.velozcastillo@kaust.edu.sa
2 Dipartimento di Neuroscienze "Rita Levi Montalcini", Università degli Studi di Torino, 10124 Torino, Italy
3 Neuroscience Institute Cavalieri Ottolenghi, 10043 Orbassano, Italy
* Correspondence: pierre.magistretti@kaust.edu.sa (P.J.M.); corrado.cali@unito.it (C.C.)

Abstract: More and more evidence shows how brain energy metabolism is the linkage between physiological and morphological synaptic plasticity and memory consolidation. Different types of memory are associated with differential inputs, each with specific inputs that are upstream diverse molecular cascades depending on the receptor activity. No matter how heterogeneous the response is, energy availability represents the lowest common denominator since all these mechanisms are energy consuming and the brain networks adapt their performance accordingly. Astrocytes exert a primary role in this sense by acting as an energy buffer; glycogen granules, a mechanism to store glucose, are redistributed at glance and conveyed to neurons via the Astrocyte–Neuron Lactate Shuttle (ANLS). Here, we review how different types of memory relate to the mechanisms of energy delivery in the brain.

Keywords: lactate; glycogen; metabolism; behavior; learning

Citation: Veloz Castillo, M.F.; Magistretti, P.J.; Calì, C. L-Lactate: Food for Thoughts, Memory and Behavior. *Metabolites* **2021**, *11*, 548. https://doi.org/10.3390/metabo11080548

Academic Editor: Maria Fuller

Received: 14 July 2021
Accepted: 11 August 2021
Published: 20 August 2021

1. Brain Energy Metabolism

The brain represents only 2% of the total body mass, yet to ensure its proper function, it uses between 20 and 25% of the energy produced by the body. This energy consumption is reflected by the use of glucose and oxygen delivered by the blood flow, which represents over 10% of the cardiac output [1]. Glucose is the major energy substrate for mammalian cells; in the brain, it is almost entirely oxidized to CO_2 and H_2O through its sequential processing by glycolysis, the tricarboxylic acid (TCA) cycle and the associated oxidative phosphorylation. First, glucose metabolism in astrocytes mainly proceeds through aerobic glycolysis, resulting in lactate production. Lactate taken up by neurons and transformed to pyruvate is then processed through the tricarboxylic acid cycle and the associated respiratory chain [2]. Glucose is also an important constituent of macromolecules and it can be incorporated in glycolipids and glycoproteins present in neural cells. Finally, it may enter the metabolic pathways that result in the synthesis of glutamate, GABA and acetylcholine, key neurotransmitters of the brain [3]. Glucose is also stored in astrocytes in the form of glycogen, a multibranched polymer consisting of thousands of glucose units assembled around a core protein called glycogenin, resulting in various sized granules.

Glycogen is commonly found in the liver, accounting for 6–8%, and skeletal muscle, representing 1–2% of its respective weight. It is also found in the brain, although it only represents about 0.1% of the total brain weight. So the commonly accepted ratio of glycogen in liver, skeletal muscle and brain is 100:10:1 and a variable size of 10 to 80 nanometers in diameter [4–6]. Despite its low abundance in the brain, glycogen is the largest cerebral energy reserve and is specifically localized in astrocytes under physiological conditions [7,8]. The glycogen granules act as an energy source under hypoglycemia or ischemia. In the first case, they are able to support energy metabolism by providing a glucose supply for up to 100 min. During ischemia, since no oxidative metabolism occurs, the glycogen stores deplete within two minutes [9,10]. Moreover, it plays a critical role in physiological brain functions such as synaptic activity and memory formation, two conditions requiring a high energy demand [3,11].

Glucose metabolism in the brain was first linked to glutamate-mediated neuronal activity through molecular mechanisms based on the role of astrocytes in coupling synaptic activity with vascular glucose intake. This is known as the Neuron-Glia Vasculature (NGV) unit [2,12–15]. These observations led to the first formalization of the Astrocyte–Neuron Lactate Shuttle (ANLS) model (Figure 1), which states that astrocytes respond to glutamate-mediated neuronal activity by enhancing their level of aerobic glycolysis [16]. Because glycogen is the largest energy reserve in the brain, one of its primary functions is to provide a metabolic buffer during neurotransmission. Under ketogenic conditions, such as breastfeeding, diabetes or starvation, ketone bodies may provide an energy source for the brain [3].

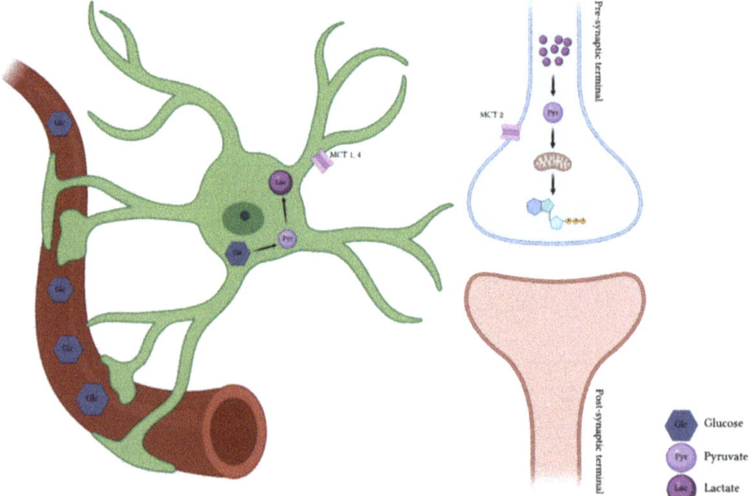

Figure 1. Representation of the astrocyte–neuron lactate shuttle which establishes that in response to glutamate-mediated neuronal activity, astrocytes take up glucose and process it through aerobic glycolysis resulting in lactate formation. Lactate can also be formed through the breakdown of glycogen by increased extracellular K^+ levels associated with increased neuronal activity or by the activation of noradrenaline and b2 adrenergic receptors [1,17]. Lactate is consequently shuttled from astrocytes via MCT1,4 and taken up by neurons via MCT2 to fuel their tricarboxylic acid cycle.

Astrocytes possess the necessary enzymatic machinery for glycogen breakdown and conversion to pyruvate/lactate, which is consequently shuttled to neurons to fuel their tricarboxylic acid (TCA) cycle [3,11]. The metabolism of glucose via glycogen, also known as glycogen shunt activity, has been demonstrated to operate in exercising muscle, as well as in the brain [18,19]. This model establishes that glial glucose flux is divided between glycolysis and glycogenolysis and that the fraction following the glycogenolytic pathway will increase with neuronal activity. This increase in the glycogenolytic pathway allows a rapid neurotransmitter clearance, which will result in a lower oxygen to glucose index, and higher lactate concentrations [19,20]. Lactate is released via monocarboxylate transporters (MCTs) 1 and 4 and taken up by active neurons through MCT 2 to satisfy their energy demands [17]. Recent studies report the crucial role of glycogen metabolism in long-term memory formation, maintenance of long-term potentiation and learning-dependent synaptic stabilization [8,21].

Initially, astrocytes were only considered as non-excitable support cells of the brain, necessary for neuronal distribution and interactions. As the field evolved, it became more evident that astrocytes are necessary to ensure optimal neuronal functioning and communication [22,23]. It is known now, that astrocytes contribute to the morphological remodeling associated with synaptic plasticity, hence acting as spatial and temporal integrators of neuronal activity and plasticity. In fact, synaptic plasticity and memory processes rely on astrocytic regulation of nutrients, i.e., glucose entry to the brain and its metabolism, as well as glycogen accumulation to fulfill high-energy demands [24]. Astrocytes synthetize TCA intermediates needed for the synthesis of glutamate formed by the rapid degradation of glycogen and stimulated by the activation of β_2-noradrenergic receptors. This, makes the learning process dependent on glycogenolysis and stimulated by noradrenaline [25]. Now, in the context of synaptic plasticity, astrocytes release certain molecules important for this process. Such is the case of TNF-a, which participates in synaptic scaling, a form of homeostatic plasticity that modulates the strength of an entire synaptic network depending on its activity history. Another example is D-serine, which acts as an endogenous NMDA receptor and plays a role in the induction of long-term potentiation in hippocampal synapses [26].

2. Memory Systems

Learning and memory are tightly related concepts. In simple terms, learning is the process of acquiring new information; some authors define it as the change in performance as a function of practice. Memory is also the persistence of learning so that it can be recalled later. It is a lasting representation that is reflected in thought, experience or behavior. Both processes demand a wide range of brain areas and involve a series of stages [27,28]. First is encoding, which is a process occurring during the presentation of the learning material. The second stage is storage (also known as retention), which results from encoding and where the information is stored within the memory system. Finally, the third stage is retrieval, which involves recovering or extracting the information previously stored in a particular memory system [29,30].

Memory can be classified, depending on the temporal availability of the information, as short-term or long-term (Figure 2). Short-term memory is the memory for information currently held in mind; it has a limited capacity and a duration of several seconds. Long-term memory refers to stored information that does not need to be presently accessed or even consciously accessible. It is considered to have an unlimited capacity essentially, and it can last throughout the lifespan [29,31–33]. Nevertheless, this classification gives little reference to the underlying molecular mechanisms or the brain areas involved in this process. For instance, memories can be divided into different categories according to how information is learned, encoded and stored. First, memories are classified either as explicit/declarative or implicit/non-declarative. Declarative memory involves episodic and semantic memory, which refers to the capacity to recollect facts, concepts or ideas, as well as recall particular life experiences or events. This type of memory is representational, providing a way to model the external world. Non-declarative memory is the capacity to recall unconsciously or retain information about automatic learned responses. This type of memory is expressed through performance rather than recollection and includes priming and perceptual learning, procedural memories, simple classical conditioning and non-associative learning (simple reflexes) [34–36]. Each of these memories is associated with a particular brain structure, as described below.

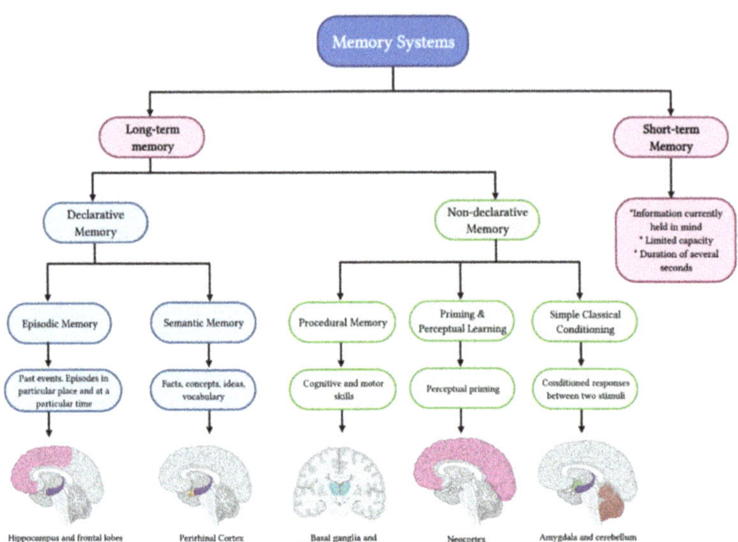

Figure 2. Types of memory and memory systems.

Episodic memory refers to the memory of past events in the life of an individual; this involves episodes in particular places at a specific time. These are also known as what, where, when and who or "wwww memories" [34,37,38]. This type of memory allows an individual to re-experience a past event in the context in which it initially occurred, which involves an association of different spatial or non-spatial clues to describe such an event. Therefore, episodic memory requires different brain regions such as the hippocampus and the frontal lobes [35,39]. Several studies show impairments in episodic memory followed by damage to the perirhinal cortex, which has connections to the hippocampus [40].

On the other hand, semantic memory represents information such as facts, concepts, ideas and vocabulary, which is explicitly known and available for recall. This type of memory is usually viewed as an associative network of concepts in which concepts similar to one another are functionally stored together. Some authors even describe it as a concrete and literal "picture memory." According to classical literature, for semantic memory, the main brain area involved is the perirhinal cortex [34,35,37,41]. More recent studies have shown, particularly analyzing PET scans and fMRI images, reveal that semantic memory is represented by spatially overlapping cortical patterns rather than anatomically segregated regions [42]. Binder and co-workers performed a meta-analysis of 120 studies and concluded that semantic processing occupies a large portion of the cortex and that it could be divided in three broad categories: Posterior heteromodal association cortex (posterior inferior parietal lobe, middle temporal gyrus and fusiform gyrus), subregions of the heteromodal prefrontal cortex (dorsal, inferior, ventromedial prefrontal cortex) and medial paralimbic regions (parahippocampus and posterior cingulate gyrus) [43].

Procedural memory is the kind of memory that stores processes, allowing the ease of performing specific activities or cognitive operations; this may include stimulus-response associations. In this case, the information is learned unconsciously as a skill, it can also be difficult to explain verbally and the memory persists for a long time [41,44]. Procedural memory can be further subdivided into motor, perceptual or cognitive. Examples of these are speech production, riding a bike, typing on a keyboard, swimming, walking, playing golf and driving a car. Procedural memory mainly depends on the basal ganglia to encode and consolidate an event; it involves complex and collective synaptic firings in the frontal–basal ganglia–thalamocortical circuits [29,33]. A study carried out in 2008 showed irregular striatum activity in an obsessive-compulsive disorder model, where procedural memory tasks are often confused, emphasizing the importance of the basal ganglia in this type of memory [45].

Priming and perceptual learning can be described as a technique by which a priming stimulus is used to sensitize the neuronal representation of the stimulus to train for a later presentation of that or a similar stimulus [29,33]. Several studies show that this type of memory is preserved in patients with amnesia [34,35] and that priming facilitates perceptual processing [46]. Priming effects can be perceptually, conceptually or semantically driven [29,47]. While this type of memory is only expressed in performance and cannot be reflected in a verbal report, its effects can be associated with declarative and procedural memory and it may depend on the neocortex. Notably, pavlovian conditioning relies on this type of memory [34,35].

The concept of classical conditioning was first mentioned in the early 1900s as a means of studying associative learning. Classical conditioning is an associative memory between a conditional stimulus and unconditioned responses, such as rewards and punishments. In this way, two stimuli (one that naturally produces a response and a neutral one) are presented together to produce a new learned response; after repeated pairings, the neutral stimulus alone will elicit the response. Therefore, classical conditioning studies the relationship between the stimuli and the environment [29,33,48]. When the conditioned stimulus triggers an emotional response, the amygdala is the brain area involved in this process [35]. Studies carried out in monkeys and rats show that amygdala lesions produce a lack of emotional response, excessive examination of objects and an incorrect pairing of food rewards, demonstrating an impairment in the processing of reward-related stimuli [40]. If the conditioned stimulus results in a skeletal response, the cerebellum is the brain area involved [35]. These findings were supported by multiple classical eyeblink conditioning studies, a valuable experiment for analyzing the behavioral and neuronal aspects of acquisition and retention of learned responses [49,50].

In recent years transgenic knockout mice have been used to carry out classical conditioning studies; this has made it possible to study changes in hippocampal synapses [51]. Additionally, neural recordings of the cerebellum during eyeblink conditioning in a rabbit show increased and decreased extracellular activity in the dentate/interpositus deep nuclei and cortex. These results are correlated with the conditioned stimulus and response, as well as unconditioned stimulus and response. It was also found that the excitability of Purkinje cells is highly correlated with the acquisition of a conditioned response [48]. Several models propose cerebellar plasticity at the synapses between the parallel fibers and the Purkinje cells, resulting from the activation of mossy and climbing fibers [50].

Molecular Mechanisms behind Memory

Memory retention is the process in which acquired information is transformed into a stored mental representation that is maintained over time without needing an active rehearsal [29]. In addition, memory consolidation is a progressive stabilization of long-term memory traces so that they become relatively resistant to decay or disruption. Memory consolidation is divided between rapid (synaptic) consolidation and system consolidation. The first one is accomplished within the first minutes to hours after learning and involves gene transcription and protein formation, leading to lasting cellular channels to support long-term memory. System consolidation can take from days to years to complete and involves the interaction between the medial temporal lobe and the neocortex [28]. These processes require neurotransmitter receptors for the acquisition and storage of new memories [36]. Examples of these are N-methyl-D-aspartate (NMDA) [52], Adenosine A2A [53], dopamine Drd1a [54], AMPA, GABA and metabotropic glutamate receptors [36], as well as acetylcholine, serotonin and norepinephrine [55,56]. It has been established that during the consolidation process, the medial temporal areas play a critical role; this system undergoes several functions related to memory, such as encoding, consolidation and retrieval. Eventually, memories will become independent of the medial temporal area and will depend on specific neocortical regions [33,57].

Changes in synaptic strength underlie memory storage and other adaptive responses, including pain control, mood stability and reward behavior. Synaptic consolidation indicates the development and stabilization of protein synthesis-dependent modifications in synaptic strength to support long-term memory formation and maintenance. This process is observed during long-term potentiation (LTP) and long-term depression (LTD). Additionally, synaptic consolidation requires brain-derived neurotrophic factor (BDNF) signaling and the immediate early gene activity-regulated cytoskeleton-associated protein Arc [28,58].

LTP is considered a neural mechanism essential for synaptic plasticity and it is the most common mechanism underlying associative learning. LTP can be evoked by high-frequency stimulation (HFS), which results in the long-lasting enhancement of synaptic efficacy. Moreover, activation of NMDA receptors is sufficient for inducing LTP, since it has been demonstrated to be the molecular substrate of the process [51,52].

BDNF triggers synaptic consolidation in mature excitatory synapses through its tyrosine kinase receptor (TrkB); this process is carried out in two stages. First, the translation stage is where high-frequency stimulation leads to the post-synaptic release of BDNF and the activation of TrkB receptors, present in pre- and post-synaptical elements of glutamatergic synapses. Particularly, post-synaptic TrkB receptors rest in the post-synaptic density (PSD) while TrkB co-immunoprecipitates are found in the NMDA receptor protein complex. Second is the Arc dependent consolidation stage; Arc encodes the only mRNA known to undergo transport to distal dendritic processes of granule cells [58].

Hippocampal studies show that Arc mRNA is enriched at stimulated synapses and Arc protein is elevated in dendrites following LTP induction. This sustained translation of Arc is crucial for cofilin phosphorylation, local F-actin expansion and the formation of stable LTP [58]. During behavioral training, Arc is expressed in principal neurons, which is necessary for long-term spatial memory [59,60]. Interestingly, a recent study in the primary visual cortex demonstrates that Arc protein in spines increases in LTD and decreases in LTP. The authors of this study conclude that Arc helps organize the distribution of potentiated and depressed spines, which underlies the plasticity of neuronal responses [61].

The synthesis of new proteins, i.e., mRNA translation, is critical for memory formation and long-lasting synaptic plasticity and for reconsolidation. This process can be triggered by gene expression changes, learning-induced activation of neuronal receptors, intracellular signaling pathways or epigenetic mechanisms [62]. Some transcription factors that participate in these tasks are cAMP Response Element–Binding Protein (CREB), C/EBP, AP1, c-Fos, Zif268, NFkB, activating transcription factor [ATF-4] [63,64]. They can bind to DNA response elements such as CRE to regulate RNA polymerase activity and determine the time and level of gene expression. Moreover, the CREB1 acts as a transcriptional activator after its phosphorylation by kinases such as PKA, MAPK, CamKIIa [65,66]. In addition to the previously mentioned transcription factors, nuclear factor kappaB (NF-kB), serum response factor, junB and neuronal Per-Arnt-Sim homology factor 4 (NPAS4) also play important roles in the memory consolidation process [64,67,68]. Protein synthesis in neurons occurs in the dendrites; this allows a rapid and precise localization of protein expression in response to synaptic activity and also provides a critical mechanism for synaptic formation, maintenance and plasticity [62].

A clear example of this is mTOR, a protein that integrates inputs from several signals such as activation of neurotransmitters, growth factor receptors and cellular metabolism changes. This protein acts together with TORC complex 1 or 2 (mTORC1, mTORC2) to regulate protein synthesis and cell growth or to control cell cycle progression and energy metabolism [62]. Several studies have shown the important role of mTORC1 in synaptic plasticity and memory reconsolidation. Findings in rat hippocampal slices, for example, show that the disruption of mTOR signaling reduces late-phase LTP expression induced by HFS without affecting early phase LTP and also blocks the synaptic potentiation induced by BDNF [69]. The idea that mTOR signaling is required to form and reconsolidate long-term memory is further supported by behavioral studies that tested spatial memory formation [70], object recognition memory [71] or involving fear-motivated tasks [72]. These studies will be discussed in detail in the following sections of this review.

Leptin receptors (LepR) in hippocampal astrocytes have also been demonstrated to play an essential role in synaptic transmission, plasticity and brain metabolism [73–75]. In one study, Naranjo et al. used a genetic mouse model that lacked the expression of LepR in GFAP-positive cells. They evaluated synaptic transmission and hippocampal plasticity using electrophysiological recordings and assessed the expression of enzymes and transporters involved in glutamate metabolism. Their findings confirmed that LepR in astrocytes are involved in maintaining glutamate homeostasis and neurotransmission since LepR depletion reduced basal synaptic transmission in CA1 cells and impaired NMDA-LTD. In addition, genetically modified mice exhibited lower glutamate uptake efficacy and upregulation of GLUT-1, GLT-1, GFAP and glutamine synthase, which could impact learning and memory processes [76].

3. Lactate: A Key Molecule for Memory

Brain energy is crucial to support the action potentials required for neuronal communication, maintenance of ionic gradients across the plasma membrane, protein synthesis, phospholipid metabolism or neurotransmitter recycling [3]. The astrocyte–neuron lactate shuttle establishes that presynaptic glutamate released from excitatory boutons is taken up by astrocytes; this glutamate is then recycled as glutamine (glutamate-glutamine cycle) and further released from astrocytes to neurons to form new glutamate for vesicle storage. Glutamate is taken up with Na^+ through specific astrocyte transporters, resulting in a dissipation of the Na^+ gradient which are reestablished through the activity of the Na/K-ATPase [77]. Both Na/K-ATPase activity and glutamine formation from glutamate are highly energy consumptive processes; astrocytes increase the glucose uptake from the bloodstream. Surprisingly, instead of using glucose through oxidative phosphorylation in mitochondria to produce ATP, they use the glycolysis pathway to produce a few ATP molecules; this process (also known as "aerobic glycolysis" or "Warburg effect") is accompanied by the synthesis of lactate, which is released via MCT1 and MCT4 and taken up by active neurons through MCT2. In neurons, lactate is transformed into pyruvate and is subsequently metabolized through oxidative phosphorylation, yielding between 14 and 17 ATPs per lactate molecule [12,78]. Lactate shuttle from astrocytes and further uptake by neurons play an essential role in learning, memory consolidation and LTP [8,21,79,80].

One such study to confirm the importance of ANLS in LTP and hippocampal memory formation was performed by Suzuki and collaborators using electrophysiological and behavioral experiments. The electrophysiological studies showed that LTP could be triggered in CA1 neurons following Schaffer collateral stimulation with the increased fEPSP slope, a classic indication and monitor of increased synaptic efficacy. Behavioral trials were conducted on rats performing the inhibitory avoidance test. Rodents received a bilateral hippocampal injection of 1,4-dideoxy-1,4-imino-D-arabinitol (DAB), which is a glycogen phosphorylase inhibitor. Researchers found that DAB prevented LTP maintenance and hypothesized that the intrahippocampal application of additional lactate could bypass it. These findings indicate that neurons require lactate uptake to meet the energy demands of LTP induction, even when displaying average concentrations of glucose. Therefore, lactate should be available for neurons during the conditioning phase of the behavioral test. Results show that the application of lactate after conditioning does not restore LTP. This indicates that ANLS plays a critical role in long-term synaptic plasticity, long-term memory, as well as molecular and synaptic changes [21].

A similar study by Duran et al. examined the learning capacities and electrophysiological properties of the hippocampal CA3-CA1 synapse using glycogen synthase knockout mice. The electrophysiological results show that paired-pulse facilitation (a form of short-term plasticity related with short-term memory) is enhanced in the mice lacking glycogen synthase. Moreover, paired-pulse stimulation (an indirect measurement of the probability of neurotransmitter release) reflects a disturbance in the release of neurotransmitters at the presynaptic terminal in the knockout mice. This confirms the role of glycogen as a precursor of glutamate and its importance in short-term memory processes. Finally, the knockout mice did not show significant LTP after the stimulation session, suggesting that glycogen is a crucial energy source to evoke this change in synaptic strength. The authors also conducted a behavioral test using the Skinner box. The results from his test reveal a significant impairment in the learning process of mice lacking glycogen synthase, which supports the previous results [81].

To support the idea that lactate regulates synaptic potentiation at central synapses and contributes to the process of memory formation, Herrera-López and co-workers carried out a series of electrophysiological experiments on hippocampal slices. They demonstrated that extracellular lactate induces glutamatergic potentiation on the recurrent collateral synapses of hippocampal CA3 pyramidal cells (CA3 PC). This potentiation occurs through a post-synaptic lactate receptor mechanism, calcium accumulation and NMDA receptor activation. The researchers found that lactate does not induce potentiation at the mossy fiber synapses of CA3 PC, concluding that lactate triggers an input-specific form of synaptic plasticity on the hippocampus and that it increases the output discharge of CA3 neurons when recurrent collaterals are repeatedly activated during lactate perfusion [82].

The degree to which long-term modifications in synaptic strength are complemented by modifications in lactate dynamics is still a matter of research. To understand it, Bingul et al. induced LTP of synapses in the dentate gyrus in freely behaving rats; this process was done through HFS of the medial perforant pathway. Before, during and up to 72 h after LTP induction, the extracellular lactate concentrations were measured using fixed potential amperometry, allowing the evaluation of how changes in synaptic strength modify local glycolytic activity. They found that synaptic potentiation was associated with persistent alterations in acute lactate dynamics following neuronal activation and observed chronic lactate availability within the dentate gyrus. These changes in lactate dynamics were only visible 24 h after HFS, whereas synaptic potentiation and altered lactate dynamics lasted up to 72 h. The authors conclude that these observations reflect a metaplastic effect that could regulate the memory consolidation process. Furthermore, these changes in extracellular lactate concentrations could support the increased energetic demands or play a neuroprotective role [83]. In order to monitor lactate dynamics Mächler and co-workers used a genetically encoded FRET sensor in combination with in vivo two-photon laser scanning microscopy. Following opening of MCTs in astrocytes and neurons using a transactivation process, they observed at first a decrease in lactate signal in astrocytes followed by an increase of it in neurons, demonstrating a lactate gradient between these two cell types that favor the flow of lactate from astrocytes to neurons, consistent with the ANLS [84].

The ANLS model establishes that lactate is released from astrocytes through MCT1 and MCT4 and taken up by neurons through MCT2, which makes these transporters critical in learning and memory formation [17]. To better understand their role, Netzahualcoyotzi and Pellerin used transgenic mice and a viral vector to decrease the expression of each transporter within the dorsal hippocampus. They demonstrate that both neuronal MCT2 and astroglial MCT4 are essential in spatial information acquisition and retention in different hippocampal-dependent tasks. After an intracerebral injection of lactate, mice with reduced levels of MCT4 exhibited improved spatial memory, but this manipulation did not affect mice with an MCT2 knockdown, supporting the idea that ANLS contributes to hippocampal-dependent learning. In contrast, MCT2 is shown to be required for long-term memory formation seven days after training, and plays an important role in mature neurons in the process of adult neurogenesis in the dentate gyrus [85].

Long-term memory formation is also affected by the release of noradrenaline and β-adrenergic signaling, which occurs in states of arousal because the coeruleo-cortical noradrenergic projection, results in noradrenaline release in the cortex. Noradrenaline has been shown to trigger glycogenolysis in astrocytes [86] resulting in aerobic glycolysis, consequently stimulating lactate production from glycogen [87]. Fink and collaborators studied single noradrenaline-stimulated astrocytes by measuring cytosolic lactate concentration using a FRET nanosensor; this process was done under different pharmacological conditions. First, they used 2-deoxy-D-glucose, a non-metabolizable form of D-glucose, to interfere with lactate metabolism; second, DAB, a potent inhibitor of glycogen phosphorylase and glycogen degradation; and finally, 3-nitropropionic acid (3-NPA), an irreversible inhibitor of succinate dehydrogenase, a Krebs cycle enzyme. Their findings reveal that D-glucose uptake is critical for the noradrenaline-induced increase in lactate concentration resulting from glycogen degradation, suggesting that most glucose molecules in the noradrenaline-stimulated cells transit through a glycogen shunt. In addition, it was observed that under these pharmacological conditions and a defined transmembrane glucose gradient, the glycolytic flux intermediates are used to produce lactate and support oxidative phosphorylation via pyruvate. This was demonstrated by an increase in lactate concentration during inhibition of the Krebs cycle [88].

To confirm the role of noradrenaline in lactate production, Zuend et al. investigated lactate dynamics in neurons and astrocytes in awake mice. They exposed the mice to isoflurane, which caused a strong arousal response, pupil dilatation and Ca^{2+} elevations in both neurons and astrocytes. These alterations in cortical activity triggered an extracellular lactate release which correlates with a fast and prominent lactate dip in astrocytes, followed by a delayed rise in neuronal and astrocytic lactate [87]. The work by Gao and collaborators also illustrates the role of adrenergic signaling in modulating long-term memory consolidation by activating glycogenolysis and subsequent lactate release [89]. These changes altogether suggest activity-dependent glycogen mobilization and further lactate release from astrocytes, which are critical in the long-term memory formation and consolidation processes [84,87,89].

Lactate also plays an important role in supporting the expression of genes such as Arc, c-Fos, Bdnf and Zif268, which involved in plasticity and neuronal activity [90]. Yang and co-workers investigated this matter in vitro in primary cultures of neurons and in vivo in the mouse sensory-motor cortex. They found that lactate stimulates the expression of genes such as Arc, c-Fos and Zif268, which are related to synaptic plasticity, and that these effects were not replicable with glucose nor pyruvate. This upregulation is carried out through a mechanism involving NMDA receptor activity and its downstream signaling cascade Erk1/2. The researchers found that lactate potentiates NMDA receptor-mediated currents, which produces elevated intracellular calcium via an increased calcium influx. Furthermore, lactate increases the intracellular levels of NADH associated with changes in the redox state of neurons. NADH mimics the effects of lactate on NMDA signaling, leading to the idea that an increase in NADH directly affects the effects of lactate [91]. In another study Margineanu and collaborators used RNA-sequencing to identify synaptic plasticity promoting genes. In addition to those found by Yang et al., they identified that Erg2, Erg3, Erg4, Npas4, Nr4a3 and Rgs4 are modulated by L-lactate in cortical neurons. Moreover, they identified ten genes associated with the MAPK signaling pathway; those are: c-Fos Bdnf, Atf4, Nr4a1, Gadd45g, Map3k11, Dusp4, Dusp6 and Dusp10 [92]. These studies lead to the conclusion that lactate can be considered a signaling molecule in neuronal plasticity, in addition to its role in energy metabolism.

The Role of L-Lactate in Disease

Lactate production in astrocytes and its sequential shuttle to neurons is an essential process in learning, memory consolidation and LTP. Accordingly, anomalies in the brain energy metabolism can result in severe pathologies or aggravate pre-existing conditions. In particular, Alzheimer's Disease (AD), amyotrophic lateral sclerosis (ALS), depression, stress and schizophrenia show disruptive lactate signaling between astrocytes and neurons [93]. For instance, Positron Emission Tomography (PET) scans have documented reduced glucose utilization in brain regions affected by patients with Alzheimer, Parkinson and Huntington's disease, as well as with ALS [94].

AD is one of the most common forms of dementia. In its preclinical stage, brain glucose hypometabolism is recognized as a prominent anomaly and some studies suggest that glycogenolysis plays a critical role in the development of the disease [95]. Impairments in glycogen synthesis could reduce glycogen levels, impeding the physiological flux of glucose units through glycogen, consequently affecting learning and memory processes [96]. Research shows reduced levels of GLUT1 and GLUT3, which correlates with less glucose uptake, which translates into a subsequent cognitive decline. Furthermore, the enzymatic activity of phosphofructokinase, phosphoglycerate mutase, aldolase, glucose-6-phosphate isomerase and lactate dehydrogenase display a loss of activity in patients with AD in comparison with age-matched controls [94]. Ryu and collaborators compared neural progenitor cells and astrocytes differentiated from late-onset AD patients. The authors found a significant downregulation of lactate dehydrogenase A in both cell types and that astrocytes from late-onset AD have a reduced metabolism of lactate [97].

In the case of Parkinson's Disease (PD), glucose hypometabolism has been documented. Key enzymes glucose-6-phosphate dehydrogenase and 6-phosphogluconate dehydrogenase are expressed in lower levels in putamen and cerebellum of PD patients [94]. Other studies show an increase in lactate levels in the striatum of patients and animal models of advanced PD [98,99].

On the other hand, ALS is characterized in patients by loss of motor neurons in the brain and spinal cord, as well as glucose intolerance, insulin resistance and hyperlipidemia. At the cellular level is common to find altered endothelial transporter proteins and astrocyte end feet degradation [94]. Nonetheless, research has shown that lactate could be used directly as cerebral uptake or indirectly as gluconeogenic precursor to improve ALS symptoms [100,101].

Schizophrenia and bipolar disorders are common and severe psychiatric disorders. They characterize by overlapping genetic background, brain abnormalities and clinical presentations. Some research suggests that alterations in brain metabolism and mitochondrial function are evident in these disorders. A set if studies ex-vivo using mouse models of schizophrenia, bipolar disorder and autism spectrum disorders showed lower pH and higher lactate levels in all the models [102]. In vivo studies in animal models and in patients confirm this evidence. Lactate concentrations are elevated and negative correlated with general cognitive function and functional capacity [103–105]. In contrast, patients suffering from depression can benefit from lactate as a treatment option. It has been proved that lactate administration produces antidepressant-like effects, promotes resilience to stress and rescues social avoidance and anxiety behaviors [106,107].

4. Behavioral Perspective

Most of the studies presented previously provide evidence that supports the importance of brain energy metabolism in learning and memory processes. This section aims to describe how behavioral studies enlighten our knowledge on brain energy metabolism in particular types of memories.

4.1. Spatial Memory

Spatial working memory is mediated by astrocytic glycogenolysis and by the expression of synaptic plasticity promoting genes [79,108,109]. To better understand this, Newman, Korol and Gold used a spontaneous alternation task using the plus-shaped maze in behaving rats. For this experiment glucose or lactate-sensitive biosensor was used to measure glucose and lactate levels in extracellular fluid in the rat hippocampus before, during and after memory tests. The recordings from the biosensors revealed a significant increase in lactate concentrations at the beginning of the behavioral test. Then, glucose levels dropped 5 min after initiating the task and 5 to 10 min while performing the test, the glucose levels raised again; these changes could correspond to an increase in blood glucose levels. After completing the task, a significant increase of lactate was again recorded; it is believed that this is a consequence of handling after removing the mice from the maze. Additionally, a pharmacological inhibition of astrocytic glycogenolysis and a pharmacological block of MCT2 by a hippocampal injection of α-cyano-4-hydroxycinnamate, resulted in memory impairment. In the first case, lactate or glucose administration was sufficient to reverse this effect; nevertheless, either glucose or lactate were able to restore the memory impairment caused by the block of MCT2 [79].

To highlight the importance of MCTs in the ANLS, Ding and co-workers established a model of long-term ketamine administration aiming to examine changes in MCTs expression that will lead to learning and memory deficits. In this case, mice were exposed to intraperitoneal administration of ketamine for six months; long-term ketamine administration is associated with abnormalities in MCTs that cause hippocampal dysfunctions. During the ketamine-administration period, mice were trained and tested for the Morris water maze (MWM) to assess their spatial memory performance and for the Radial arm maze (RAM) to evaluate their spatial working memory performance. The authors report that mice exhibited learning and memory deficits. When quantifying hippocampal proteins, the membrane fraction showed a significant decline of MCT1 and MCT4 proteins, whereas the cytoplasmic fraction showed increased levels of MCT1 and MCT4. Moreover, the global expression of MCT2 was enhanced. Finally, mRNA analysis showed that the expression on MCT2 mRNA was significantly increased, whereas MCT1 and MCT4 transcripts displayed no changes. Supposedly, cognitive deficits observed in the behavioral tests were related to the reduced levels of hippocampal membrane MCT1 and MCT4 [110].

As previously mentioned, mTOR plays an important role in regulating protein-synthesis-dependent synaptic plasticity and memory formation [62]. Dash, Orsi and Moore investigated the role of mTOR in long-term spatial memory formation using the MWM. The researchers administered either rapamycin (mTOR inhibitor), glucose, 5-amonoimidazole-4-carboxamide-1-B-4-ribonucleoside (AICAR; AMP kinase activator) or a mix of glucose and rapamycin into the dorsal hippocampus of Long-Evans rats after training in the MWM. The results suggest that AICAR and rapamycin impair long-term spatial memory, whereas glucose improves it. Moreover, the authors aimed to examine a potential mechanism to restore memory impairment by the co-administration of glucose and rapamycin; however, in this case, memory impairment was not reversed [70].

Learning and memory retrieval are both energetically demanding processes. In order to explore the role of lactate production in these processes Harris and colleagues injected dichloroacetate (DCA) into the frontal cortex and hippocampus of mice. DCA is a chemical inhibitor of lactate production; it inhibits pyruvate dehydrogenase (PDH) kinase by enhancing the activity of PDH, which further attenuates the conversion of pyruvate to lactate. The authors examined the effect of DCA on spatial learning and memory, which requires communication between the frontal cortex and the hippocampus. For this, they used the MWM as a behavioral task. The results were obtained by in vivo 13C-pyruvate magnetic resonance spectroscopy, revealing a decrease in pyruvate conversion to lactate after the DCA administration, which was accompanied by a reduction in the phosphorylation of PDH. The behavioral studies showed impaired learning in those mice injected with DCA 30 min before training, which resulted in memory impairment during the probe trial. In contrast, mice that received the DCA injection before the probe trial and not before training exhibited a standard memory. When testing memory retrieval using the MWM, the researchers found that DCA administration does not significantly affect the recall of established memories, even four days after training. These findings suggest that aerobic glycolysis, and hence lactate production, are required for memory acquisition but not for retrieval [111].

4.2. Object Recognition Memory

To understand the role of BDNF in the rat hippocampus Radiske and collaborators used the novel object recognition (NOR) test. For this, rats were trained in NOR with two different stimuli objects; 24 h later, they were undergoing a refresh session for five minutes using a familiar and a novel object. After memory reactivation, rats received a bilateral injection in the CA1 area of the hippocampus; they received either a vehicle or an anti-BDNF antibody. Then, the animals were exposed to a familiar and a novel object for five minutes to evaluate long-term memory retention. The authors found that reactivation in the presence of a novel object destabilizes object memory recognition to initiate reconsolidation in the hippocampus. These results indicate that BDNF is sufficient for controlling the integration of new information into the memory system and that object recognition memory retrieval increases BDNF levels in dorsal CA1. Finally, the amnesia caused by mRNA and protein synthesis inhibitors can be reversed by BDNF signaling reactivation following memory refresh [112].

L-Lactate plays a role as a metabolic and signaling molecule, accordingly, Vaccari-Cardoso and co-workers developed a viral vector to express a modified version of lactate oxidase (LOx) originating from the bacteria Aerococcus viridans. Their results in vitro show that LOx expression in astrocytes reduced their intracellular lactate levels and its release to the extracellular space. The researchers used the hole board test to measure exploratory behavior and they observed that mice expressing LOx in hippocampal astrocytes manifested an increased activity compared to control mice. Mice expressing LOx exhibited improved performance in the Y-maze task, which tests spatial recognition memory, but not in the Y-maze spontaneous alternation task or the NOR test. They concluded that a selective decrease in intracellular lactate pool in hippocampal astrocytes contributes to increased responsiveness to novel stimuli [113].

To elucidate the role of the basolateral complex of the amygdala (BLA) in recognition memory, Jobim et al. used the NOR task in Wistar rats. The researchers compared the effects of mTOR inhibition by rapamycin infusion into the BLA or dorsal hippocampus; this was done before or after training or reactivation. Results show that rapamycin infusion, either before or after training, impairs NOR retention tested 24 h after training. In particular, memory retention is impaired when the infusion is given before reactivation on BLA or dorsal hippocampus and measured 24 h after the reactivation, but this does not occur if measured six hours after reactivation. These findings indicate that mTOR signaling is crucial for the consolidation and stabilization of object recognition memory, either in the hippocampus or BLA. mTOR acts as a regulator of glucose uptake, glycolysis, lipid synthesis and mitochondrial metabolism; consequently, its inhibition might influence neuronal metabolism, which will affect memory-modulatory function [71].

Continuing with the exploration of different brain areas involved in object recognition memory, Korol and colleagues evaluated the involvement of lactate in the hippocampus and striatum. This test was performed with Long-Evans rats and recognition memory was assessed using the double object location (DOL) task and the double object replacement (DOR) task. Rats received an injection of lidocaine (Na^+ channel blocker) and 4-CIN (MCT2 inhibitor) into either hippocampus or striatum after three training sessions and before the test trial. Findings demonstrate that both lidocaine and 4-CIN impair recognition memory for objects and their relative location; this only occurs when the substance is administered to the particular brain area necessary for that type of recognition. Infusion into the hippocampus impairs the recognition in the DOL task, whereas the ones in the striatum impaired the recognition in the DOR task. In conclusion, neuronal lactate uptake in both hippocampus and striatum is necessary for object recognition memory [80].

4.3. *Fear Conditioned Memory*

Noradrenaline acts through adrenergic receptors, of which β-adrenergic receptors (β -AR) in the amygdala or hippocampus play crucial roles in encoding and consolidating memories, particularly fear related memories. Noradrenaline activates glycogenolysis and consequently lactate release, which is critical in memory processing [88,89,114,115]. With this in mind, Gao et al. used the hippocampus-dependent IA task to determine if astrocytic or neuronal β-ARs in the hippocampus mediate memory consolidation. Their results show that astrocytic ARs (β2-AR) play a critical role in the consolidation of a fear-based contextual memory. Moreover, β2-AR mediates learning-dependent lactate release from astrocytes, which is necessary to support the molecular changes needed for long-term memory formation [89].

Noradrenergic innervation to the cortex originates from neurons in the locus coeruleus and this system plays a key role in the sleep wake cycle, arousal, respiration, learning and memory [115]. To investigate how noradrenergic activity modulates Ca^{2+} and cAMP dynamic during fear conditioning, Oe and co-workers imaged astrocytes in the auditory cortex of behaving mice. First, they tested if a startle response increases Ca^{2+} and cAMP levels; to do so, mice received unpredictable air puffs on the right side of the face. As a result, Ca^{2+} was elevated, but no significant cAMP increase was recorded. Then, the researchers examined astrocytic activity during fear memory acquisition. In this case, the mice had their head fixed to allow imaging of the cortex throughout the experiment and they received a foot shock after a sound cue. During recall on the next day, only the sound cue was presented and the establishment of the fear memory was manifested by increased immobility times. The results of this second experiment showed that foot shock induces and elevates both astrocytic Ca^{2+} and cAMP, although it is attenuated with repeated shocks. Finally, the authors conclude that these changes might be involved in the modulation of synaptic transmission and memory consolidation [116].

To test if the supply of glycolytic metabolites such as pyruvate or β-hydroxybutyrate can functionally replace lactate in a memory impairment model Descalzi and collaborators performed a series of experiments. First, they used the IA task in adult rats and injected DAB into the dorsal hippocampus to generate memory impairment. After the DAB injection, they observed that it reduced the latency of rat entry into the shock compartment postconditioning; nevertheless, this latency was increased by a co-injection of DAB and either pyruvate or β-hydroxybutyrate. Then, they performed an expression knockdown of MCT1, MCT2 and MCT4 that resulted in a reduced latency post-conditioning. The supply of pyruvate and β-hydroxybutyrate counteracted this effect and rescued the memory loss caused by the knockdown of MCT1 and MCT4, but it did not affect the decrease in latency in the MCT2 knockdown. In conclusion lactate is critical in providing energy for neuronal responses required in long-term memory. The authors suggested that learning and training increase mRNA translation expressed in excitatory and inhibitory neurons, which can be blocked by inhibiting glycogenolysis and rescued with a co-injection of DAB and lactate [117].

To assess the role of mTOR signaling in long-term memory, Beckinschtein and co-workers performed a one-trial IA test in rats. First, the authors demonstrate that IA training is associated with rapid increases in the phosphorylation state of mTOR and its downstream substrate p70S6K in the hippocampus. Then, the animals received a bilateral infusion of rapamycin (mTOR inhibitor) in the CA1 region of the hippocampus. Their results show that rats who received the infusion 15 min before training showed impairment in long-term memory without affecting short-term memory and that these rats were capable of learning the IA task in a second training session. In contrast, rats that received the infusion immediately after training showed no effect on their long-term memory retention scores. The authors also found an increase in the activation of p70S6K 15 min after training and conclude that the mTOR-p70S6K cascade is required for protein synthesis required for memory processing [72].

4.4. *Drug-Associated Memories*

Drug-associated memories persist for a long time after abstinence and this represents a core symptom of addiction. Re-exposure to drug-associated cues reactivates drug memories and triggers neuroplastic changes that promote drug-seeking behaviors. Since memory and addiction share a common neural circuitry and molecular mechanisms, clinical and laboratory studies conclude that addiction represents the pathologic hijacking of neural processes that would typically account for reward-related learning [118,119]. With this in mind, we can say that drug addiction depends on the remodeling of synapses that shape long-term memory.

In the last few years, several research studies have been published showing the role of ANLS or lactate itself in the reconsolidation of drug memories. For instance,

Zang and collaborators conducted a series of experiments to determine the role of lactate transport in the reconsolidation of drug memories. They trained a group of rats for cocaine-induced conditioned place preference (CPP) or self-administration and injected DAB into the BLA immediately after retrieval. Results show that DAB injection into the BLA prevented cocaine-induced CPP expression for up to 14 days and reduced drug-seeking behavior in rats trained to self-administer cocaine. The scientists measure the lactate concentration immediately after retrieval and found a lower concentration of lactate in the BLA. The reason to choose BLA as an injection site is because the amygdala controls emotional responses, and therefore it plays a key role in encoding conditioned drug-related information. Additionally, the authors used antisense oligonucleotides to disrupt the expression of MCTs and observed that the disruption of MCT1 and MCT2 in the BLA caused deficits in the expression of cocaine-induced CPP. While lactate co-administration can rescue the effects in MCT1 it does not occur in MCT2. Finally, it is demonstrated that glycogenolysis inhibition and its consequent reduction in lactate release decreases the gene expression of pCREB, pERK and pcoffilin, which are associated with synaptic plasticity and memory reconsolidation [120].

A similar work involving DAB administration into the BLA was reported by Boury-Jamot and co-workers. They were aiming to explore the role of the ANLS for the acquisition and maintenance of cocaine-induced memories. Their findings demonstrate that inhibition of glycogenolysis prevents the acquisition of cocaine-induced CPP in a lactate-reversible manner. This manipulation also disrupts the expression of BDNF and Zif-268, which are involved in the modulation of synaptic morphology and plasticity underlying the learning processes that strengthen conditioned responses to cocaine. Moreover, the co-administration of lactate rescues drug-associated memory through a mechanism that requires Zif-268 and the ERK signaling pathway, but not BDNF. Finally, the authors conclude that the storage and retrieval of drug-associated memories require astrocyte-derived lactate [121].

Another study that reflects the role of BLA in the reconsolidation of cocaine-associated memories was carried out by Wu et al. They aim to explore the role of glycogen synthase kinase 3β (GSK-3β) in this process. Their results show increased GSK-3β activity in the BLA in rats that acquired cocaine-induced CPP after a memory reactivation process. Some rats received a systemic injection of lithium chloride or SB216763 (GSK-3β inhibitors); this resulted in an impaired reconsolidation of cocaine cue memories and the consequent GSK-3β activity in the BLA. These findings indicate the importance of GSK-3β in the BLA in the consolidation of drug-associated memories [122].

Other molecules that play a key role in synaptic plasticity and memory consolidation are eukaryotic initiation factors. In a series of experiments, Jian and collaborators elucidated the role of eIF2a dephosphorylation in the BLA to reconsolidate drug-associated memories; this was done using the CPP task and self-administration procedures in rats. Their results display decreased levels of eIF2a phosphorylation and the activation transcription factor 4 (ATF4) in the BLA after memory retrieval procedure in a morphine- and cocaine-paired context. A group of animals received an intra-BLA infusion of Sal003 (eIF2a dephosphorylation inhibitor) immediately after retrieval; they showed a disruption in the reconsolidation of drug-induced CPP, leading to the suppression of stimulus-induced craving. This disruption in reconsolidation was blocked by the knockdown of ATF4 expression in the BLA [123].

5. Morphological Changes Associated with Memory Consolidation: Role of Lactate

The most common brain-imaging techniques to study brain energy metabolism in vivo are positron emission tomography (PET) and functional magnetic resonance imaging (fMRI). PET monitors changes in the blood flow, oxygen consumption and glucose utilization, whereas fMRI tracks the degree of blood oxygenation and flow. Furthermore, nuclear magnetic resonance (NMR) studies provide an insight into the relationship between glucose consumption and glutamate-glutamine cycling [3,78,124]. In the past few years, neuroscientists have been using high-resolution imaging techniques such as scanning electron microscopy (SEM) and transmission electron microscopy (TEM). These techniques can be adapted to novel set-ups that allow the acquisition of high-resolution image stacks. The segmentation of these image stacks results in accurate 3D models of brain structures which are further analyzed using 3D visualization tools or virtual reality set-ups. These approaches facilitate brain morphology analysis at the nanoscale and the study of relationships between energy consumption and glycogen storage [6,7,14,15,125,126].

To exhibit the extent of high-resolution imaging techniques and 3D morphological reconstructions, Vezzoli et al. recently published a work where they demonstrate the role of lactate to rescue memory in mice treated with DAB. Their findings show that mice injected with DAB have fewer synaptic spines per unit volume than control mice and that a co-injection can reverse the memory loss effect with lactate. They also observed that spine density increased after learning and that DAB prevented this increase. Finally, the authors concluded that a co-administration of lactate is sufficient to rescue the memory but not to increase the number of spines or post-synaptic density [8].

6. Concluding Remarks

This article has reviewed the interconnection between energy metabolism, synaptic plasticity and memory consolidation. The main message is that energy substrates are not only necessary to fluke the energy-consuming process associated with synaptic plasticity, but that lactate in particular has an additional function as a signaling molecule regulating the levels of expression of plasticity-associated genes and processes. These findings suggest that pharmacological interventions aimed at promoting lactate production by astrocytes may be useful in clinical conditions characterized by cognitive impairments such as Alzheimer's disease.

Author Contributions: C.C., Conceptualization of the review; C.C. and P.J.M., supervision; M.F.V.C., writing, original draft preparation; P.J.M., writing, review and editing, funding acquisition. All authors have read and agreed to the published manuscript.

Funding: This research was funded by KAUST baseline to P.J.M.

Acknowledgments: We thank Matthew Fechtmeyer (English Language and Communication Program, KAUST) for critical review and grammatical editing of the manuscript.

Conflicts of Interest: The authors declare no conflicts of interest.

References

1. Magistretti, P.J.; Allaman, I. A cellular perspective on brain energy metabolism and functional imaging. *Neuron* **2015**, *86*, 883–901. [CrossRef]
2. Calì, C.; Agus, M.; Kare, K.; Boges, D.J.; Lehväslaiho, H.; Hadwiger, M.; Magistretti, P.J. 3D cellular reconstruction of cortical glia and parenchymal morphometric analysis from serial block-face electron microscopy of juvenile rat. *Prog. Neurobiol.* **2019**, *183*, 101696. [CrossRef] [PubMed]
3. Allaman, I.; Magistretti, P.J. Brain Energy Metabolism. In *Fundamental Neuroscience*, 3rd ed.; Squire, L.R., Berg, D., Bloom, F.E., Du Lac, S., Ghosh, A., Spitzer, N.C., Eds.; Elsevier Inc.: Amsterdam, The Netherlands, 2013; pp. 261–284. [CrossRef]
4. Nelson, S.R.; Schulz, D.W.; Passonneau, J.V.; Lowry, O.H. Control of glycogen levels in brain. *J. Neurochem.* **1968**, *15*, 1271–1279. [CrossRef] [PubMed]
5. Brown, A.M. Brain glycogen re-awakened. *J. Neurochem.* **2004**, *89*, 537–552. [CrossRef] [PubMed]
6. Agus, M.; Boges, D.; Gagnon, N.; Magistretti, P.J.; Hadwiger, M.; Calí, C. GLAM: Glycogen-derived Lactate Absorption Map for visual analysis of dense and sparse surface reconstructions of rodent brain structures on desktop systems and virtual environments. *Comput. Graph.* **2018**, *74*, 85–98. [CrossRef]
7. Calì, C.; Baghabra, J.; Boges, D.J.; Holst, G.R.; Kreshuk, A.; Hamprecht, F.A.; Srinivasan, M.; Lehväslaiho, H.; Magistretti, P.J. Three-dimensional immersive virtual reality for studying cellular compartments in 3D models from EM preparations of neural tissues. *J. Comp. Neurol.* **2016**, *524*, 23–38. [CrossRef]
8. Vezzoli, E.; Calì, C.; De Roo, M.; Ponzoni, L.; Sogne, E.; Gagnon, N.; Francolini, M.; Braida, D.; Sala, M.; Muller, D.; et al. Ultrastructural evidence for a role of astrocytes and glycogen-derived lactate in learning-dependent synaptic stabilization. *Cere.l Cortex* **2020**, *30*, 2114–2127. [CrossRef]
9. Swanson, R.A.; Sagar, S.M.; Sharp, F.R. Regional brain glycogen stores and metabolism during complete global ischaemia. *Neurol. Res.* **1989**, *11*, 24–28. [CrossRef]

10. Gruetter, R. Glycogen: The forgotten cerebral energy store. *J. Neurosci. Res.* **2003**, *74*, 179–183. [CrossRef]
11. Obel, L.F.; Müller, M.S.; Walls, A.B.; Sickmann, H.M.; Bak, L.K.; Waagepetersen, H.S.; Schousboe, A. Brain glycogen—New perspectives on its metabolic function and regulation at the subcellular level. *Front. Neuroenergetics* **2012**, *4*, 1–15. [CrossRef]
12. Magistretti, P.J.; Pellerin, L. Astrocytes couple synaptic activity to glucose utilization in the brain. *Physiology* **1999**, *14*, 177–182. [CrossRef]
13. Magistretti, P.J. Neuron-glia metabolic coupling and plasticity. *J. Exp. Biol.* **2006**, *209*, 2304–2311. [CrossRef]
14. Coggan, J.S.; Cali, C.; Keller, D.; Agus, M.; Boges, D.; Abdellah, M.; Kare, K.; Lehväslaiho, H.; Eilemann, S.; Jolivet, R.; et al. A process for digitizing and simulating biologically realistic oligocellular networks demonstrated for the neuro-glio-vascular ensemble. *Front. Neurosci.* **2018**, *12*, 1–23. [CrossRef] [PubMed]
15. Calì, C.; Kare, K.; Agus, M.; Veloz Castillo, M.F.; Boges, D.; Hadwiger, M.; Magistretti, P. A method for 3D reconstruction and virtual reality analysis of glial and neuronal cells. *JoVE J. Vis. Exp.* **2019**, e59444. [CrossRef] [PubMed]
16. Pellerin, L.; Magistretti, P.J. Glutamate uptake into astrocytes stimulates aerobic glycolysis: A mechanism coupling neuronal activity to glucose utilization. *Proc. Natl. Acad. Sci. USA* **1994**, *91*, 10625–10629. [CrossRef] [PubMed]
17. Magistretti, P.J.; Allaman, I. Lactate in the brain: From metabolic end-product to signalling molecule. *Nat. Rev. Neurosci.* **2018**, *19*, 235–249. [CrossRef] [PubMed]
18. Shulman, R.G.; Rothman, D.L. The 'glycogen shunt' in exercising muscle: A role for glycogen in muscle energetics and fatigue. *Proc. Natl. Acad. Sci. USA* **2001**, *98*, 457–461. [CrossRef] [PubMed]
19. Walls, A.B.; Heimbürger, C.M.; Bouman, S.D.; Schousboe, A.; Waagepetersen, H.S. Robust glycogen shunt activity in astrocytes: Effects of glutamatergic and adrenergic agents. *Neuroscience* **2009**, *158*, 284–292. [CrossRef]
20. Shulman, R.G.; Hyder, F.; Rothman, D.L. Cerebral energetics and the glycogen shunt: Neurochemical basis of functional imaging. *Proc. Natl. Acad. Sci. USA* **2001**, *98*, 6417–6422. [CrossRef]
21. Suzuki, A.; Stern, S.A.; Bozdagi, O.; Huntley, G.W.; Walker, R.H.; Magistretti, P.J.; Alberini, C.M. Astrocyte-neuron lactate transport is required for long-term memory formation. *Cell* **2011**, *144*, 810–823. [CrossRef]
22. Volterra, A.; Meldolesi, J. Astrocytes, from brain glue to communication elements: The revolution continues. *Nat. Rev. Neurosci.* **2005**, *6*, 626–640. [CrossRef] [PubMed]
23. Kol, A.; Adamsky, A.; Groysman, M.; Kreisel, T.; London, M.; Goshen, I. Astrocytes contribute to remote memory formation by modulating hippocampal–cortical communication during learning. *Nat. Neurosci.* **2020**, *23*, 1229–1239. [CrossRef]
24. Steinman, M.Q.; Gao, V.; Alberini, C.M. The role of lactate-mediated metabolic coupling between astrocytes and neurons in long-term memory formation. *Front. Integr. Neurosci.* **2016**, *10*, 1–14. [CrossRef]
25. Gibbs, M.E.; Hutchinson, D.; Hertz, L. Astrocytic involvement in learning and memory consolidation. *Neurosci. Biobehav. Rev.* **2008**, *32*, 927–944. [CrossRef]
26. Bezzi, P.; Volterra, A. Astrocytes: Powering memory. *Cell* **2011**, *144*, 644–645. [CrossRef]
27. Squire, L.R. Definitions: From Synapses to Behavior. In *Memory and Brain*; Oxford University Press Inc.: New York, NY, USA, 1987; pp. 3–9.
28. Baars, B.J.; Gage, N.M. Learning and Memory. In *Cognition, Brain, and Consciousness*; Elsevier: Amsterdam, The Netherlands, 2010; pp. 304–343. [CrossRef]
29. Lucas, J.A. Memory, Overview. In *Encyclopedia of the Human Brain*; Elsevier: Amsterdam, The Netherlands, 2002; pp. 817–833. [CrossRef]
30. Eysenck, M.W.; Keane, M.T. Learning Ans Memory. In *Cognitive Psychology: A Student's Handbook*, 5th ed.; Psychology Press: New York, NY, USA, 2005; pp. 189–228.
31. Squire, L.R. Short-Term and Long-Term Memory Processes. In *Memory and Brain*; Oxford University Press Inc.: New York, NY, USA, 1987; pp. 134–150. Available online: https://ebookcentral.proquest.com/lib/kaust-ebooks/detail.action?docID=272279 (accessed on 28 February 2021).
32. Ward, J. The Remembering Brain. In *The Student's Guide to Cognitive Neuroscience*, 1st ed.; Psychology Press: New York, NY, USA, 2006; pp. 176–201.
33. Kropotov, J.D. Memory Systems. In *Quantitative EEG, Event-Related Potentials and Neurotherapy*; Elsevier: Amsterdam, The Netherlands, 2009; pp. 310–324. [CrossRef]
34. Squire, L.R. Divisions of Long-Term Memory. In *Memory and Brain*; Oxford University Press Inc.: New York, NY, USA, 1987; pp. 151–174. Available online: https://ebookcentral.proquest.com/lib/kaust-ebooks/detail.action?docID=272279 (accessed on 28 February 2021).
35. Squire, L.R. Memory systems of the brain: A brief history and current perspective. *Neurobiol. Learn. Mem.* **2004**, *82*, 171–177. [CrossRef]
36. Alberini, C.M.; Cruz, E.; Descalzi, G.; Bessières, B.; Gao, V. Astrocyte glycogen and lactate: New insights into learning and memory mechanisms. *Glia* **2018**, *66*, 1244–1262. [CrossRef] [PubMed]
37. Kinsbourne, M. Brain mechanisms and memory. *Hum. Neurobiol.* **1987**, *6*, 81–92. [CrossRef]
38. Tulving, E. Episodic memory: From mind to brain. *Annu. Rev. Psychol.* **2002**, *53*, 1–25. [CrossRef]
39. Eichenbaum, H. How does the brain organize memories? *Science* **1997**, *277*, 330–332. [CrossRef] [PubMed]
40. Rolls, E.T. Memory systems in the brain. *Annu. Rev. Psychol.* **2000**, *51*, 599–630. [CrossRef]
41. Radvansky, G.; Tamplin, A. Types of Memory. In *Encyclopedia of Human Behavior*, 2nd ed.; Ramachandran, V.S., Ed.; Elsevier Science & Technology: Berkeley, CA, USA, 2012; pp. 585–592. Available online: http://waraqa.org/wp-content/uploads/2021/04/Encyclopedia-of-Human-Behavior-Second-Edition.pdf (accessed on 28 February 2021).

42. Zhang, Y.; Han, K.; Worth, R.; Liu, Z. Connecting concepts in the brain by mapping cortical representations of semantic relations. *Nat. Commun.* **2020**, *11*, 1–13. [CrossRef]
43. Binder, J.R.; Desai, R.H.; Graves, W.W.; Conant, L.L. Where is the semantic system? A critical review and meta-analysis of 120 functional neuroimaging studies. *Cereb. Cortex* **2009**, *19*, 2767–2796. [CrossRef]
44. Zichlin, M. Procedural Memory. In *Encyclopedia of Clinical Neuropsychology*; Kreutzer, J.S., DeLuca, J., Caplan, B., Eds.; Springer: New York, NY, USA, 2011; pp. 2033–2034. [CrossRef]
45. Rueda-Orozco, P.E.; Montes-Rodriguez, C.J.; Soria-Gomez, E.; Méndez-Díaz, M.; Prospéro-García, O. Impairment of endocannabinoids activity in the dorsolateral striatum delays extinction of behavior in a procedural memory task in rats. *Neuropharmacology* **2008**, *55*, 55–62. [CrossRef] [PubMed]
46. Kim, J.I.; Humphreys, G.W. Working memory, perceptual priming, and the perception of hierarchical forms: Opposite effects of priming and working memory without memory refreshing. *Atten. Percept. Psychophys.* **2010**, *72*, 1533–1555. [CrossRef]
47. Woltz, D.J. Perceptual and conceptual priming in a semantic reprocessing task. *Mem. Cogn.* **1996**, *24*, 429–440. [CrossRef] [PubMed]
48. Schreurs, B.G.; Alkon, D.L. Imaging learning and memory: Classical conditioning. *Anat. Rec.* **2001**, *265*, 257–273. [CrossRef] [PubMed]
49. Robleto, K. Brain mechanisms of extinction of the classically conditioned eyeblink response. *Learn. Mem.* **2004**, *11*, 517–524. [CrossRef] [PubMed]
50. Villarreal, R.P.; Steinmetz, J.E. Neuroscience and learning: Lessons from studying the involvement of a region of cerebellar cortex in eyeblink classical conditioning. *J. Exp. Anal. Behav.* **2005**, *84*, 631–652. [CrossRef] [PubMed]
51. Gruart, A.; Delgado-García, J.M. Activity-dependent changes of the hippocampal CA3-CA1 synapse during the acquisition of associative learning in conscious mice. *Genes Brain Behav.* **2007**, *6*, 24–31. [CrossRef]
52. Gruart, A.; Leal-Campanario, R.; López-Ramos, J.C.; Delgado-García, J.M. Functional basis of associative learning and their relationships with long-term potentiation evoked in the involved neural circuits: Lessons from studies in behaving mammals. *Neurobiol. Learn. Mem.* **2015**, *124*, 3–18. [CrossRef]
53. Fontinha, B.M.; Delgado-García, J.M.; Madroñal, N.; Ribeiro, J.A.; Sebastião, A.M.; Gruart, A. Adenosine A(2A) receptor modulation of hippocampal CA3-CA1 synapse plasticity during associative learning in behaving mice. *Neuropsychopharmacology* **2009**, *34*, 1865–1874. [CrossRef]
54. Ortiz, O.; Delgado-García, J.M.; Espadas, I.; Bahí, A.; Trullas, R.; Dreyer, J.L.; Gruart, A.; Moratalla, R. Associative learning and CA3-CA1 synaptic plasticity are impaired in D 1R Null, Drd1a-/- mice and in hippocampal SiRNA silenced Drd1a mice. *J. Neurosci.* **2010**, *30*, 12288–12300. [CrossRef] [PubMed]
55. Myhrer, T. Neurotransmitter systems involved in learning and memory in the rat: A meta-analysis based on studies of four behavioral tasks. *Brain Res. Rev.* **2003**, *41*, 268–287. [CrossRef]
56. Meneses, A. Neurotransmitters and Memory: Cholinergic, Glutamatergic, GaBAergic, Dopaminergic, Serotonergic, Signaling, and Memory. In *Identification of Neural Markers Accompanying Memory*; Elsevier: Amsterdam, The Netherlands, 2014; pp. 5–45. [CrossRef]
57. Squire, L.R. Amnesia and the Functional Organization of Memory. In *Memory and Brain*; Oxford University Press Inc.: New York, NY, USA, 1986; pp. 202–223.
58. Bramham, C.R. Control of synaptic consolidation in the dentate gyrus: Mechanisms, functions, and therapeutic implications. *Prog. Brain Res.* **2007**, *163*, 453–471. [CrossRef]
59. Plath, N.; Ohana, O.; Dammermann, B.; Errington, M.L.; Schmitz, D.; Gross, C.; Mao, X.; Engelsberg, A.; Mahlke, C.; Welzl, H.; et al. Arc/Arg3.1 is essential for the consolidation of synaptic plasticity and memories. *Neuron* **2006**, *52*, 437–444. [CrossRef]
60. Vazdarjanova, A.; Ramirez-Amaya, V.; Insel, N.; Plummer, T.K.; Rosi, S.; Chowdhury, S.; Mikhael, D.; Worley, P.F.; Guzowski, J.F.; Barnes, C.A. Spatial exploration induces ARC, a plasticity-related immediate-early gene, only in calcium/calmodulin-dependent protein kinase II-positive principal excitatory and inhibitory neurons of the rat forebrain. *J. Comp. Neurol.* **2006**, *498*, 317–329. [CrossRef] [PubMed]
61. El-Boustani, S.; Ip, J.P.K.; Breton-Provencher, V.; Knott, G.W.; Okuno, H.; Bito, H.; Sur, M. Locally coordinated synaptic plasticity of visual cortex neurons in vivo. *Science* **2018**, *360*, 1349–1354. [CrossRef]
62. Roesler, R. Molecular mechanisms controlling protein synthesis in memory reconsolidation. *Neurobiol. Learn. Mem.* **2017**, *142*, 30–40. [CrossRef]
63. Alberini, C.M. Transcription factors in long-term memory and synaptic plasticity. *Physiol. Rev.* **2009**, *89*, 121–145. [CrossRef]
64. Alberini, C.M.; Kandel, E.R. The regulation of transcription in memory consolidation. *Cold Spring Harb. Perspect. Biol.* **2015**, *7*, a021741. [CrossRef]
65. Kandel, E.R. The molecular biology of memory storage: A dialogue between genes and synapses. *Science* **2001**, *294*, 1030–1038. [CrossRef]
66. Kandel, E.R. The molecular biology of memory: CAMP, PKA, CRE, CREB-1, CREB-2, and CPEB. *Mol. Brain* **2012**, *5*, 1–12. [CrossRef]
67. Kaltschmidt, B.; Kaltschmidt, C. NF-KappaB in long-term memory and structural plasticity in the adult mammalian brain. *Front. Mol. Neurosci.* **2015**, *8*, 1–11. [CrossRef]
68. Abraham, W.C.; Jones, O.D.; Glanzman, D.L. Is plasticity of synapses the mechanism of long-term memory storage? *Npj Sci. Learn.* **2019**, *4*, 1–10. [CrossRef]
69. Tang, S.J.; Reis, G.; Kang, H.; Gingras, A.C.; Sonenberg, N.; Schuman, E.M. A rapamycin-sensitive signaling pathway contributes to long-term synaptic plasticity in the hippocampus. *Proc. Natl. Acad. Sci. USA* **2002**, *99*, 467–472. [CrossRef] [PubMed]
70. Dash, P.K.; Orsi, S.A.; Moore, A.N. Spatial memory formation and memory-enhancing effect of glucose involves activation of the tuberous sclerosis complex-mammalian target of rapamycin pathway. *J. Neurosci.* **2006**, *26*, 8048–8056. [CrossRef] [PubMed]

71. Jobim, P.F.C.; Pedroso, T.R.; Werenicz, A.; Christoff, R.R.; Maurmann, N.; Reolon, G.K.; Schröder, N.; Roesler, R. Impairment of object recognition memory by rapamycin inhibition of mTOR in the amygdala or hippocampus around the time of learning or reactivation. *Behav. Brain Res.* **2012**, *228*, 151–158. [CrossRef] [PubMed]
72. Bekinschtein, P.; Katche, C.; Slipczuk, L.N.; Igaz, L.M.; Cammarota, M.; Izquierdo, I.; Medina, J.H. mTOR signaling in the hippocampus is necessary for memory formation. *Neurobiol. Learn. Mem.* **2007**, *87*, 303–307. [CrossRef]
73. Funahashi, H.; Yada, T.; Suzuki, R.; Shioda, S. Distribution, function, and properties of leptin receptors in the brain. *Int. Rev. Cytol.* **2003**, *224*, 1–27. [CrossRef]
74. Harvey, J.; Solovyova, N.; Irving, A. Leptin and its role in hippocampal synaptic plasticity. *Prog. Lipid Res.* **2006**, *45*, 369–378. [CrossRef]
75. Irving, A.J.; Harvey, J. Leptin regulation of hippocampal synaptic function in health and disease. *Philos. Trans. R. Soc. B Biol. Sci.* **2014**, *369*, 20130155. [CrossRef] [PubMed]
76. Naranjo, V.; Contreras, A.; Merino, B.; Plaza, A.; Lorenzo, M.P.; García-Cáceres, C.; García, A.; Chowen, J.A.; Ruiz-Gayo, M.; Del Olmo, N.; et al. Specific deletion of the astrocyte leptin receptor induces changes in hippocampus glutamate metabolism, synaptic transmission and plasticity. *Neuroscience* **2020**, *447*, 182–190. [CrossRef] [PubMed]
77. Pellerin, L.; Magistretti, P.J. Glutamate uptake stimulates Na+,K+-ATPase activity in astrocytes via activation of a distinct subunit highly sensitive to ouabain. *J. Neurochem.* **1997**, *69*, 2132–2137. [CrossRef] [PubMed]
78. Magistretti, P.J.; Pellerin, L.; Rothman, D.L.; Shulman, R.G. Energy on demand. *Science* **1999**, *283*, 496–497. [CrossRef] [PubMed]
79. Newman, L.A.; Korol, D.L.; Gold, P.E. Lactate produced by glycogenolysis in astrocytes regulates memory processing. Edited by Darrell Brann. *PLoS ONE* **2011**, *6*, e28427. [CrossRef]
80. Korol, D.L.; Gardner, R.S.; Tunur, T.; Gold, P.E. Involvement of lactate transport in two object recognition tasks that require either the hippocampus or striatum. *Behav. Neurosci.* **2019**, *133*, 176–187. [CrossRef] [PubMed]
81. Duran, J.; Saez, I.; Gruart, A.; Guinovart, J.J.; Delgado-García, J.M. Impairment in long-term memory formation and learning-dependent synaptic plasticity in mice lacking glycogen synthase in the brain. *J. Cereb. Blood Flow Metab.* **2013**, *33*, 550–556. [CrossRef]
82. Herrera-López, G.; Griego, E.; Galván, E.J. Lactate induces synapse-specific potentiation on CA3 pyramidal cells of rat hippocampus. *PLoS ONE* **2020**, *15*, e0242309. [CrossRef]
83. Bingul, D.; Kalra, K.; Murata, E.M.; Belser, A.; Dash, M.B. Persistent changes in extracellular lactate dynamics following synaptic potentiation. *Neurobiol. Learn. Mem.* **2020**, *175*, 107314. [CrossRef] [PubMed]
84. Mächler, P.; Wyss, M.T.; Elsayed, M.; Stobart, J.; Gutierrez, R.; Von Faber-Castell, A.; Kaelin, V.; Zuend, M.; Martiín, A.S.; Romero-Gómez, I.; et al. In vivo evidence for a lactate gradient from astrocytes to neurons. *Cell Metab.* **2016**, *23*, 94–102. [CrossRef] [PubMed]
85. Netzahualcoyotzi, C.; Pellerin, L. Neuronal and astroglial monocarboxylate transporters play key but distinct roles in hippocampus-dependent learning and memory formation. *Prog. Neurobiol.* **2020**, *19*, 101888. [CrossRef] [PubMed]
86. Magistretti, P.J.; Morrison, J.H. Noradrenaline- and vasoactive intestinal peptide-containing neuronal systems in neocortex: Functional convergence with contrasting morphology. *Neuroscience* **1988**, *24*, 367–378. [CrossRef]
87. Zuend, M.; Saab, A.S.; Wyss, M.T.; Ferrari, K.D.; Hösli, L.; Looser, Z.J.; Stobart, J.L.; Duran, J.; Guinovart, J.J.; Barros, L.F.; et al. Arousal-induced cortical activity triggers lactate release from astrocytes. *Nat. Metab.* **2020**, *2*, 179–191. [CrossRef] [PubMed]
88. Fink, K.; Velebit, J.; Vardjan, N.; Zorec, R.; Kreft, M. Noradrenaline-induced l-lactate production requires d-glucose entry and transit through the glycogen shunt in single-cultured rat astrocytes. *J. Neurosci. Res.* **2021**, *99*, 1084–1098. [CrossRef] [PubMed]
89. Gao, V.; Suzuki, A.; Magistretti, P.J.; Lengacher, S.; Pollonini, G.; Steinman, M.Q.; Alberini, C.M. Astrocytic B2- adrenergic receptors mediate hippocampal long- term memory consolidation. *Proc. Natl. Acad. Sci. USA* **2016**, *113*, 8526–8531. [CrossRef] [PubMed]
90. Flavell, S.W.; Greenberg, M.E. Signaling mechanisms linking neuronal activity to gene expression and plasticity of the nervous system. *Annu. Rev. Neurosci.* **2008**, *31*, 563–590. [CrossRef]
91. Yang, J.; Ruchti, E.; Petit, J.M.; Jourdain, P.; Grenningloh, G.; Allaman, I.; Magistretti, P.J. Lactate promotes plasticity gene expression by potentiating NMDA signaling in neurons. *Proc. Natl. Acad. Sci. USA* **2014**, *111*, 12228–12233. [CrossRef]
92. Margineanu, M.B.; Mahmood, H.; Fiumelli, H.; Magistretti, P.J. L-lactate regulates the expression of synaptic plasticity and neuroprotection genes in cortical neurons: A transcriptome analysis. *Front. Mol. Neurosci.* **2018**, *11*, 1–17. [CrossRef]
93. Powell, C.L.; Davidson, A.R.; Brown, A.M. Universal glia to neurone lactate transfer in the nervous system: Physiological functions and pathological consequences. *Biosensors* **2020**, *10*, 183. [CrossRef]
94. Camandola, S.; Mattson, M.P. Brain metabolism in health, aging, and neurodegeneration. *EMBO J.* **2017**, *36*, 1474–1492. [CrossRef]
95. Zhang, X.; Alshakhshir, N.; Zhao, L. Glycolytic metabolism, brain resilience, and Alzheimer's disease. *Front. Neurosci.* **2021**, *15*, 476. [CrossRef] [PubMed]
96. Bak, L.K.; Walls, A.B.; Schousboe, A.; Waagepetersen, H.S. Astrocytic glycogen metabolism in the healthy and diseased brain. *J. Biol. Chem.* **2018**, *293*, 7108–7116. [CrossRef]
97. Ryu, W.-I.; Bormann, M.K.; Shen, M.; Kim, D.; Forester, B.; Park, Y.; So, J.; Seo, H.; Sonntag, K.-C.; Cohen, B.M. Brain cells derived from Alzheimer's disease patients have multiple specific innate abnormalities in energy metabolism. *Mol. Psychiatry* **2021**, 1–13. [CrossRef]
98. Emamzadeh, F.N.; Surguchov, A. Parkinson's disease: Biomarkers, treatment, and risk factors. *Front. Neurosci.* **2018**, *12*, 1–14. [CrossRef] [PubMed]
99. Yang, C.; Zhang, T.; Wang, W.; Xiang, Y.; Huang, Q.; Xie, C.; Zhao, L.; Zheng, H.; Yang, Y.; Gao, H. Brain-region specific metabolic abnormalities in Parkinson's disease and levodopa-induced dyskinesia. *Front. Aging Neurosci.* **2020**, *12*, 1–11. [CrossRef]

100. Tefera, T.W.; Borges, K. Metabolic dysfunctions in amyotrophic lateral sclerosis pathogenesis and potential metabolic treatments. *Front. Neurosci.* **2017**, *10*, 1–15. [CrossRef] [PubMed]
101. Tefera, T.W.; Steyn, F.J.; Ngo, S.T.; Borges, K. CNS glucose metabolism in amyotrophic lateral sclerosis: A therapeutic target? *Cell Biosci.* **2021**, *11*, 1–17. [CrossRef]
102. Dogan, A.E.; Yuksel, C.; Du, F.; Chouinard, V.-A.; Öngür, D. Brain lactate and PH in schizophrenia and bipolar disorder: A systematic review of findings from magnetic resonance studies. *Neuropsychopharmacology* **2018**, *43*, 1681–1690. [CrossRef]
103. Rowland, L.M.; Pradhan, S.; Korenic, S.; Wijtenburg, S.A.; Hong, L.E.; Edden, R.A.; Barker, P.B. Elevated brain lactate in schizophrenia: A 7 T magnetic resonance spectroscopy study. *Transl. Psychiatry* **2016**, *6*, 1–7. [CrossRef]
104. Kuang, H.; Duong, A.; Jeong, H.; Zachos, K.; Andreazza, A.C. Lactate in bipolar disorder: A systematic review and meta-analysis. *Psychiatry Clin. Neurosci.* **2018**, *72*, 546–555. [CrossRef]
105. Sullivan, C.R.; Mielnik, C.A.; Funk, A.; O'Donovan, S.M.; Bentea, E.; Pletnikov, M.; Ramsey, A.J.; Wen, Z.; Rowland, L.M.; McCullumsmith, R.E. Measurement of lactate levels in postmortem brain, IPSCs, and animal models of schizophrenia. *Sci. Rep.* **2019**, *9*, 1–7. [CrossRef]
106. Carrard, A.; Elsayed, M.; Margineanu, M.; Boury-Jamot, B.; Fragnière, L.; Meylan, E.M.; Petit, J.-M.; Fiumelli, H.; Magistretti, P.J.; Martin, J.-L. Peripheral administration of lactate produces antidepressant-like effects. *Mol. Psychiatry* **2016**, *23*, 392–399. [CrossRef] [PubMed]
107. Karnib, N.; El-Ghandour, R.; El Hayek, L.; Nasrallah, P.; Khalifeh, M.; Barmo, N.; Jabre, V.; Ibrahim, P.; Bilen, M.; Stephan, J.; et al. Lactate is an antidepressant that mediates resilience to stress by modulating the hippocampal levels and activity of histone deacetylases. *Neuropsychopharmacology* **2019**, *44*, 1152–1162. [CrossRef] [PubMed]
108. Guzowski, J.F.; Setlow, B.; Wagner, E.K.; McGaugh, J.L. Experience-dependent gene expression in the rat hippocampus after spatial learning: A comparison of the immediate-early genesArc, c-fos, and zif268. *J. Neurosci.* **2001**, *21*, 5089–5098. [CrossRef] [PubMed]
109. El Hayek, L.; Khalifeh, M.; Zibara, V.; Abi Assaad, R.; Emmanuel, N.; Karnib, N.; El-Ghandour, R.; Nasrallah, P.; Bilen, M.; Ibrahim, P.; et al. Lactate mediates the effects of exercise on learning and memory through SIRT1-dependent activation of hippocampal brain-derived neurotrophic factor (BDNF). *J. Neurosci.* **2019**, *39*, 2369–2382. [CrossRef]
110. Ding, R.; Tan, Y.; Du, A.; Wen, G.; Ren, X.; Yao, H.; Ren, W.; Liu, H.; Wang, X.; Yu, H.; et al. Redistribution of monocarboxylate 1 and 4 in hippocampus and spatial memory impairment induced by long-term ketamine administration. *Front. Behav. Neurosci.* **2020**, *14*, 1–10. [CrossRef]
111. Harris, R.A.; Lone, A.; Lim, H.; Martinez, F.; Frame, A.K.; Scholl, T.J.; Cumming, R.C. Aerobic glycolysis is required for spatial memory acquisition but not memory retrieval in mice. *ENeuro* **2019**, *6*. [CrossRef]
112. Radiske, A.; Rossato, J.I.; Gonzalez, M.C.; Köhler, C.A.; Bevilaqua, L.R.; Cammarota, M. BDNF controls object recognition memory reconsolidation. *Neurobiol. Learn. Mem.* **2017**, *142*, 79–84. [CrossRef] [PubMed]
113. Vaccari Cardoso, B.; Shevelkin, A.V.; Terrillion, S.; Mychko, O.; Mosienko, V.; Kasparov, S.; Pletnikov, M.V.; Teschemacher, A.G. Reducing L-lactate release from hippocampal astrocytes by intracellular oxidation increases novelty induced activity in mice. *Glia* **2021**, *69*, 1241–1250. [CrossRef]
114. Hott, S.C.; Gomes, F.V.; Fabri, D.R.S.; Reis, D.G.; Crestani, C.C.; Côrrea, F.M.A.; Resstel, L.B.M. Both A1- and B1-adrenoceptors in the bed nucleus of the stria terminalis are involved in the expression of conditioned contextual fear. *Br. J. Pharmacol.* **2012**, *167*, 207–221. [CrossRef]
115. Giustino, T.F.; Maren, S. Noradrenergic modulation of fear conditioning and extinction. *Front. Behav. Neurosci.* **2018**, *12*, 1–20. [CrossRef]
116. Oe, Y.; Wang, X.; Patriarchi, T.; Konno, A.; Ozawa, K.; Yahagi, K.; Hirai, H.; Tsuboi, T.; Kitaguchi, T.; Tian, L.; et al. Distinct temporal integration of noradrenaline signaling by astrocytic second messengers during vigilance. *Nat. Commun.* **2020**, *11*, 1–15. [CrossRef]
117. Descalzi, G.; Gao, V.; Steinman, M.Q.; Suzuki, A.; Alberini, C.M. Lactate from astrocytes fuels learning-induced MRNA translation in excitatory and inhibitory neurons. *Commun. Biol.* **2019**, *2*, 247. [CrossRef] [PubMed]
118. Boury-Jamot, B.; Halfon, O.; Magistretti, P.J.; Boutrel, B. Lactate release from astrocytes to neurons contributes to cocaine memory formation. *BioEssays* **2016**, *38*, 1266–1273. [CrossRef] [PubMed]
119. Boutrel, B.; Magistretti, P.J. A role for lactate in the consolidation of drug-related associative memories. *Biol. Psychiatry* **2016**, *79*, 875–877. [CrossRef]
120. Zhang, Y.; Xue, Y.; Meng, S.; Luo, Y.; Liang, J.; Li, J.; Ai, S.; Sun, C.; Shen, H.-W.; Zhu, W.; et al. Inhibition of lactate transport erases drug memory and prevents drug relapse. *Biol. Psychiatry.* **2016**, *79*, 928–939. [CrossRef]
121. Boury-Jamot, B.; Carrard, A.; Martin, J.L.; Halfon, O.; Magistretti, P.J.; Boutrel, B. Disrupting astrocyte-neuron lactate transfer persistently reduces conditioned responses to cocaine. *Mol. Psychiatry* **2016**, *21*, 1070–1076. [CrossRef] [PubMed]
122. Wu, P.; Xue, Y.-X.; Ding, Z.-B.; Xue, L.-F.; Xu, C.-M.; Lu, L. Glycogen Synthase Kinase 3β in the Baso...Ocaine Reward Memory_Enhanced Reader.Pdf. *J. Neurochem.* **2011**, *118*, 113–125. [CrossRef]
123. Jian, M.; Luo, Y.X.; Xue, Y.X.; Han, Y.; Shi, H.S.; Liu, J.F.; Yan, W.; Wu, P.; Meng, S.-Q.; Deng, J.-H.; et al. EIF2α dephosphorylation in basolateral amygdala mediates reconsolidation of drug memory. *J. Neurosci.* **2014**, *34*, 10010–10021. [CrossRef]
124. Pellerin, L.; Magistretti, P.J. The Central Role of Astrocytes in Neuroenergetics. In *Neuroglia*, 2nd ed.; Kettenmann, H., Ransom, B.R., Eds.; Oxford University Press: Oxford, UK, 2013. [CrossRef]

125. Agus, M.; Calì, C.; Al-Awami, A.; Gobbetti, E.; Magistretti, P.; Hadwiger, M. Interactive volumetric visual analysis of glycogen-derived energy absorption in nanometric brain structures. *Comput. Graph. Forum* **2019**, *38*, 427–439. [CrossRef]
126. Boges, D.J.; Agus, M.; Magistretti, P.J.; Calì, C. Forget About Electron Micrographs: A Novel Guide for Using 3D Models for Quantitative Analysis of Dense Reconstructions. In *Neuromethods*; Humana Press: New York, NY, USA, 2020; Volume 155, pp. 263–304. [CrossRef]

 metabolites

Review

Control of Adipose Cell Browning and Its Therapeutic Potential

Fernando Lizcano [1,*] and Felipe Arroyave [2]

[1] Center of Biomedical Investigation, (CIBUS), Universidad de La Sabana, 250008 Chia, Colombia
[2] Doctoral Program in Biociencias, Universidad de La Sabana, 250008 Chia, Colombia; felipe.arroyave@unisabana.edu.co
* Correspondence: Fernando.lizcano@unisabana.edu.co

Received: 28 August 2020; Accepted: 2 November 2020; Published: 19 November 2020

Abstract: Adipose tissue is the largest endocrine organ in humans and has an important influence on many physiological processes throughout life. An increasing number of studies have described the different phenotypic characteristics of fat cells in adults. Perhaps one of the most important properties of fat cells is their ability to adapt to different environmental and nutritional conditions. Hypothalamic neural circuits receive peripheral signals from temperature, physical activity or nutrients and stimulate the metabolism of white fat cells. During this process, changes in lipid inclusion occur, and the number of mitochondria increases, giving these cells functional properties similar to those of brown fat cells. Recently, beige fat cells have been studied for their potential role in the regulation of obesity and insulin resistance. In this context, it is important to understand the embryonic origin of beige adipocytes, the response of adipocyte to environmental changes or modifications within the body and their ability to transdifferentiate to elucidate the roles of these cells for their potential use in therapeutic strategies for obesity and metabolic diseases. In this review, we discuss the origins of the different fat cells and the possible therapeutic properties of beige fat cells.

Keywords: beige adipocyte; white adipocyte; brown adipocyte; obesity; diabetes mellitus; differentiation

1. Introduction

Obesity is a disease that induces a series of cardiovascular, metabolic and osteoarticular complications that reduce life expectancy. It is prevalent on a global scale, with multiple factors contributing to its development. The treatment possibilities for metabolic diseases have increased dramatically. However, in the circumstance of obesity, many medications have been withdrawn from the market due to undesirable side effects [1,2]. Other drugs have not presented the desired efficacy and the projections of therapeutic efficiency are low with high costs. Furthermore, one of the difficulties in many countries is that obesity has not been declared a disease. Obesity is a highly stigmatized condition that has long been generally regarded by the public as a reversible consequence of personal choices. For this reason, obesity is seen even in some countries, as a circumstance of the person, without policies for its prevention, adequate therapy and of course the risk of the appearance of complications is not supported [3,4]. Recent observations have proven that a variety of types of adipose tissue dysfunction clearly play a role in the genesis of many obesity-related diseases. These include impairments in adipocyte storage and release of fatty acids, overproduction or underproduction of "adipokines" and cytokines, hormonal conversion, and the adverse mechanical effects of greater tissue mass [5,6]. Additionally, adipose tissue has been shown to be dynamically more active than initially considered [7]. Adipose tissue is constituted by white adipose tissue (WAT), which has the property of storing energy in the form of triglycerides and is useful in preventing deficiencies of energy during periods of prolonged starvation. In comparison, there is the brown adipose tissue (BAT), which is more metabolically active and has the property of producing heat through the activation of uncoupling proteins (UCP1). BAT controls energy homeostasis during periods of low temperature and hibernation [8]. In adult humans, WAT is believed to predominate, while BAT has a predominant role in the first few months of life when heat production by adipose tissue is necessary to maintain body temperature [9]. However, another type of adipose tissue has been described in adults, whose functional characteristics may be similar to those observed in BAT. This type of adipose tissue is observed in circumstances of low temperatures or after sympathetic activation [10]. Although this adipose tissue is functionally very similar to BAT, it has specific characteristics and for this reason it has been referred to as beige adipose tissue [11,12]. Although adipose tissue has a mesenchymal origin, in the differentiation process manifest differences are established and BAT has more similarities with muscle cells of mesenchymal origin than with WAT [13,14]. Additionally, in the process of differentiation, some external factors can modify the cells that give rise to WAT and change the phenotype of these cells towards cells that are more metabolically active, such as beige adipocyte [15,16]. In the human body, energy homeostasis is regulated in the central nervous system by the hypothalamus, which has several widely interconnected hypothalamic neural circuits. The hypothalamus is involved in the physiological control of many functions, including metabolic homeostasis. The chemical signals that can modulate the function of the hypothalamus are the signals that come from the immune system, those that come from the activation of the sympathetic nervous system and other signals that can induce epigenetic modifications, which influence the expression of specific genes that influence the transcription of specific genes [17,18]. This review discusses the metabolic control of organisms, the effect of the environment on the regulation of thermogenesis and the different adipocytes identified to date, emphasizing the therapeutic potential of beige adipocytes.

2. Hypothalamic Control of Energy Homeostasis

The control of energy in the body is based on the balance between the intake of energy through food and the expenditure of calories. Energy homeostasis is regulated in the central nervous system by the hypothalamus, which has several widely interconnected hypothalamic neural circuits. The hypothalamus is involved in the physiological control of many functions, including the regulation of hormonal axes, autonomic nervous system activity, and metabolic homeostasis [19]. Hypothalamic nuclei play a major role in the transmission of information from peripheral signals on energy availability, including hormone and nutrient signals, integrating them and generating an appropriate response in terms of food intake and energy expenditure [20,21]. Due to its involvement in numerous metabolic processes, the hypothalamus is considered to be the master regulator of energy homeostasis [22].

Energy expenditure is the sum of the thermic effect of food, locomotor activity and thermogenesis [23,24]. Interestingly, basal thermogenesis, which is the heat produced by metabolism, is sufficient to preserve body temperature at adequate levels without involving thermoregulatory mechanisms. This range of temperatures in which the body is maintained in a harmonic state with the environmental temperature is called thermoneutrality [25,26], with temperatures below thermoneutrality inducing an immediate response through peripheral vasoconstriction [24]. However, this primary response only provides limited effects in maintaining body temperature. Therefore, the body uses additional thermogenic mechanisms, referred to as adaptive thermogenesis [27], that can be induced by shivering or involve non-shivering thermogenesis. Whereas shivering produces additional heat from movement, mammals have developed a non-shivering adaptive thermogenic mechanism that is carried out by BAT [28]. Core temperature control is essential to preserve energy homeostasis. The preoptic area (POA) is located in the anterior region of the hypothalamus and controls body temperature. The POA has the ability to receive thermosensitive peripheral signals from the skin and intestinal organs and trigger the activation of efferent signals that can promote BAT thermogenesis [29,30].

Moreover, changes in behavior, including the waking state, the response of the immune system and stress are characterized by elevated body temperature. Although the neural circuits and transmitters involved in the response to behavioral changes are not well understood, the activity of orexins may have an important role in these behavioral events [31]. Additionally, the high metabolic rate of beige adipose tissue and BAT during thermogenesis cannot be maintained without a reliable supply of metabolic fuels, particularly oxygen, lipolytic products, and glucose [32,33]. A great deal of information has been obtained from the study of neuronal regulators, specifically, those that function at the hypothalamic level that are responsible for these events. The arcuate nucleus (ARC) of the hypothalamus is the region that is most implicated in the control of eating habits. The ARC is made up of two primary neural populations: (a) orexigenic (feeding-promoting) neurons and co-expressing neuropeptide Y (NPY) and related protein agouti (AgRP) and (b) anorexigenic (feeding-inhibiting) neurons that co-express a transcription factor related to cocaine and amphetamines (CART) and pro-opiomelanocortin (POMC), the precursor of alpha-melanocyte-stimulating hormone (α-MSH) and adrenocorticotrophic hormone (ACTH) [22,34]. After integration by the ARC, peripheral signals are transmitted by neural projections to hypothalamic areas of the dorsomedial nucleus (DMH), paraventricular nucleus (HPV), lateral hypothalamic area (LHA), and ventromedial nucleus of the hypothalamus (VMH) [35]. VMH neurons can communicate with other hypothalamic areas, such as the DMH, LHA, and ARC, in addition to other brain regions, such as the vagus motor dorsal nucleus (DMV), the nucleus of the solitary tract (NTS), the pale raphe (RPa) and the lower olive (IO) [36].

In addition to the established biochemical neurotransmission mechanisms, functional studies have identified specific areas of the brain that generate WAT browning. For example, the role of neuropeptide-Y (NPY) in DMH nuclei, in addition to oxidative stress and the administration of CART to the paraventricular nucleus (PVN), has been shown to induce an increase in uncoupling protein1 (UCP1) levels in WAT [37]. The regulation of WAT metabolic activity by thyroid hormones, bone morphogenic protein 8B (BMP8b) and the incretin glucagon-like peptide-1 (GLP-1) can be reduced by blocking AMP activated protein kinase (AMPK) in the VMH nucleus [28]. Increased expression of the endoplasmic reticulum (ER) chaperone protein GRP78, which improves ER stress, induces signal activation mediated by β-3-adrenergic receptors, increasing browning and reducing weight [38].

In recent years, several studies have shown the important role that the stimulation of LHA nucleus neurons has in BAT activation. The nerve terminals that reach beige adipocytes promote the browning process. Under cold exposure, the central neural circuits in the hypothalamic (HPV and LHA) and brain stem of RPa and locus coeruleus are rearranged with higher proportions of neurons projecting into BAT and beige adipocytes [39]. These data provide strong evidence indicating a likely reorganization of nervous system connectivity after WAT browning [40] (Figure 1).

The results of pharmacological experiments and animal studies have shown that orexins induce increased energy expenditure. Mice with reduced orexin expression show a reduced ability to maintain temperature after exposure to cold [41,42]. The administration of orexin A and B to the VMH and LHA nuclei of rats induces the thermogenic activity of BAT [43–45]. There is evidence to indicate that orexins can induce the thermogenic activity of BAT by regulating the energy sensor AMPK and ER stress [46]. Some groups have observed that BMP8b can affect both BAT and the browning process of WAT [42,47]. Despite multiple lines of evidence showing the role of orexins in rodents, it has not been possible to elucidate their roles in stimulating BAT activity in humans. Treatment with Orexin A alone or in combination with an adrenergic stimulation does not affect thermogenesis [48,49]. These observations are closely related to findings in patients with narcolepsy, a disease that involves the selective deterioration of neurons that produce orexin. Patients with narcolepsy present an abnormal distribution of body fat but retain BAT deposits at the supraclavicular level [50]. Additionally, BAT has been shown to be functional in these patients after cold exposure. These observations show that the role of orexins in humans is controversial and that the control of thermogenesis and fat cell activity at the hypothalamic level in humans requires additional studies [51]. In humans, the control of thermogenesis may be influenced by multiple hormones that are secreted in different tissues, such as insulin and glucagon by the pancreas; leptin by adipose tissue; incretins, gastric inhibitory peptide (GIP), GLP-1, and ghrelin produced in gastric fundus of the gastrointestinal system; thyroid hormones and estrogens. All of these hormones may affect the regulation of adaptive thermogenesis in humans [52–55].

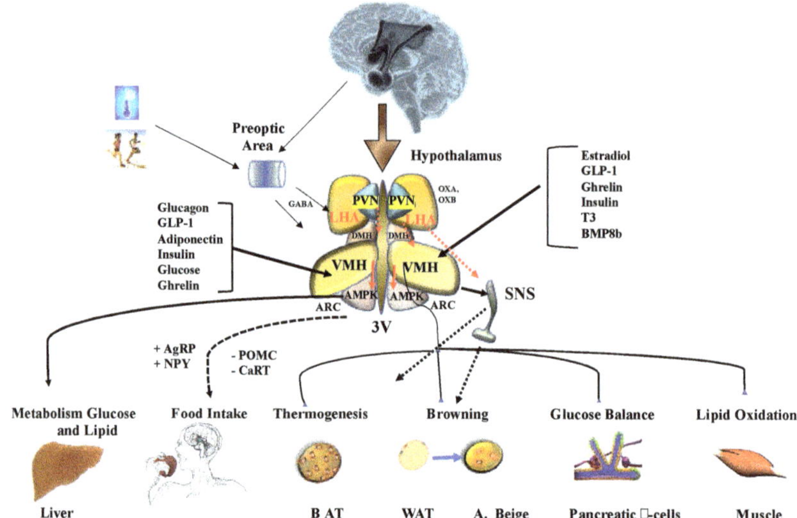

Figure 1. Energy homeostasis controlled by hypothalamus. Cold and physical peripheral signals reach the central nervous system where they interact with their specific receptors in the preoptic area. Adenine monophosphate activated protein kinase (AMPK) can regulate food intake, liver glucose production, lipid metabolism, brown adipose tissue thermogenesis (BAT), white adipose tissue (WAT) browning, and lipid and glycogen synthesis in skeletal muscle. AMPK activity in peripheral tissues is mediated by the activity of the sympathetic nervous system (SNS). Abbreviations: 3V, third ventricle; AgRP, agouti-related peptide; ARC, arcuate nucleus of the hypothalamus; BMP8b, bone morphogenetic protein 8B; CART, cocaine and amphetamine-regulated transcript; DMH, dorsomedial nucleus of the hypothalamus; GLP-1, glucagon-like peptide-1; LHA, lateral hypothalamic area; NPY, neuropeptide Y; POMC: pro-opiomelanocortin; PVN, paraventricular nucleus of the hypothalamus; T3, 3,3′,5-triiodothyronine; VMH, ventromedial nucleus of the hypothalamus. (+) increase intake; (-) reduce intake.

Energy consumption and food composition can influence the thermogenic activity of BAT. In addition to the heat released during the digestive process and nutrient transport, food has a postprandial thermal effect. Glucose and insulin can contribute to the activation of the sympathetic system and b-adrenergic receptors [56]. However, the activation of thermogenesis by glucose may be influenced by other hormones such as cholecystokinin (CCK), which acts as a sensor of the caloric value of other nutrients [57]. Hypoglycemia induces hypothermia at least in part by reducing the thermogenic activity of BAT. The reduction in glucose in the VLM nuclei completely inhibits BAT activity [58]. The contribution of lipids in the activity of postprandial thermogenesis is characterized by the activation of sympathetic nerve activity. In rodents, diets with a high lipid content stimulate the thermogenic activity of BAT and the activation of b-adrenergic receptors [59]. However, this effect diminishes after a few weeks [60]. Other nutrients that can stimulate thermogenesis are proteins. Although the effect of proteins on BAT is not yet fully elucidated, ketogenic diets with a high content of lipids and proteins and no carbohydrates induce activation of the sympathetic system [61]. High levels of ketone bodies increase BAT and energy expenditure. β-Hydroxybutyrate increases BAT activity and norepinephrine secretion, and the most likely mechanism for the effect of (beta) β-Hydroxybutyrate is through the VMH and HPV [62].

3. Environment and Obesity

Endocrine-disrupting chemicals (EDCs) are substances that are located in the environment and interfere with normal hormone action, thereby increasing the risk of developing diabetes, cancer, reproductive impairment, behavior disorders and obesity. These chemical substances have been shown to potentially cause alterations in metabolic activity in organisms and increase lipogenic activity [63]. Adipocytes are functionally classified as endocrine-active and are sensitive to changes induced by endocrine disruptors, which have been termed "obesogens" [64]. These substances may trigger an increase in adiposity by altering the cellular development program in adipocytes, increasing triglyceride storage and interfering with the neuroendocrine control of hunger and satiety centers [65,66]. Some of the chemicals we typically encountered tend to cause changes in fat cell metabolism. Bisphenol A (BPA), phytoestrogens, and tributyltin (TBT), among others, have been shown in experimental studies to induce weight increases in animal models. Although there are multiple mechanisms by which endocrine disruptors can induce obesity, the modifications described so far suggest that epigenetic modifications of gene expression through DNA methylation and covalent modifications of chromatin are common and affect subsequent generations [67,68].

Numerous potentially obesogenic compounds have been identified using in vitro assays that evaluate the ability of candidate chemicals to promote the differentiation of established cell lines, such as mesenchymal stem cells from humans and mice, and 3T3-L1 preadipocytes [69,70]. Many of the chemicals shown to promote the differentiation of white adipocytes in these analyses activate PPARγ and/or RXR [71]. Based on the central role of the PPARγ:RXR heterodimer as the master regulator of adipogenesis, it is not surprising that these nuclear transcription factors are one of the most important targets of obesogens [72–75] (Figure 2).

Human exposure to organotin can occur through dietary intake, such as by seafood contaminated with TBT, or after the consumption of products that have been in contact with pesticides such as triphenyltin [76,77].

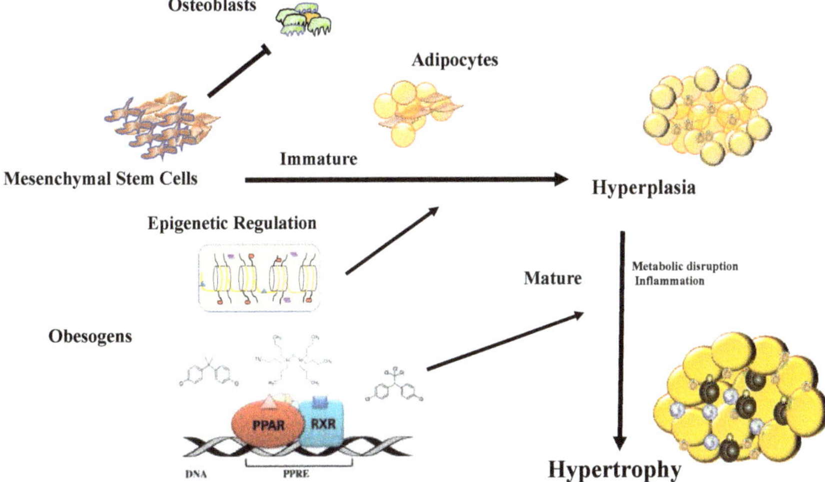

Figure 2. Chemical substances with the ability to modify thermogenesis primarily act through PPARγ:RXR receptors, the masters of adipocyte differentiation. Other changes that obesogenic substances can induce are epigenetic modifications (DNA methylation, covalent modifications of histones, and noncoding RNA). Obesogens can induce mesenchymal cells to direct differentiation towards the adipocyte line and reduce osteoblastic activity. Then, the obesogens induce hyperplasia of adipocytes and later produce hypertrophy with an accumulation of inflammatory cells that produce the secondary effects of obesity.

TBT has the ability to bind and activate PPARg, promoting adipogenesis and lipid accumulation. The results of studies using 3T3-L1 cells and human mesenchymal cells have shown that TBT at nanomolar concentrations can induce their differentiation into adipocytes [78]. Studies in humans have revealed that elevated levels of TBT in urine are associated with metabolic diseases such as diabetes and obesity. In a recent study, individuals with high levels of perfluorinated chemicals exhibited a low metabolic rate and tended to gain weight easily [79].

Similar observations have been made with other EDCs, such as phthalates, plastic components, and epoxy resins. The phthalate MEHP (mono-2-ethylhexyl phthalate) can induce adipogenesis in 3T3-L1 cells by activating PPARγ. Prenatal exposure to bisphenol A (BPA) has been linked to several adverse effects, including weight gain, reproductive disorders, and behavioral changes in mice and rats. BPA can interfere with the activity of estrogen receptors (ERs) and with PPARγ [80].

Although several studies have observed that obesogens primarily influence PPARγ receptor activity, recent studies have shown that other functional variations can be attributed to obesogens, such as modifications to the retinoid acid receptor and other nuclear receptors, such as glucocorticoid or thyroid hormone receptors [81,82]. Other functional variations have been observed after epigenetic modifications in WAT, including structural changes in chromatin or modifications in the gut microbiota [83,84].

Epigenetic modifications mediated by TBT include a reduction in the trimethylation of histone 3 lysine 27 (H3K27me3), which increases the expression of genes involved in the development of adipogenesis [85]. Other modifications have also been observed, such as a reduction in DNA methylation [86], and it has even been argued that DNA methylation modifications may have a transgenerational impact [87,88]. However, both the evidence and the mechanisms for this epigenetic modification have not been fully elucidated. In the induction of adipogenesis by TBT, these adipocytes have been shown to exhibit an altered functionality due to changes in oxidative respiration, difficulty in expressing thermogenic proteins after cold stimulation or activation of b-adrenergic receptors. These adipocytes also tend to be insulin resistant and express profibrotic proteins [89,90].

Recently, other chemicals with potential obesogenic effects have been described. The TBT metabolite dibutyltin (DBT), the prevalence of which is higher than TBT in the environment, can also induce 3T3-L1 preadipocyte differentiation. In mouse studies, DBT has been observed to cause insulin resistance [91]. Analogs of bisphenol A, bisphenol S (BPS) and bisphenol F (BPF) are able to activate PPARγ and induce adipogenesis. A longitudinal study of a cohort of children demonstrated that exposure to BPS and BPF was significantly related to obesity in children [92,93]. Acrylamide is widely used as a colorant in the manufacture of paper and other industrial products and can be generated after cooking foods with a high carbohydrate content at high temperatures. Interestingly, acrylamide increases the activity of the adenosine 5′-monophosphate-activated protein kinase-acetyl-CoA carboxylase (AMPK-ACC) pathway [94]. The results of two European studies, one in France and one in Norway, showed that children exposed to high levels of polyacrylamide during the prenatal period were more likely to be short for their gestational age and tended to be obese after 3 years of age [95,96]. Surfactants such as dioctyl sodium sulfosuccinate (DOOS) and Span 80 can activate the PPARγ and RXR receptors, and the combination of these products can induce an increase in adipogenesis. Other chemical compounds used as preservatives, such as 3-tertbutyl-4-hydroxyanisole (3-BHA), induce the differentiation of 3T3-L1 preadipocytes [97], while the flavoring monosodium glutamate (MSG) can promote adipogenesis by modifying the secretion of GLP-1. In addition, some herbicides that remain in use in some countries, such as glyphosate, can induce obesity in the F2 and F3 offspring of females exposed to during gestation [98].

4. Functional Variation of Adipocytes

Adipose tissue has allowed mammals to adapt to changes due to energy demand, environmental conditions, and nutrient availability. In recent years, knowledge of adipose tissue has been ostensibly expanded, which is partly due to increases in metabolic diseases and cardiovascular risk [99–101]. Although fat cells are the primary component of adipose tissue, almost 40% comprises vascular components, macrophages, fibroblasts, endothelial cells and adipocyte precursor cells [89,102,103]. The size of the adipocytes can vary considerably from 20 to 200 μm in diameter, showing that they have great plasticity and a high capacity to modify their volume [104,105]. Adipose tissue pathologies can result as a consequence of lipid accumulation and adipocyte cell hypertrophy. WAT secretes endocrine and inflammatory signals under hypertrophic conditions, inducing a prothrombotic state and generating a chronic inflammation disorder [106,107]. This condition causes many of the cardiovascular complications presented by patients with metabolic diseases [108,109]. Most of the adipose tissue in adults is WAT, which has the functional characteristic of saving energy for periods of famine. Phenotypically, white adipocytes have a single cytoplasmic lipid droplet that is responsible for storing triglycerides as a consequence of lipogenic processes. Recent observations have shown that WAT exhibits great phenotypic plasticity [7,110]. These adipocytes, under the regulation of the sympathetic nervous system, can release fatty acids and secrete substances with endocrine effects [111]. Since the discovery of leptin in the 1990s [112,113], adipose tissue has not only been seen as an exclusive energy storage organ but also as a dynamic organ with an endocrine function, and several WAT-secreted adipokines with various biological activities have been described in recent years. Leptin regulates the energy homeostasis of the body and interferes with various neuroendocrine and immune functions, regulating food intake and increasing energy utilization through hypothalamic signals [114]. Leptin administration reduces the weight of mice or humans with congenital leptin deficiency [115]. However, the function of leptin in humans with diet-induced obesity is subject to the control of receptors and transporters that make pharmacological therapy difficult [116,117]. Adiponectin is another adipokine that is predominantly secreted by WAT, although it can also be secreted by skeletal muscle and cardiomyocytes. Adiponectin plays a crucial role in glucose and lipid metabolism, inflammation and oxidative stress. Adiponectin levels increase with exposure to insulin-sensitizing drugs, with adiponectin plasma levels observed that are inversely proportional to insulin resistance. Adiponectin also has anti-inflammatory properties with anti-atherogenic effects and promotes angiogenesis [118,119]. Resistin is secreted by WAT and macrophages and has an important role in inflammatory processes that trigger insulin resistance, with some studies having determined that elevated plasma levels of resistin are a predictor of the future development of type 2 diabetes mellitus [120,121]. The mechanisms by which resistin can induce insulin resistance are not fully elucidated in humans. However, the results of in vitro studies have shown that the activities of pro-inflammatory cytokines such as tumor necrosis factor-α (TNF-α) and interleukin 6 (IL-6) in addition to the functional modification of 5′AMP-activated protein kinase (AMPK) may be involved in resistin-mediated insulin resistance [122]. Visfatin, also known as nicotinamide phosphoribosyl-transferase (Nampt), is an adipocytokine secreted by adipocytes, macrophages, and inflamed endothelial tissue. High levels of visfatin are observed in patients with obesity, type 2 diabetes mellitus, chronic inflammatory conditions and cancer, and an association between serum visfatin levels and cardiovascular disease has recently been observed in patients with type 2 diabetes [123–125]. Moreover, BAT exhibits multiple cytoplasmic lipid inclusions and numerous mitochondria. Compared to WAT, BAT is highly vascularized and rapidly metabolizes fatty acids, favoring optimal oxygen consumption and heat production [23]. Many environmental or molecular stimuli can increase the appearance of BAT [126]. Brown adipocytes are primarily observed in small mammals and in the newborn, with the embryological formation of BAT preceding that of WAT due to its thermogenic function in newborns. BAT originates from a subpopulation of the dermomyotome that expresses molecular markers such as paired box 7 (Pax7), engreiled-1 (En1), and myogenic factor 5 (Myf5) [7,127–129]. BAT can secrete cytokines that have an effect on different tissues and prevent diet-induced obesity. Follistatin is a soluble glycoprotein secreted by BAT that can blockade the activities of some members of the transforming growth factor (TGF) family, induce insulin sensitivity and prevent diet-induced obesity [130,131]. The c-terminal fragment of slit guidance ligand 2 (SLIT-C) belongs to the Slit family of secreted proteins that play an important role in various physiological and pathological activities, including inflammatory cell chemotaxis. SLIT-C is secreted by BAT and induces thermogenic WAT browning and metabolic processes associated with substrate supply to fuel thermogenesis [128,132]. Growth differentiation factor 8 (GDF8, also known as myostatin) and growth differentiation factor 15 (GDF15) are members of the transforming growth factor family, which are involved in the control of hunger-related neural circuits. GDF15 overexpression has been shown to prevent obesity and insulin resistance by increasing the expression of thermogenic genes [133]. Fibroblast growth factor 21 (FGF21) is a regulator of energy homeostasis that is primarily secreted by the liver. BAT-secreted FGF21 prevents hyperglycemia and hyperlipidemia in mice [134], and FGF21 analogues tested in overweight/obese patients with type 2 diabetes mellitus have been shown to reduce dyslipidemia and hepatic steatosis, although they do not lead to improvements in glucose control and body weight [135]. Although FGF21 was reported to have anti-inflammatory effects on white adipocytes, it remains to be determined if FGF21 has a similar action in BAT [136,137].

Beige adipose tissue is the newest of these adipose tissues and has some morphological characteristics in common with WAT and BAT. The nature of these cells is controversial, although it is believed that their origin is secondary to the differentiation of WAT, and their differentiation from cell precursors has been observed [30,138]. Beige adipocytes have simple lipid inclusions similar to WAT, but when faced with stimuli such as cold exposure, their behavior is similar to that of BAT cells. The thermogenic capacity and potential role of beige adipose tissue in the regulation of obesity and insulin resistance are currently being studied [139,140]. Beige adipocyte biogenesis, also called beige adipogenesis or (browning/beigeing), is induced by chronic exposure to external cues such as cold, adrenergic stimulation, and long-term treatment with peroxisome proliferator-activated receptor gamma (PPARγ) agonists, among others [141] (Table 1).

Table 1. Characteristics of different adipocyte tissues.

White Adipocytes (WAT)	Brown Adipocytes (BAT)	Beige Adipocytes * (Cold, TZD, FGF21, IL-4, IL-6)		
Fatty Ac Oxidation (+)	Fatty Ac Oxidation (+++)	Fatty Ac Oxidation (+)	→	Fatty Ac Oxidation (+++)
Lipid Storage (+++)	Lipid Storage (+)	Lipid Storage (+++)	→	Lipid Storage (+)
Mitochondria (+)	Mitochondria (+++)	Mitochondria (+)	→	Mitochondria (+++)
PGC-1α (+)	PGC-1α (+++)	PGC-1α (+)	→	PGC-1α (+++)
Respiratory Chain (+)	Respiratory Chain (+++)	Respiratory Chain (+)	→	Respiratory Chain (+++)
Succinate (+)	Succinate (+++)	Succinate (+)	→	Succinate (+++)
UCP1 (-)	UCP1 (+++)	UCP1 (-)	→	UCP1 (+++)
Unilocular	Multilocular	Unilocular	→	Multilocular
Markers: ASC-1, Resistin, Leptin	Markers: Eva1, Lhx8, Zic1	Markers: CD137, CIDEA, Cited, SLIT2, Tbx, Tmem26,		

* Some conditions that can induce thermogenic activity of beige adipocytes. ASC-1, adipocyte-specific cell surface protein-1; CIDEA, cell death-inducing DFFA-like effector; CD 137, cluster of differentiation 137; Cited1, Cbp/P300-interacting transactivator 1; Eva1, epithelial V-like antigen 1; FABP4, fatty acid binding protein 4; FGF21, fibroblast growth factor 21; Lhx8, LIM/homeobox protein; PGC-1α, peroxisome proliferator-activated receptor-gamma coactivator alpha 1; Pdk4, pyruvate dehydrogenase kinase 4; SLIT2, slit guidance ligand 2; Tbx1,T-box transcription factor 1; Tmem2, transmembrane protein 26; TZD, thiazolidinedione; UCP1, uncoupling protein 1; Zic1, zinc finger protein 1. (+) increase; (+++) highest increase; (-) reduce; (→) Changes in Beige adipocytes after browning.

Browning is a temporary adaptive response that lasts even after the dissipation of external environmental signals [140,142]. Beige adipocytes have an origin that is not yet fully clarified. Some are believed to arise from WAT from cell precursors that express CD34, in addition to platelet-derived growth factor receptor alpha (PDGFRα), and spinocerebellar ataxia type 1 (SCA1) proteins [143–145].

Beige adipocytes may also be derived from the Myf5-negative precursors of inguinal WAT [30], and the results of a number of studies suggest that beige adipocyte precursors, such as WAT precursors, reside in adipose tissue vasculature [141,146,147]. Recently, some beige adipocytes have been shown to express myosin heavy chain 11 (Myh11), which is a selective marker of smooth muscle cells [141]. These observations may indicate that during embryonic development beige adipocytes have a different cellular origin from that seen with classic brown adipocytes [148–150].

Another of the functional characteristics of beige adipocytes in relation to other adipocytes is their functional flexibility. Although the differentiation process of beige adipocytes from precursor cells is highly inducible, there is clear evidence that mature white adipocytes could be transdifferentiated into beige adipocytes by specific exogenous factors [105].

Whether this change is manifestation of a real transdifferentiation from white adipocytes, a direct transformation of white adipocytes to beige adipocytes, or resembles beige adipocytes that previously remained hidden among white adipocytes is a matter of debate [151]. One of the considerations that are still under evaluation is the amount of BAT in adult humans, because it is considered that most of the thermogenic adipose tissue in adults corresponds to beige adipocytes [148,152–154]. However, BAT can be observed in adults in specific areas, such as the posterior neck and perirenal area [155–158]. The adipose progenitor cells (APC) maintain a continuous supply of adipocytes in the different tissues in the body. In this way, the number of adipocytes in the body remains constant in adults despite the fact that the individual is obese or thin. This indicates that the amount of adipocytes in the body is established during childhood and adolescence in a correspondent way with the size of the different organs of the body [150]. According to recent studies, adult APC cells are considered to reside in the stromal vascular fraction (SVF). Specifically, studies in mice have identified cells that have APC characteristics in SVF and express PPARγ [149,159]. Using genetic-tracing methodologies, PPARγ-expressing APCs were shown to be crucial for adipocyte formation in vitro and in vivo [160]. In vivo tracking of PPARγ cells has indicated that these cells reside within blood vessel walls. In line with a vascular residency, these APCs resemble mural cells (pericytes and vascular smooth muscle cells) due to their expression of several mural cell markers, such as platelet-derived growth factor receptor-beta (PDGFRβ) and alpha-smooth muscle actin (α-SMA). The results of smooth muscle genetic fate-mapping studies have suggested that cells expressing Myh11, PDGFRβ, and SMA can generate beige adipocytes in response to cold exposure [12,161]. SMA perivascular cells have been shown to generate 50–70% of new beige adipocytes after 1 week of cold exposure. Remarkably, blocking adipogenesis within SMA cells or ablating SMA positive cells led to the failure to generate cold-induced beige adipocytes, and mice were unable to either preserve their temperature or lower plasma glucose levels [162,163].

5. Thermogenesis by Brown and Beige Adipose Tissue

Free energy for life-sustaining biochemical processes in mammals is provided by reduced substrates. The energy in the cells is subject to the oxidation of substrates in the inner membrane of the mitochondria through oxidative phosphorylation. In this electrochemical pathway, proton conductance is established by the mitochondrial respiratory chain producing energy [164]. Thus, cell metabolism is carried out in the membranes of the mitochondria through oxidative phosphorylation. The protonmotive force (Dp) generated by mitochondrial respiration drives protons back into the mitochondrial matrix through ATP synthase, providing energy for the reaction ADP+Pi/ATP [165]. The hydrolysis of ATP into ADP and inorganic phosphate releases 30.5 kJ / mol, with a change in free energy of 3.4 kJ/mol [166]. The energy released by the division of a unit of phosphate (Pi) or pyrophosphate (PPi) of ATP in the standard 1 M state is: $ATP + H_2O \rightarrow ADP + Pi\ \Delta G = -30.5$ kJ/mol (−7.3 kcal/mol). Interestingly, most of the thermal energy that is produced from the oxidation of substrates is conserved in a small fraction, and most of this energy is released in the form of heat. Thereby, in cells that maintain oxidative metabolism from the mitochondrial respiratory chain, the production of heat is subject to the rate of mitochondrial respiration. Thermogenesis in brown and beige adipocytes is effectively controlled by modulating the steps that modify the mitochondrial respiratory chain [167,168]. UCP1, previously referred to as thermogenin, is responsible for the conductance of protons in brown and beige adipocytes. Taking into account the laws of thermodynamics, most of the energy generated by the electrochemical potential in the oxidation of brown and beige adipocytes is dissipated as heat and is not used for the phosphorylation of ADP. Thus, the activation of UCP1 acts as a small radiator in the brown and beige adipocytes [169,170]. Experiments performed with mitochondria from brown adipocyte have shown that free fatty acids can increase UCP1-mediated proton conductance. In these experiments it was observed that purine nucleotides can inhibit proton translocation by binding to the cytosolic face of UCP1. Under basal conditions, proton leakage is maintained by a predominant role of purine nucleotides on UCP1 [25]. Studies in rats observed that while the basal proton conductance represents 20 to 30% of the metabolic rate in hepatocytes, it increases to up to 50% in skeletal muscle. Taking into account the large proportion of skeletal muscle and the high metabolic activity of the liver, the metabolic rate in a mammal at rest is governed by proton conductance under thermoneutrality conditions and in the postabsorptive state [171]. However, experiments both in vivo and in cell cultures have shown that the thermogenic capacity of beige and brown adipocytes can be subject to external stimuli. In these experiments, it was observed that exposition of brown and beige adipocytes to cold induced a strong increase in mitochondrial respiration [8]. In humans, cold stimulation activates cold-sensitive thermoreceptors in the skin or viscera and transmits afferent signals to the hypothalamus and brain stem. Centrally, the release of noradrenaline from sympathetic nerves is stimulated, leading to the stimulation of postganglionic sympathetic nerves supplying brown adipocytes [105,172,173]. Noradrenaline acts on β-adrenergic receptors on the adipocyte surface, which generate a chain of stimulation that causes liberation of free fatty acids from stored triglycerides. Upon adrenergic stimulus that results in the activation of the brown (and white) adipocyte lipolytic cascade, respiration increases in a UCP1-dependent manner [174]. However, thermogenesis has been shown to be independent of lipolysis, and it has been demonstrated that stimulating lipolysis of cytosolic lipid droplets in brown adipocytes is not required for cold-induced non-shivering thermogenesis [175]. Recently, it has been observed in adipocytes that genetic or pharmacological elevation of levels in reactive oxygen species (ROS) is sufficient to drive thermogenesis [21]. Furthermore, studies in mice showed that application of heat stress (4 °C) or a β-adrenergic stimulus induces the activation of thermogenesis in BAT and results in an elevation of mitochondrial superoxide, mitochondrial hydrogen peroxide and lipid hydroperoxides. Oxidation of cysteine thiols by ROS can initiate the thermogenic activity of mitochondria in brown and beige adipocytes in a UCP1-dependent manner [176,177]. Similarly, it was identified that the accumulation of succinate, an intermediate metabolite of the tricarboxylic acid cycle in the mitochondria, rises independently of adrenergic stimulation in brown fat cells, and was sufficient to increase thermogenesis [178]. In this study, it was observed that the oxidation of succinate by the enzyme succinate dehydrogenase induces the production of ROS and manages thermogenic respiration [179,180]. The various pathways that thermogenesis may have in adipocytes were studied in UCP1 knockout mice (UCP1-KO). It was observed that UCP1-KO can be sensitive to cold, after being crossed with transgenic mice that express the PR domain containing 16 (PRDM16) that have the fatty acid binding protein 4 (Fabp4/aP2) promoter, which is expressed primarily in adipocytes [181]. Remarkably, UCP1-KO mice were resistant to diet-induced obesity at low temperatures, presumably by alternate activation pathways of energy loss, which are not yet well described. However, the possibility of thermogenesis being UCP1-independent remains controversial, and it is likely that thermogenesis in mice with UCP-1KO is induced by muscular activity that promotes shivering thermogenesis [182]. Creatine has been shown to be involved in metabolism and mitochondrial heat production, with recent observations suggesting the existence of a mitochondrial substrate cycle that is regulated by creatine to drive thermogenic respiration [176,183,184]. The thermogenic activity of creatine appears to only occur when ADP is limiting, which is expected during this physiological cellular state. However, the mechanism by which creatine influences the mitochondrial metabolism has yet to be established. Experiments using different animal models with genetic modifications have shown that reducing creatine may predispose animals to obesity. Interestingly, in a recent analysis of 18F-FDG PET/CT scans in human subjects, it was shown that renal creatinine clearance is a good predictor of activated human BAT [185]. Based on these observations and the fact that creatinine is a metabolite of the muscle energy store phosphocreatine, we can infer that creatine can be an activator of thermogenesis of BAT in humans and creatinine could be used as a biomarker of BAT activity [186].

6. Transcriptional Control of Browning

The phenotypic change in WAT into more energetically active cells indicates the involvement of gene expression regulation, in which internal or external regulators of WAT physiology must modify the chromatin structure or the DNA promoter methylation pattern of the target genes [74,187,188]. Recently, modifications of noncoding RNAs have been shown to act as an additional level of gene expression control [189,190]. Notably, a number of regulators function by modifying four transcriptional or coregulator factors: PPARγ, CCAAT enhancer binding protein beta (C/EBPβ), PPARγ co-activator-1α (PGC1α) and PRDM16. PPARγ and C/EBPβ act as transcription factors and directly bind DNA [191,192]. PRDM16 and PGC1α function as transcriptional coregulators, with PRDM16 forming a transcriptional complex with the canonical DNA binding transcription factors PPARγ and C/EBPβ through its zinc finger domains to activate the selective gene program for browning [193,194]. Similarly, it was observed in an analysis of chromatin immunoprecipitation sequencing (ChIP-seq) that PRDM16 co-localized with PPARγ and C/EBP in a large number of genes binding sites, further supporting their co-regulatory functions [195]. Although still not completely defined, beige adipocytes can be detected after the generation of a pre-adipocyte population of cells that are positive for platelet derived growth factor receptor-alfa (PDGFRα) and stem cell antigen 1 (SCA1) or precursors of MYH11. Its appearance occurs as a response to a variety of internal or external stimuli, including chronic exposure to cold, PPARγ agonists, cancer cachexia, exercise and various endocrine hormones [153,154]. Some factors that control the differentiation of BAT adipocytes also regulate the differentiation of beige adipocytes. Early beta-Cell transcription factor 2 (EBF2) is a key factor for the differentiation of BAT and has an important role in inducing the development of beige adipocytes [147,196]. EBF2 is highly expressed in PDGFRα positive cells, and the overexpression of EBF2 in primary white adipocytes or WAT induces the expression of thermogenic genes, increases oxygen consumption and suppresses high-fat diet-induced weight gain [197]. There are several proteins that can control beige differentiation through the functional control of PRDM16. The differentiation of beige adipocytes can be promoted by the activating or repressive activity of PRDM16. The formation of a repressor complex of PRDM16 with CtBP1 and CtBP2 reduces WAT adipogenesis [127]. On the other hand, the family of retinoblastoma proteins (pRb) antagonize the activity of PPARγ and PRDM16 [198]. pRb is a determinant of the choice of mesenchymal cells towards the osteoblastic lineage, thus in vivo experiments have shown that a pRb deficiency increases the development of mesenchymal precursor cells towards the brown adipocyte lineage. [199]. We previously observed that pRb inhibition by EP300 interacting inhibitor of differentiation-1 (EID-1) can induce the differentiation of beige adipocytes in humans [112]. Additionally, EID-1 can control adipogenesis through the transcriptional regulation of glycerol-3-phosphate dehydrogenases (GPDH), a key enzyme in the synthesis of triglycerides [200].

Several studies revealed peroxisome proliferator activated receptor γ coactivator 1 alpha (PGC1α) can regulate thermogenesis by directly inducing the expression of UCP1. PGC1α was first discovered as an interacting partner of PPARγ in brown adipocytes [201]. Pgc1α gene expression is greatly induced by cold exposure and is further activated following phosphorylation by the cAMP-PKA-p38/MAPK signaling pathway. PGC1α increases the transcription of specific genes through coactivation of by binding to transcription factors belonging to the nuclear receptor superfamily and recruitment of histone acetyltransferases such as CBP/p300 and GCN5 [202]. PGC1α binds to nuclear respiratory factors 1 and 2 (NRF-1 and NRF-2) to promote the activation of many mitochondrial genes. PGC1α mainly co-activates the nuclear hormone receptors, including PPARγ, PPARα, and estrogen related receptor (ERRα/β/γ), all of which participate in the transcription of brown fat genes [203]. PGC1α overexpression in adipocytes, myotubes, or cardiomyocytes promotes mitochondrial biogenesis and increases oxygen consumption [204,205]. Although PGC1α is a regulator of UCP1 expression, BAT Pgc1α-deficient mice display mildly increased lipid droplet accumulation but express normal levels of Ucp1 and other brown fat-selective genes [206]. Pgc1α-deficient BAT in culture fails to efficiently activate the thermogenic machinery in response to adrenergic stimulation [207]. These results demonstrate that PGC1α is required for the acute transcriptional activation of thermogenesis. Interestingly, the deletion of Pgc1α in adipocytes severely impairs the development of beige adipocytes in WAT [208].

7. Therapy with Inductors of Beige Cells

Every day the search for factors that can generate a metabolic change in the body is steadily expanding. Some of these experimental drugs have been used as possible obesity therapies by regulating BAT activity or adipogenesis of beige adipocytes [209,210]. Some regulatory peptides may have an effect on BAT and browning by stimulating hypothalamic nuclei. Dopamine agonists can induce BAT activity by stimulation of dopamine receptors 2 (D2R) in the VMH nuclei of the hypothalamus. It has been observed that patients treated with cabergoline, a D2R agonist, for 12 months showed a reduction in body mass index (BMI)and body fat together with an increase in resting energy expenditure (REE) and an improvement in glucose and lipid metabolism [211].

Polyphenols are substances characterized by the presence of several phenolic rings. They originate mainly in plants, which synthesize them in large quantities, as a product of their secondary metabolism. Some polyphenols are essential for plant physiological functions, others participate in defense functions in circumstances of stress. Among the polyphenols, flavonoids, catechins, capsaicin, and resveratrol stand out [212]. Capsaicin and capsinoids can trigger the activation of BAT and the browning of WAT. Studies with animal models of obesity have shown that capsinoids exert their mechanism of action by selectively activating the channels of the transient receptor potential vanilloid 1 (TRPV1) [213,214]. The effect of amplification of TRPV1 channels by capsinoid treatment ultimately results in activation of the vagal afferent nerves that project into the VMH hypothalamus [215]. Therefore, the mechanisms by which capsinoids induce BAT activity involve activation of β-adrenergic receptors. Human studies have shown that capsinoids can exponentially increase BAT activity after exposure to cold [216]. Resveratrol in elevated doses can reduce weight by increasing the activity of sirtuina 1 (SIRT1), a histone deacetylase dependent on NAD+. SIRT1 increases the function of AMPK-PGC1α and can trigger browning [217,218]. Recent studies have observed that the function of resveratrol can be mediated by the modification of the intestinal microbiome. The action of resveratrol on the gut microbiome is mediated by three potential ways: it can alter the composition of obesity-related gut microbiota, improve gut function and barrier integrity, or undergo gut microflora mediated-biotransformation to active metabolites in the intestinal tract [219,220]. Other polyphenols have been shown to regulate the activity of BAT; however, these observations are preliminary and have not been shown to have clinical relevance [212]. Gut microbiota can be considered as a contributing factor to the pathophysiology of obesity and may have potential therapeutic implications. Obesity is associated with elevated levels of Firmicutes such as Ruminococcaceae and depleted levels of Bacteroidetes such as Bacteroidaceae and Bacteroides. After exposure to cold in obese patients, an increase in the Firmicutes Ruminococcaceae family has been observed [221]. In fact, it was associated with high levels of acetate in plasma and was positively related to the expression of PRDM16 [222]. In support of these findings, several studies have shown that acetate increases the activity of brown fat and induces the formation of beige adipocytes [223,224]. Given the evidence from some studies, it seems that the use of prebiotics could improve adaptive thermogenic capacity. However, assumed that it is a new field, more studies are needed that can evidence preclinical observations. Furthermore, the underlying mechanisms that explain the relationship between prebiotic supplementation and the induction of thermogenesis and browning of white adipocytes are still indeterminate. An important point of connection between thermogenic capacity and prebiotics is the production of secondary metabolites by resident bacteria after the fermentation of prebiotics. Thus, the profile of short chain fatty acids (SCFA) and secondary bile acids could be important metabolites derived from the metabolism of bacteria capable of connecting prebiotic supplementation with the regulation of thermogenic capacity in BAT and WAT [225]. In recent years, the development of incretin effect mediations, especially glucagon-like peptide receptor agonists (GLP-1R), has had a beneficial effect, not only for the control of glucose levels in the blood but also for the reduction in body weight. The most prominent effects of GLP-1R are the anorectic effect and the modification of gastric emptying [226]. However, the GLP-1R effect on BAT energy expenditure, and possibly on WAT browning or hepatic lipid oxidation, will lead to a reduction in the weight and depletion in endogenous lipid stores [55]. The most likely effect of GLP-1R is mediated by modulation of AMPK activity located in the VMH nucleus of the hypothalamus. However, the possibility that extrahypothalamic areas are also involved in the effects of GLP-1R agonists on BAT thermogenesis and energy expenditure cannot be ruled out [227].

Considerable attention has been recognized to the secretory capacity of BAT and beige fat cells, and the substances secreted by these cells have been called "batokines", which can have a paracrine or endocrine effect. These batokines play an important role in contributing to metabolic health by improving glucose and lipid homeostasis. The results of various BAT transplant studies have shown improvements in conditions associated with obesity, such as body weight and insulin sensitivity [228,229] (Figure 3).

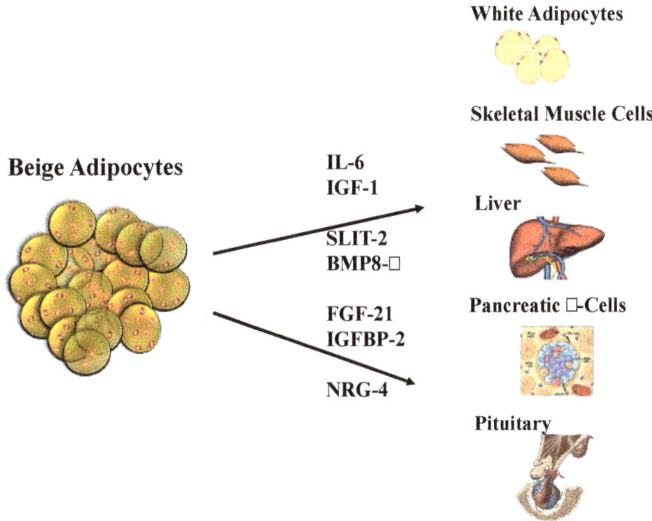

Figure 3. Adipokines secreted by beige adipocytes, which are also known as "batokines". Beige adipocytes secrete molecules that have endocrine effects on some of the most important tissues involved in the regulation of body weight and lipid and carbohydrate metabolism. BMP8-β, bone morphogenetic protein 8b; IGF-1, insulin-like growth factor 1; IGFBP-2, insulin-like growth factor binding protein 2; IL-6, interleukin 6; NRG-4, neuregulin 4; SLIT-2, slit homolog protein 2. Adapted from Arroyave et al. [230].

The activation of β3-adrenergic receptors is perhaps the most widely used pharmacological method for the development of browning. Recently, a study of the β3-adrenergic receptor agonist mirabegron, an food and drug administration (FDA) drug approved for overactive bladder treatment, was used in patients with prediabetes. Treatment with mirabegron for 3 months decreased insulin resistance without causing a reduction in weight or any cardiovascular effects [231]. However, the development of adrenergic ligands for obesity and metabolic disease applications has the drawback of producing unwanted cardiovascular, autonomic, and bone effects over time [232–234]. Similarly, bone morphogenetic proteins 4, 7, and 8b (BMP4, BMP7, and BMP8-β), atrial and brain-type natriuretic peptides (ANP and BNP), FGF21, vascular endothelial growth factor-α (VEGF-α), and prostaglandins have all been shown to promote browning in vivo [89,235–237]. However, the use of these peptides as medications for obesity or metabolic diseases has induced undesirable effects that have restricted their clinical use. FGF21 is an activator of thermogenesis that has gained particular interest from various pharmaceutical companies. However, studies conducted in humans using FGF21 analogs have not shown a significant effect of these compounds on body weight or glycemic control [238,239], although a beneficial effect on hepatic steatosis has been observed [240]. Another experimental molecule as possible therapeutic target is SLIT2, a factor secreted by beige adipocytes. The expression of the SLIT2 gene is regulated by PRDM16. SLIT2 induces a PKA-dependent thermogenic pathway in adipocytes and improves general metabolic parameters in response to a high-fat diet. [128]. The c-terminal fibrinogen-like domain of angiopoietin-like 4 (FLD of Angptl4) induces cAMP-PKA-dependent lipolysis in white adipocytes and reduces diet-induced obesity [241]. Angptl4 increases the thermogenic program and promotes subsequent protection against weight gain and improved glucose tolerance in high-fat diet-fed mice [242]. Recently, kynurenic acid was observed to increase energy expenditure by activating G-protein-coupled receptor 35 (Gpr35), which in turn stimulates a thermogenic program in adipose tissue and increases regulator of G protein signaling 14 (Rgs14) levels in adipocytes, leading to enhanced β-adrenergic receptor signaling [243]. The results of many clinical studies in adult humans have suggested the beneficial effect of activating browning from the WAT. In an interesting study it was observed that a reduction in temperature (17 °C) for two hours a day for 6 weeks induced BAT activity and a progressive increase in energy expenditure [216]. Prolonged cold exposure for 5 to 8 h increased resting REE by 15%, increased glucose clearance in brown/ beige adipocytes, and significantly increased whole body glucose clearance [244]. In subjects with type 2 diabetes mellitus, 10 days of cold acclimation increased peripheral insulin sensitivity by 43% [245,246]. In another study, cold exposure produced a translocation in transporter type 4 (GLUT4) and increased glucose uptake in skeletal muscle [247]. Adipose-derived stem cells can be induced to differentiate to beige adipocytes in many ways, and the results of a large amount of research in this area suggest that a substantial number of potential drugs will become available in the next few years [248–254]. Although many effects remains to be clarified, it is evident that the increase in the activity of beige adipocytes is a consistent mechanism to increase glucose metabolism, supporting its role in treating obesity and related metabolic disorders in humans, especially for type 2 diabetes mellitus [255], which has two key pathophysiological components: peripheral resistance to the action of insulin especially in muscular cells and adipocytes and reduction in insulin secretion in β-pancreatic cells. A change in the sensitivity to insulin mediated by an increase in the number of beige adipocytes can be a highly valuable therapeutic strategy [256,257]. Additionally, a reduction in body weight may lead to a reduced load on pancreatic activity. Experiments carried out in humans have shown that a reduction in glucose levels combined with an increase in insulin sensitivity can be obtained by inducing beige/brown fat [244]. Despite the encouraging observations, from a therapeutic point of view some biochemical aspects should be evaluated. First, it should be determined whether the metabolic effects of beige adipocytes are subject to increased UCP1 levels or if there are alternative metabolic pathways that could improve the condition of adipocytes [183]. The effects of beige and brown adipocytes, and the UCP1-independent pathways, on the metabolic benefits in glucose and lipids, should be determined [89]. Another important aspect that should be taken into account is the detection of the volume of beige adipocytes in organisms. Although F-FDG-PET has considerably improved, it is desirable to develop new tools or instruments that can quantify the amount of beige adipocyte tissue in the body. Due to the current lack of understanding of beige adipocyte biology, it is necessary to gain a comprehension of the relevant physiological conditions, the number of days that the cells can survive, and the elements necessary to maintain the functionality of these cells.

Finally, it is important to recognize the specific factors that could induce plasticity in adipocytes, which may allow white adipocytes to be converted into beige adipocytes. The browning process from both adipocyte precursor cells and white adipocytes may be desirable for weight reduction and for the treatment of metabolic diseases.

Author Contributions: Conceptualization, F.L. and F.A.; Methodology, F.L.; Software, F.A.; Validation, F.L.; Formal Analysis, F.L.; Investigation, F.L. and F.A.; Resources, F.L.; Writing-Original Draft Preparation, F.A. and F.L.; Writing-Review & Editing, F.L.; Visualization, F.L.; Supervision, F.L.; Project Administration, F.L.; Funding Acquisition, F.L. All authors have read and agreed to the published version of the manuscript.

Funding: The work of the authors was supported in part by MED-255-2019 from Direction of Investigation (DIN) Universidad de La Sabana.

Acknowledgments: The authors thank the members of the laboratory for helpful suggestions. This work was supported by the School of Medicine and DIN (Research Department) from Universidad de La Sabana.

Conflicts of Interest: The authors declare no conflict of interest.

References

1. Saunders, K.H.; Umashanker, D.; Igel, L.I.; Kumar, R.B.; Aronne, L.J. Obesity Pharmacotherapy. *Med. Clin.* **2018**, *102*, 135–148. [CrossRef] [PubMed]
2. Khera, R.; Murad, M.H.; Chandar, A.K.; Dulai, P.S.; Wang, Z.; Prokop, L.J.; Loomba, R.; Camilleri, M.; Singh, S. Association of Pharmacological Treatments for Obesity With Weight Loss and Adverse Events: A Systematic Review and Meta-analysis. *JAMA* **2016**, *315*, 2424–2434. [CrossRef] [PubMed]
3. Kyle, T.K.; Dhurandhar, E.J.; Allison, D.B. Regarding Obesity as a Disease: Evolving Policies and Their Implications. *Endocrinol. Metab. Clin.* **2016**, *45*, 511–520. [CrossRef] [PubMed]
4. Bray, G.A.; Heisel, W.E.; Afshin, A.; Jensen, M.D.; Dietz, W.H.; Long, M.; Kushner, R.F.; Daniels, S.R.; Wadden, T.A.; Tsai, A.G.; et al. The Science of Obesity Management: An Endocrine Society Scientific Statement. *Endocr. Rev.* **2018**, *39*, 79–132. [CrossRef]
5. Heymsfield, S.B.; Wadden, T.A. Mechanisms, Pathophysiology, and Management of Obesity. *N. Engl. J. Med.* **2017**, *376*, 1492. [CrossRef]
6. Halberg, N.; Wernstedt-Asterholm, I.; Scherer, P.E. The adipocyte as an endocrine cell. *Endocrinol. Metab. Clin.* **2008**, *37*, 753–768. [CrossRef]
7. Cohen, P.; Spiegelman, B.M. Cell biology of fat storage. *Mol. Biol. Cell* **2016**, *27*, 2523–2527. [CrossRef] [PubMed]
8. Ye, L.; Wu, J.; Cohen, P.; Kazak, L.; Khandekar, M.J.; Jedrychowski, M.P.; Zeng, X.; Gygi, S.P.; Spiegelman, B.M. Fat cells directly sense temperature to activate thermogenesis. *Proc. Natl. Acad. Sci. USA* **2013**, *110*, 12480–12485. [CrossRef] [PubMed]
9. Wu, J.; Cohen, P.; Spiegelman, B.M. Adaptive thermogenesis in adipocytes: Is beige the new brown? *Genes Dev.* **2013**, *27*, 234–250. [CrossRef]
10. Kajimura, S.; Spiegelman, B.M.; Seale, P. Brown and Beige Fat: Physiological Roles beyond Heat Generation. *Cell Metab.* **2015**, *22*, 546–559. [CrossRef] [PubMed]
11. Hussain, M.F.; Roesler, A.; Kazak, L. Regulation of adipocyte thermogenesis: Mechanisms controlling obesity. *FEBS J.* **2020**, *2020*, 1–16. [CrossRef]
12. Ikeda, K.; Maretich, P.; Kajimura, S. The Common and Distinct Features of Brown and Beige Adipocytes. *Trends Endocrinol. Metab.* **2018**, *29*, 191–200. [CrossRef] [PubMed]
13. Keipert, S.; Jastroch, M. Brite/beige fat and UCP1—Is it thermogenesis? *Biochim. Biophys. Acta* **2014**, *1837*, 1075–1082. [CrossRef]
14. Harms, M.; Seale, P. Brown and beige fat: Development, function and therapeutic potential. *Nat. Med.* **2013**, *19*, 1252–1263. [CrossRef]
15. Lee, P.; Werner, C.D.; Kebebew, E.; Celi, F.S. Functional thermogenic beige adipogenesis is inducible in human neck fat. *Int. J. Obes.* **2014**, *38*, 170–176. [CrossRef]
16. Rahman, S.; Lu, Y.; Czernik, P.J.; Rosen, C.J.; Enerback, S.; Lecka-Czernik, B. Inducible brown adipose tissue, or beige fat, is anabolic for the skeleton. *Endocrinology* **2013**, *154*, 2687–2701. [CrossRef] [PubMed]
17. Kajimura, S. Engineering Fat Cell Fate to Fight Obesity and Metabolic Diseases. *Keio J. Med.* **2015**, *64*, 65. [CrossRef]
18. Holmes, D. Epigenetics: On-off switch for obesity. *Nat. Rev. Endocrinol.* **2016**, *12*, 125. [CrossRef]
19. Saper, C.B.; Lowell, B.B. The hypothalamus. *Curr. Biol.* **2014**, *24*, R1111–R1116. [CrossRef]

20. Lee, Y.H.; Petkova, A.P.; Konkar, A.A.; Granneman, J.G. Cellular origins of cold-induced brown adipocytes in adult mice. *FASEB J.* **2015**, *29*, 286–299. [CrossRef]
21. Cakir, I.; Nillni, E.A. Endoplasmic Reticulum Stress, the Hypothalamus, and Energy Balance. *Trends Endocrinol. Metab.* **2019**, *30*, 163–176. [CrossRef]
22. Waterson, M.J.; Horvath, T.L. Neuronal Regulation of Energy Homeostasis: Beyond the Hypothalamus and Feeding. *Cell Metab.* **2015**, *22*, 962–970. [CrossRef]
23. Cannon, B.; Nedergaard, J. Brown adipose tissue: Function and physiological significance. *Physiol. Rev.* **2004**, *84*, 277–359. [CrossRef] [PubMed]
24. Silva, J.E. Thermogenic mechanisms and their hormonal regulation. *Physiol. Rev.* **2006**, *86*, 435–464. [CrossRef]
25. Chouchani, E.T.; Kazak, L.; Spiegelman, B.M. New Advances in Adaptive Thermogenesis: UCP1 and Beyond. *Cell Metab.* **2019**, *29*, 27–37. [CrossRef]
26. Silva, J.E. Physiological importance and control of non-shivering facultative thermogenesis. *Front BioSci* **2011**, *3*, 352–371. [CrossRef]
27. Contreras, C.; Nogueiras, R.; Dieguez, C.; Medina-Gomez, G.; Lopez, M. Hypothalamus and thermogenesis: Heating the BAT, browning the WAT. *Mol. Cell Endocrinol.* **2016**, *438*, 107–115. [CrossRef]
28. Gonzalez-Garcia, I.; Milbank, E.; Martinez-Ordonez, A.; Dieguez, C.; Lopez, M.; Contreras, C. HYPOTHesizing about central comBAT against obesity. *J. Physiol. Biochem.* **2020**, *76*, 193–211. [CrossRef]
29. McKemy, D.D.; Neuhausser, W.M.; Julius, D. Identification of a cold receptor reveals a general role for TRP channels in thermosensation. *Nature* **2002**, *416*, 52–58. [CrossRef]
30. Min, S.Y.; Kady, J.; Nam, M.; Rojas-Rodriguez, R.; Berkenwald, A.; Kim, J.H.; Noh, H.L.; Kim, J.K.; Cooper, M.P.; Fitzgibbons, T.; et al. Human 'brite/beige' adipocytes develop from capillary networks, and their implantation improves metabolic homeostasis in mice. *Nat. Med.* **2016**, *22*, 312–318. [CrossRef]
31. Milbank, E.; Lopez, M. Orexins/Hypocretins: Key Regulators of Energy Homeostasis. *Front. Endocrinol.* **2019**, *10*, 830. [CrossRef]
32. Venner, A.; Karnani, M.M.; Gonzalez, J.A.; Jensen, L.T.; Fugger, L.; Burdakov, D. Orexin neurons as conditional glucosensors: Paradoxical regulation of sugar sensing by intracellular fuels. *J. Physiol.* **2011**, *589*, 5701–5708. [CrossRef]
33. Karnani, M.M.; Apergis-Schoute, J.; Adamantidis, A.; Jensen, L.T.; de Lecea, L.; Fugger, L.; Burdakov, D. Activation of central orexin/hypocretin neurons by dietary amino acids. *Neuron* **2011**, *72*, 616–629. [CrossRef] [PubMed]
34. Gautron, L.; Elmquist, J.K.; Williams, K.W. Neural control of energy balance: Translating circuits to therapies. *Cell* **2015**, *161*, 133–145. [CrossRef]
35. Everitt, B.J.; Hokfelt, T. Neuroendocrine anatomy of the hypothalamus. *Acta Neurochir. Suppl.* **1990**, *47*, 1–15. [CrossRef]
36. Schneeberger, M.; Gomis, R.; Claret, M. Hypothalamic and brainstem neuronal circuits controlling homeostatic energy balance. *J. Endocrinol.* **2014**, *220*, T25–T46. [CrossRef]
37. Stanley, S.; Pinto, S.; Segal, J.; Perez, C.A.; Viale, A.; DeFalco, J.; Cai, X.; Heisler, L.K.; Friedman, J.M. Identification of neuronal subpopulations that project from hypothalamus to both liver and adipose tissue polysynaptically. *Proc. Natl. Acad. Sci. USA* **2010**, *107*, 7024–7029. [CrossRef] [PubMed]
38. Sidossis, L.S.; Porter, C.; Saraf, M.K.; Borsheim, E.; Radhakrishnan, R.S.; Chao, T.; Ali, A.; Chondronikola, M.; Mlcak, R.; Finnerty, C.C.; et al. Browning of Subcutaneous White Adipose Tissue in Humans after Severe Adrenergic Stress. *Cell Metab.* **2015**, *22*, 219–227. [CrossRef]
39. Downs, J.L.; Dunn, M.R.; Borok, E.; Shanabrough, M.; Horvath, T.L.; Kohama, S.G.; Urbanski, H.F. Orexin neuronal changes in the locus coeruleus of the aging rhesus macaque. *NeuroBiol. Aging* **2007**, *28*, 1286–1295. [CrossRef]
40. Wiedmann, N.M.; Stefanidis, A.; Oldfield, B.J. Characterization of the central neural projections to brown, white, and beige adipose tissue. *FASEB J.* **2017**, *31*, 4879–4890. [CrossRef] [PubMed]
41. Sellayah, D.; Bharaj, P.; Sikder, D. Orexin is required for brown adipose tissue development, differentiation, and function. *Cell Metab.* **2011**, *14*, 478–490. [CrossRef]
42. Martins, L.; Seoane-Collazo, P.; Contreras, C.; Gonzalez-Garcia, I.; Martinez-Sanchez, N.; Gonzalez, F.; Zalvide, J.; Gallego, R.; Dieguez, C.; Nogueiras, R.; et al. A Functional Link between AMPK and Orexin Mediates the Effect of BMP8B on Energy Balance. *Cell Rep.* **2016**, *16*, 2231–2242. [CrossRef] [PubMed]

43. Russell, S.H.; Small, C.J.; Sunter, D.; Morgan, I.; Dakin, C.L.; Cohen, M.A.; Bloom, S.R. Chronic intraparaventricular nuclear administration of orexin A in male rats does not alter thyroid axis or uncoupling protein-1 in brown adipose tissue. *Regul. Pept.* **2002**, *104*, 61–68. [CrossRef]
44. Yasuda, T.; Masaki, T.; Kakuma, T.; Hara, M.; Nawata, T.; Katsuragi, I.; Yoshimatsu, H. Dual regulatory effects of orexins on sympathetic nerve activity innervating brown adipose tissue in rats. *Endocrinology* **2005**, *146*, 2744–2748. [CrossRef]
45. Tupone, D.; Madden, C.J.; Cano, G.; Morrison, S.F. An orexinergic projection from perifornical hypothalamus to raphe pallidus increases rat brown adipose tissue thermogenesis. *J. NeuroSci.* **2011**, *31*, 15944–15955. [CrossRef]
46. Yavari, A.; Stocker, C.J.; Ghaffari, S.; Wargent, E.T.; Steeples, V.; Czibik, G.; Pinter, K.; Bellahcene, M.; Woods, A.; Martinez de Morentin, P.B.; et al. Chronic Activation of gamma2 AMPK Induces Obesity and Reduces beta Cell Function. *Cell Metab.* **2016**, *23*, 821–836. [CrossRef] [PubMed]
47. Whittle, A.J.; Carobbio, S.; Martins, L.; Slawik, M.; Hondares, E.; Vazquez, M.J.; Morgan, D.; Csikasz, R.I.; Gallego, R.; Rodriguez-Cuenca, S.; et al. BMP8B increases brown adipose tissue thermogenesis through both central and peripheral actions. *Cell* **2012**, *149*, 871–885. [CrossRef]
48. Monda, M.; Viggiano, A.; Viggiano, A.; Viggiano, E.; Messina, G.; Tafuri, D.; De Luca, V. Sympathetic and hyperthermic reactions by orexin A: Role of cerebral catecholaminergic neurons. *Regul. Pept.* **2007**, *139*, 39–44. [CrossRef]
49. Monda, M.; Viggiano, A.; Viggiano, A.; Viggiano, E.; De Luca, V. Risperidone potentiates the sympathetic and hyperthermic reactions induced by orexin A in the rat. *Physiol. Res.* **2006**, *55*, 73–78.
50. Bassetti, C.L.A.; Adamantidis, A.; Burdakov, D.; Han, F.; Gay, S.; Kallweit, U.; Khatami, R.; Koning, F.; Kornum, B.R.; Lammers, G.J.; et al. Narcolepsy—Clinical spectrum, aetiopathophysiology, diagnosis and treatment. *Nat. Rev. Neurol* **2019**, *15*, 519–539. [CrossRef]
51. Enevoldsen, L.H.; Tindborg, M.; Hovmand, N.L.; Christoffersen, C.; Ellingsgaard, H.; Suetta, C.; Stallknecht, B.M.; Jennum, P.J.; Kjaer, A.; Gammeltoft, S. Functional brown adipose tissue and sympathetic activity after cold exposure in humans with type 1 narcolepsy. *Sleep* **2018**, *41*, zsy092. [CrossRef]
52. Lopez, M.; Alvarez, C.V.; Nogueiras, R.; Dieguez, C. Energy balance regulation by thyroid hormones at central level. *Trends Mol. Med.* **2013**, *19*, 418–427. [CrossRef]
53. Lizcano, F.; Guzman, G. Estrogen Deficiency and the Origin of Obesity during Menopause. *Biomed. Res. Int.* **2014**, *2014*, 757461. [CrossRef]
54. Martinez de Morentin, P.B.; Gonzalez-Garcia, I.; Martins, L.; Lage, R.; Fernandez-Mallo, D.; Martinez-Sanchez, N.; Ruiz-Pino, F.; Liu, J.; Morgan, D.A.; Pinilla, L.; et al. Estradiol regulates brown adipose tissue thermogenesis via hypothalamic AMPK. *Cell Metab.* **2014**, *20*, 41–53. [CrossRef]
55. Gonzalez-Garcia, I.; Milbank, E.; Dieguez, C.; Lopez, M.; Contreras, C. Glucagon, GLP-1 and Thermogenesis. *Int. J. Mol. Sci.* **2019**, *20*, 3445. [CrossRef] [PubMed]
56. Madden, C.J. Glucoprivation in the ventrolateral medulla decreases brown adipose tissue sympathetic nerve activity by decreasing the activity of neurons in raphe pallidus. *Am. J. Physiol. Regul. Integr. Comp. Physiol.* **2012**, *302*, R224–R232. [CrossRef]
57. Blouet, C.; Schwartz, G.J. Duodenal lipid sensing activates vagal afferents to regulate non-shivering brown fat thermogenesis in rats. *PLoS ONE* **2012**, *7*, e51898. [CrossRef]
58. Acheson, K.; Jequier, E.; Wahren, J. Influence of beta-adrenergic blockade on glucose-induced thermogenesis in man. *J. Clin. Investig.* **1983**, *72*, 981–986. [CrossRef] [PubMed]
59. Fromme, T.; Klingenspor, M. Uncoupling protein 1 expression and high-fat diets. *Am. J. Physiol. Regul. Integr. Comp. Physiol.* **2011**, *300*, R1–R8. [CrossRef]
60. Levin, B.E.; Triscari, J.; Sullivan, A.C. Altered sympathetic activity during development of diet-induced obesity in rat. *Am. J. Physiol.* **1983**, *244*, R347–R355. [CrossRef]
61. Cannon, B.; Nedergaard, J. Yes, even human brown fat is on fire! *J. Clin. Investig.* **2012**, *122*, 486–489. [CrossRef]
62. Srivastava, S.; Baxa, U.; Niu, G.; Chen, X.; Veech, R.L. A ketogenic diet increases brown adipose tissue mitochondrial proteins and UCP1 levels in mice. *IUBMB Life* **2013**, *65*, 58–66. [CrossRef]
63. La Merrill, M.A.; Vandenberg, L.N.; Smith, M.T.; Goodson, W.; Browne, P.; Patisaul, H.B.; Guyton, K.Z.; Kortenkamp, A.; Cogliano, V.J.; Woodruff, T.J.; et al. Consensus on the key characteristics of endocrine-disrupting chemicals as a basis for hazard identification. *Nat. Rev. Endocrinol.* **2020**, *16*, 45–57. [CrossRef]
64. Grun, F.; Blumberg, B. Minireview: The case for obesogens. *Mol. Endocrinol.* **2009**, *23*, 1127–1134. [CrossRef]

65. Gonzalez-Casanova, J.E.; Pertuz-Cruz, S.L.; Caicedo-Ortega, N.H.; Rojas-Gomez, D.M. Adipogenesis Regulation and Endocrine Disruptors: Emerging Insights in Obesity. *Biomed. Res. Int.* **2020**, *2020*, 7453786. [CrossRef]
66. Heindel, J.J.; Blumberg, B. Environmental Obesogens: Mechanisms and Controversies. *Annu. Rev. Pharm. Toxicol.* **2019**, *59*, 89–106. [CrossRef]
67. Majnik, A.; Gunn, V.; Fu, Q.; Lane, R.H. Epigenetics: An accessible mechanism through which to track and respond to an obesogenic environment. *Expert Rev. Endocrinol. Metab.* **2014**, *9*, 605–614. [CrossRef]
68. Janesick, A.S.; Shioda, T.; Blumberg, B. Transgenerational inheritance of prenatal obesogen exposure. *Mol. Cell Endocrinol.* **2014**, *398*, 31–35. [CrossRef]
69. Janesick, A.S.; Dimastrogiovanni, G.; Vanek, L.; Boulos, C.; Chamorro-Garcia, R.; Tang, W.; Blumberg, B. On the Utility of ToxCast and ToxPi as Methods for Identifying New Obesogens. *Envrion. Health Perspect.* **2016**, *124*, 1214–1226. [CrossRef]
70. Darbre, P.D. Endocrine Disruptors and Obesity. *Curr. Obes. Rep.* **2017**, *6*, 18–27. [CrossRef]
71. le Maire, A.; Grimaldi, M.; Roecklin, D.; Dagnino, S.; Vivat-Hannah, V.; Balaguer, P.; Bourguet, W. Activation of RXR-PPAR heterodimers by organotin environmental endocrine disruptors. *EMBO Rep.* **2009**, *10*, 367–373. [CrossRef]
72. Grun, F.; Blumberg, B. Environmental obesogens: Organotins and endocrine disruption via nuclear receptor signaling. *Endocrinology* **2006**, *147*, S50–S55. [CrossRef]
73. Punzon, I.; Latapie, V.; Le Mevel, S.; Hagneau, A.; Jolivet, P.; Palmier, K.; Fini, J.B.; Demeneix, B.A. Towards a humanized PPARgamma reporter system for in vivo screening of obesogens. *Mol. Cell Endocrinol.* **2013**, *374*, 1–9. [CrossRef]
74. Lizcano, F.; Romero, C.; Vargas, D. Regulation of adipogenesis by nuclear receptor PPARgamma is modulated by the histone demethylase JMJD2C. *Genet. Mol. Biol.* **2011**, *34*, 19–24. [CrossRef]
75. Lizcano, F.; Vargas, D. EID1-induces brown-like adipocyte traits in white 3T3-L1 pre-adipocytes. *Biochem. Biophys. Res. Commun.* **2010**, *398*, 160–165. [CrossRef]
76. Golub, M.; Doherty, J. Triphenyltin as a potential human endocrine disruptor. *J. Toxicol. Envrion. Health B Crit. Rev.* **2004**, *7*, 281–295. [CrossRef]
77. Li, X.; Ycaza, J.; Blumberg, B. The environmental obesogen tributyltin chloride acts via peroxisome proliferator activated receptor gamma to induce adipogenesis in murine 3T3-L1 preadipocytes. *J. Steroid Biochem. Mol. Biol.* **2011**, *127*, 9–15. [CrossRef]
78. Grun, F.; Watanabe, H.; Zamanian, Z.; Maeda, L.; Arima, K.; Cubacha, R.; Gardiner, D.M.; Kanno, J.; Iguchi, T.; Blumberg, B. Endocrine-disrupting organotin compounds are potent inducers of adipogenesis in vertebrates. *Mol. Endocrinol.* **2006**, *20*, 2141–2155. [CrossRef]
79. Kanayama, T.; Kobayashi, N.; Mamiya, S.; Nakanishi, T.; Nishikawa, J. Organotin compounds promote adipocyte differentiation as agonists of the peroxisome proliferator-activated receptor gamma/retinoid X receptor pathway. *Mol. Pharm.* **2005**, *67*, 766–774. [CrossRef]
80. Rochester, J.R. Bisphenol A and human health: A review of the literature. *Reprod. Toxicol.* **2013**, *42*, 132–155. [CrossRef]
81. Egusquiza, R.J.; Ambrosio, M.E.; Wang, S.G.; Kay, K.M.; Zhang, C.; Lehmler, H.J.; Blumberg, B. Evaluating the Role of the Steroid and Xenobiotic Receptor (SXR/PXR) in PCB-153 Metabolism and Protection against Associated Adverse Effects during Perinatal and Chronic Exposure in Mice. *Envrion. Health Perspect.* **2020**, *128*, 47011. [CrossRef]
82. Evans, R.M.; Mangelsdorf, D.J. Nuclear Receptors, RXR, and the Big Bang. *Cell* **2014**, *157*, 255–266. [CrossRef]
83. Egusquiza, R.J.; Blumberg, B. Environmental Obesogens and Their Impact on Susceptibility to Obesity: New Mechanisms and Chemicals. *Endocrinology* **2020**, *161*, bqaa024. [CrossRef]
84. Guo, H.; Yan, H.; Cheng, D.; Wei, X.; Kou, R.; Si, J. Tributyltin exposure induces gut microbiome dysbiosis with increased body weight gain and dyslipidemia in mice. *Envrion. Toxicol. Pharm.* **2018**, *60*, 202–208. [CrossRef] [PubMed]
85. Shoucri, B.M.; Martinez, E.S.; Abreo, T.J.; Hung, V.T.; Moosova, Z.; Shioda, T.; Blumberg, B. Retinoid X Receptor Activation Alters the Chromatin Landscape To Commit Mesenchymal Stem Cells to the Adipose Lineage. *Endocrinology* **2017**, *158*, 3109–3125. [CrossRef]
86. Anway, M.D.; Cupp, A.S.; Uzumcu, M.; Skinner, M.K. Epigenetic transgenerational actions of endocrine disruptors and male fertility. *Science* **2005**, *308*, 1466–1469. [CrossRef]

87. Skinner, M.K.; Manikkam, M.; Tracey, R.; Guerrero-Bosagna, C.; Haque, M.; Nilsson, E.E. Ancestral dichlorodiphenyltrichloroethane (DDT) exposure promotes epigenetic transgenerational inheritance of obesity. *BMC Med.* **2013**, *11*, 228. [CrossRef]
88. Nilsson, E.E.; Sadler-Riggleman, I.; Skinner, M.K. Environmentally induced epigenetic transgenerational inheritance of disease. *Envrion. Epigenet.* **2018**, *4*, dvy016. [CrossRef]
89. Kusminski, C.M.; Bickel, P.E.; Scherer, P.E. Targeting adipose tissue in the treatment of obesity-associated diabetes. *Nat. Rev. Drug Discov.* **2016**, *15*, 639–660. [CrossRef] [PubMed]
90. Regnier, S.M.; El-Hashani, E.; Kamau, W.; Zhang, X.; Massad, N.L.; Sargis, R.M. Tributyltin differentially promotes development of a phenotypically distinct adipocyte. *Obesity* **2015**, *23*, 1864–1871. [CrossRef]
91. Chamorro-Garcia, R.; Shoucri, B.M.; Willner, S.; Kach, H.; Janesick, A.; Blumberg, B. Effects of Perinatal Exposure to Dibutyltin Chloride on Fat and Glucose Metabolism in Mice, and Molecular Mechanisms, in Vitro. *Envrion. Health Perspect.* **2018**, *126*, 057006. [CrossRef]
92. Ivry Del Moral, L.; Le Corre, L.; Poirier, H.; Niot, I.; Truntzer, T.; Merlin, J.F.; Rouimi, P.; Besnard, P.; Rahmani, R.; Chagnon, M.C. Obesogen effects after perinatal exposure of 4,4′-sulfonyldiphenol (Bisphenol S) in C57BL/6 mice. *Toxicology* **2016**, *357–358*, 11–20. [CrossRef]
93. Ahmed, S.; Atlas, E. Bisphenol S- and bisphenol A-induced adipogenesis of murine preadipocytes occurs through direct peroxisome proliferator-activated receptor gamma activation. *Int. J. Obes.* **2016**, *40*, 1566–1573. [CrossRef]
94. Lee, H.W.; Pyo, S. Acrylamide induces adipocyte differentiation and obesity in mice. *Chem. Biol. Interact.* **2019**, *298*, 24–34. [CrossRef] [PubMed]
95. Kadawathagedara, M.; Tong, A.C.H.; Heude, B.; Forhan, A.; Charles, M.A.; Sirot, V.; Botton, J.; The Eden Mother-Child Cohort Study, G. Dietary acrylamide intake during pregnancy and anthropometry at birth in the French EDEN mother-child cohort study. *Envrion. Res.* **2016**, *149*, 189–196. [CrossRef] [PubMed]
96. Kadawathagedara, M.; Botton, J.; de Lauzon-Guillain, B.; Meltzer, H.M.; Alexander, J.; Brantsaeter, A.L.; Haugen, M.; Papadopoulou, E. Dietary acrylamide intake during pregnancy and postnatal growth and obesity: Results from the Norwegian Mother and Child Cohort Study (MoBa). *Envrion. Int.* **2018**, *113*, 325–334. [CrossRef]
97. Sun, Z.; Tang, Z.; Yang, X.; Liu, Q.S.; Liang, Y.; Fiedler, H.; Zhang, J.; Zhou, Q.; Jiang, G. Perturbation of 3-tert-butyl-4-hydroxyanisole in adipogenesis of male mice with normal and high fat diets. *Sci. Total Envrion.* **2020**, *703*, 135608. [CrossRef] [PubMed]
98. Shannon, M.; Green, B.; Willars, G.; Wilson, J.; Matthews, N.; Lamb, J.; Gillespie, A.; Connolly, L. The endocrine disrupting potential of monosodium glutamate (MSG) on secretion of the glucagon-like peptide-1 (GLP-1) gut hormone and GLP-1 receptor interaction. *Toxicol. Lett.* **2017**, *265*, 97–105. [CrossRef]
99. Rawshani, A.; Rawshani, A.; Franzen, S.; Sattar, N.; Eliasson, B.; Svensson, A.M.; Zethelius, B.; Miftaraj, M.; McGuire, D.K.; Rosengren, A.; et al. Risk Factors, Mortality, and Cardiovascular Outcomes in Patients with Type 2 Diabetes. *N. Engl. J. Med.* **2018**, *379*, 633–644. [CrossRef]
100. Rosengren, A.; Edqvist, J.; Rawshani, A.; Sattar, N.; Franzen, S.; Adiels, M.; Svensson, A.M.; Lind, M.; Gudbjornsdottir, S. Excess risk of hospitalisation for heart failure among people with type 2 diabetes. *Diabetologia* **2018**, *61*, 2300–2309. [CrossRef]
101. Edqvist, J.; Rawshani, A.; Adiels, M.; Bjorck, L.; Lind, M.; Svensson, A.M.; Gudbjornsdottir, S.; Sattar, N.; Rosengren, A. BMI and Mortality in Patients With New-Onset Type 2 Diabetes: A Comparison With Age- and Sex-Matched Control Subjects From the General Population. *Diabetes Care* **2018**, *41*, 485–493. [CrossRef]
102. Sun, K.; Kusminski, C.M.; Scherer, P.E. Adipose tissue remodeling and obesity. *J. Clin. Investig.* **2011**, *121*, 2094–2101. [CrossRef]
103. Khan, T.; Muise, E.S.; Iyengar, P.; Wang, Z.V.; Chandalia, M.; Abate, N.; Zhang, B.B.; Bonaldo, P.; Chua, S.; Scherer, P.E. Metabolic dysregulation and adipose tissue fibrosis: Role of collagen VI. *Mol. Cell Biol.* **2009**, *29*, 1575–1591. [CrossRef]
104. Rosenwald, M.; Perdikari, A.; Rulicke, T.; Wolfrum, C. Bi-directional interconversion of brite and white adipocytes. *Nat. Cell Biol.* **2013**, *15*, 659–667. [CrossRef] [PubMed]
105. Barbatelli, G.; Murano, I.; Madsen, L.; Hao, Q.; Jimenez, M.; Kristiansen, K.; Giacobino, J.P.; De Matteis, R.; Cinti, S. The emergence of cold-induced brown adipocytes in mouse white fat depots is determined predominantly by white to brown adipocyte transdifferentiation. *Am. J. Physiol. Endocrinol. Metab.* **2010**, *298*, E1244–E1253. [CrossRef]

106. Galic, S.; Oakhill, J.S.; Steinberg, G.R. Adipose tissue as an endocrine organ. *Mol. Cell Endocrinol.* **2010**, *316*, 129–139. [CrossRef]
107. Armani, A.; Berry, A.; Cirulli, F.; Caprio, M. Molecular mechanisms underlying metabolic syndrome: The expanding role of the adipocyte. *FASEB J.* **2017**, *31*, 4240–4255. [CrossRef]
108. Vargas, D.; Lopez, C.; Acero, E.; Benitez, E.; Wintaco, A.; Camacho, J.; Carreno, M.; Umana, J.; Jimenez, D.; Diaz, S.; et al. Thermogenic capacity of human periaortic adipose tissue is transformed by body weight. *PLoS ONE* **2018**, *13*, e0194269. [CrossRef] [PubMed]
109. Davidson, L.E.; Hunt, S.C.; Adams, T.D. Fitness versus adiposity in cardiovascular disease risk. *Eur. J. Clin. Nutr.* **2019**, *73*, 225–230. [CrossRef]
110. Wu, Z.; Puigserver, P.; Spiegelman, B.M. Transcriptional activation of adipogenesis. *Curr. Opin. Cell Biol.* **1999**, *11*, 689–694. [CrossRef]
111. Lizcano, F.; Vargas, D. Biology of Beige Adipocyte and Possible Therapy for Type 2 Diabetes and Obesity. *Int. J. Endocrinol.* **2016**, *2016*, 9542061. [CrossRef]
112. Vargas, D.; Shimokawa, N.; Kaneko, R.; Rosales, W.; Parra, A.; Castellanos, A.; Koibuchi, N.; Lizcano, F. Regulation of human subcutaneous adipocyte differentiation by EID1. *J. Mol. Endocrinol.* **2016**, *56*, 113–122. [CrossRef] [PubMed]
113. Bornstein, S.R.; Abu-Asab, M.; Glasow, A.; Path, G.; Hauner, H.; Tsokos, M.; Chrousos, G.P.; Scherbaum, W.A. Immunohistochemical and ultrastructural localization of leptin and leptin receptor in human white adipose tissue and differentiating human adipose cells in primary culture. *Diabetes* **2000**, *49*, 532–538. [CrossRef]
114. Tilg, H.; Moschen, A.R. Adipocytokines: Mediators linking adipose tissue, inflammation and immunity. *Nat. Rev. Immunol.* **2006**, *6*, 772–783. [CrossRef]
115. Farooqi, I.S.; Jebb, S.A.; Langmack, G.; Lawrence, E.; Cheetham, C.H.; Prentice, A.M.; Hughes, I.A.; McCamish, M.A.; O'Rahilly, S. Effects of recombinant leptin therapy in a child with congenital leptin deficiency. *N. Engl. J. Med.* **1999**, *341*, 879–884. [CrossRef]
116. Crujeiras, A.B.; Carreira, M.C.; Cabia, B.; Andrade, S.; Amil, M.; Casanueva, F.F. Leptin resistance in obesity: An epigenetic landscape. *Life Sci.* **2015**, *140*, 57–63. [CrossRef]
117. Gruzdeva, O.; Borodkina, D.; Uchasova, E.; Dyleva, Y.; Barbarash, O. Leptin resistance: Underlying mechanisms and diagnosis. *Diabetes Metab. Syndr. Obes.* **2019**, *12*, 191–198. [CrossRef] [PubMed]
118. Esteve, E.; Ricart, W.; Fernandez-Real, J.M. Adipocytokines and insulin resistance: The possible role of lipocalin-2, retinol binding protein-4, and adiponectin. *Diabetes Care* **2009**, *32* (Suppl. S2), S362–S367. [CrossRef]
119. Ahima, R.S. Central actions of adipocyte hormones. *Trends Endocrinol. Metab.* **2005**, *16*, 307–313. [CrossRef] [PubMed]
120. Steppan, C.M.; Bailey, S.T.; Bhat, S.; Brown, E.J.; Banerjee, R.R.; Wright, C.M.; Patel, H.R.; Ahima, R.S.; Lazar, M.A. The hormone resistin links obesity to diabetes. *Nature* **2001**, *409*, 307–312. [CrossRef]
121. Benomar, Y.; Taouis, M. Molecular Mechanisms Underlying Obesity-Induced Hypothalamic Inflammation and Insulin Resistance: Pivotal Role of Resistin/TLR4 Pathways. *Front. Endocrinol.* **2019**, *10*, 140. [CrossRef]
122. Su, K.Z.; Li, Y.R.; Zhang, D.; Yuan, J.H.; Zhang, C.S.; Liu, Y.; Song, L.M.; Lin, Q.; Li, M.W.; Dong, J. Relation of Circulating Resistin to Insulin Resistance in Type 2 Diabetes and Obesity: A Systematic Review and Meta-Analysis. *Front. Physiol.* **2019**, *10*, 1399. [CrossRef]
123. Landecho, M.F.; Tuero, C.; Valenti, V.; Bilbao, I.; de la Higuera, M.; Fruhbeck, G. Relevance of Leptin and Other Adipokines in Obesity-Associated Cardiovascular Risk. *Nutrients* **2019**, *11*, 2664. [CrossRef]
124. Saddi-Rosa, P.; Oliveira, C.S.; Giuffrida, F.M.; Reis, A.F. Visfatin, glucose metabolism and vascular disease: A review of evidence. *Diabetol Metab. Syndr.* **2010**, *2*, 21. [CrossRef] [PubMed]
125. Zheng, L.Y.; Xu, X.; Wan, R.H.; Xia, S.; Lu, J.; Huang, Q. Association between serum visfatin levels and atherosclerotic plaque in patients with type 2 diabetes. *Diabetol Metab. Syndr.* **2019**, *11*, 60. [CrossRef]
126. Shapira, S.N.; Seale, P. Transcriptional Control of Brown and Beige Fat Development and Function. *Obesity* **2019**, *27*, 13–21. [CrossRef]
127. Kajimura, S.; Seale, P.; Tomaru, T.; Erdjument-Bromage, H.; Cooper, M.P.; Ruas, J.L.; Chin, S.; Tempst, P.; Lazar, M.A.; Spiegelman, B.M. Regulation of the brown and white fat gene programs through a PRDM16/CtBP transcriptional complex. *Genes Dev.* **2008**, *22*, 1397–1409. [CrossRef]
128. Svensson, K.J.; Long, J.Z.; Jedrychowski, M.P.; Cohen, P.; Lo, J.C.; Serag, S.; Kir, S.; Shinoda, K.; Tartaglia, J.A.; Rao, R.R.; et al. A Secreted Slit2 Fragment Regulates Adipose Tissue Thermogenesis and Metabolic Function. *Cell Metab.* **2016**, *23*, 454–466. [CrossRef]

129. Seale, P.; Bjork, B.; Yang, W.; Kajimura, S.; Chin, S.; Kuang, S.; Scime, A.; Devarakonda, S.; Conroe, H.M.; Erdjument-Bromage, H.; et al. PRDM16 controls a brown fat/skeletal muscle switch. *Nature* **2008**, *454*, 961–967. [CrossRef]
130. Koncarevic, A.; Kajimura, S.; Cornwall-Brady, M.; Andreucci, A.; Pullen, A.; Sako, D.; Kumar, R.; Grinberg, A.V.; Liharska, K.; Ucran, J.A.; et al. A novel therapeutic approach to treating obesity through modulation of TGFbeta signaling. *Endocrinology* **2012**, *153*, 3133–3146. [CrossRef]
131. Yadav, H.; Quijano, C.; Kamaraju, A.K.; Gavrilova, O.; Malek, R.; Chen, W.; Zerfas, P.; Zhigang, D.; Wright, E.C.; Stuelten, C.; et al. Protection from obesity and diabetes by blockade of TGF-beta/Smad3 signaling. *Cell Metab.* **2011**, *14*, 67–79. [CrossRef] [PubMed]
132. Tong, M.; Jun, T.; Nie, Y.; Hao, J.; Fan, D. The Role of the Slit/Robo Signaling Pathway. *J. Cancer* **2019**, *10*, 2694–2705. [CrossRef]
133. Chrysovergis, K.; Wang, X.; Kosak, J.; Lee, S.H.; Kim, J.S.; Foley, J.F.; Travlos, G.; Singh, S.; Baek, S.J.; Eling, T.E. NAG-1/GDF-15 prevents obesity by increasing thermogenesis, lipolysis and oxidative metabolism. *Int. J. Obes.* **2014**, *38*, 1555–1564. [CrossRef] [PubMed]
134. Villarroya, F.; Cereijo, R.; Villarroya, J.; Giralt, M. Brown adipose tissue as a secretory organ. *Nat. Rev. Endocrinol.* **2017**, *13*, 26–35. [CrossRef] [PubMed]
135. Struik, D.; Dommerholt, M.B.; Jonker, J.W. Fibroblast growth factors in control of lipid metabolism: From biological function to clinical application. *Curr. Opin. Lipidol.* **2019**, *30*, 235–243. [CrossRef]
136. Veniant, M.M.; Sivits, G.; Helmering, J.; Komorowski, R.; Lee, J.; Fan, W.; Moyer, C.; Lloyd, D.J. Pharmacologic Effects of FGF21 Are Independent of the "Browning" of White Adipose Tissue. *Cell Metab.* **2015**, *21*, 731–738. [CrossRef]
137. Wang, N.; Zhao, T.T.; Li, S.M.; Sun, X.; Li, Z.C.; Li, Y.H.; Li, D.S.; Wang, W.F. Fibroblast Growth Factor 21 Exerts its Anti-inflammatory Effects on Multiple Cell Types of Adipose Tissue in Obesity. *Obesity* **2019**, *27*, 399–408. [CrossRef]
138. Wang, Q.A.; Tao, C.; Jiang, L.; Shao, M.; Ye, R.; Zhu, Y.; Gordillo, R.; Ali, A.; Lian, Y.; Holland, W.L.; et al. Distinct regulatory mechanisms governing embryonic versus adult adipocyte maturation. *Nat. Cell Biol.* **2015**, *17*, 1099–1111. [CrossRef]
139. Phillips, K.J. Beige Fat, Adaptive Thermogenesis, and Its Regulation by Exercise and Thyroid Hormone. *Biology* **2019**, *8*, 57. [CrossRef]
140. Roh, H.C.; Tsai, L.T.Y.; Shao, M.; Tenen, D.; Shen, Y.; Kumari, M.; Lyubetskaya, A.; Jacobs, C.; Dawes, B.; Gupta, R.K.; et al. Warming Induces Significant Reprogramming of Beige, but Not Brown, Adipocyte Cellular Identity. *Cell Metab.* **2018**, *27*, 1121–1137.e1125. [CrossRef]
141. Long, J.Z.; Svensson, K.J.; Tsai, L.; Zeng, X.; Roh, H.C.; Kong, X.; Rao, R.R.; Lou, J.; Lokurkar, I.; Baur, W.; et al. A smooth muscle-like origin for beige adipocytes. *Cell Metab.* **2014**, *19*, 810–820. [CrossRef] [PubMed]
142. Altshuler-Keylin, S.; Shinoda, K.; Hasegawa, Y.; Ikeda, K.; Hong, H.; Kang, Q.; Yang, Y.; Perera, R.M.; Debnath, J.; Kajimura, S. Beige Adipocyte Maintenance Is Regulated by Autophagy-Induced Mitochondrial Clearance. *Cell Metab.* **2016**, *24*, 402–419. [CrossRef]
143. Bonet, M.L.; Oliver, P.; Palou, A. Pharmacological and nutritional agents promoting browning of white adipose tissue. *Biochim. Biophys. Acta* **2013**, *1831*, 969–985. [CrossRef]
144. Shan, T.; Liang, X.; Bi, P.; Zhang, P.; Liu, W.; Kuang, S. Distinct populations of adipogenic and myogenic Myf5-lineage progenitors in white adipose tissues. *J. Lipid Res.* **2013**, *54*, 2214–2224. [CrossRef]
145. Sanchez-Gurmaches, J.; Hung, C.M.; Sparks, C.A.; Tang, Y.; Li, H.; Guertin, D.A. PTEN loss in the Myf5 lineage redistributes body fat and reveals subsets of white adipocytes that arise from Myf5 precursors. *Cell Metab.* **2012**, *16*, 348–362. [CrossRef]
146. Vishvanath, L.; MacPherson, K.A.; Hepler, C.; Wang, Q.A.; Shao, M.; Spurgin, S.B.; Wang, M.Y.; Kusminski, C.M.; Morley, T.S.; Gupta, R.K. Pdgfrbeta+ Mural Preadipocytes Contribute to Adipocyte Hyperplasia Induced by High-Fat-Diet Feeding and Prolonged Cold Exposure in Adult Mice. *Cell Metab.* **2016**, *23*, 350–359. [CrossRef] [PubMed]
147. Wang, W.; Kissig, M.; Rajakumari, S.; Huang, L.; Lim, H.W.; Won, K.J.; Seale, P. Ebf2 is a selective marker of brown and beige adipogenic precursor cells. *Proc. Natl. Acad. Sci. USA* **2014**, *111*, 14466–14471. [CrossRef]
148. Seale, P. Transcriptional Regulatory Circuits Controlling Brown Fat Development and Activation. *Diabetes* **2015**, *64*, 2369–2375. [CrossRef]
149. Tang, W.; Zeve, D.; Suh, J.M.; Bosnakovski, D.; Kyba, M.; Hammer, R.E.; Tallquist, M.D.; Graff, J.M. White fat progenitor cells reside in the adipose vasculature. *Science* **2008**, *322*, 583–586. [CrossRef]

150. Spalding, K.L.; Arner, E.; Westermark, P.O.; Bernard, S.; Buchholz, B.A.; Bergmann, O.; Blomqvist, L.; Hoffstedt, J.; Naslund, E.; Britton, T.; et al. Dynamics of fat cell turnover in humans. *Nature* **2008**, *453*, 783–787. [CrossRef]
151. Fu, L.; Zhu, X.; Yi, F.; Liu, G.H.; Izpisua Belmonte, J.C. Regenerative medicine: Transdifferentiation in vivo. *Cell Res.* **2014**, *24*, 141–142. [CrossRef]
152. Rosenwald, M.; Wolfrum, C. The origin and definition of brite versus white and classical brown adipocytes. *Adipocyte* **2014**, *3*, 4–9. [CrossRef] [PubMed]
153. Zhang, Z.; Shao, M.; Hepler, C.; Zi, Z.; Zhao, S.; An, Y.A.; Zhu, Y.; Ghaben, A.L.; Wang, M.Y.; Li, N.; et al. Dermal adipose tissue has high plasticity and undergoes reversible dedifferentiation in mice. *J. Clin. Investig.* **2019**, *129*, 5327–5342. [CrossRef]
154. Shao, M.; Wang, Q.A.; Song, A.; Vishvanath, L.; Busbuso, N.C.; Scherer, P.E.; Gupta, R.K. Cellular Origins of Beige Fat Cells Revisited. *Diabetes* **2019**, *68*, 1874–1885. [CrossRef]
155. Wu, J.; Jun, H.; McDermott, J.R. Formation and activation of thermogenic fat. *Trends Genet.* **2015**, *31*, 232–238. [CrossRef]
156. Bartesaghi, S.; Hallen, S.; Huang, L.; Svensson, P.A.; Momo, R.A.; Wallin, S.; Carlsson, E.K.; Forslow, A.; Seale, P.; Peng, X.R. Thermogenic activity of UCP1 in human white fat-derived beige adipocytes. *Mol. Endocrinol.* **2015**, *29*, 130–139. [CrossRef]
157. Qian, S.; Huang, H.; Tang, Q. Brown and beige fat: The metabolic function, induction, and therapeutic potential. *Front. Med.* **2015**, *9*, 162–172. [CrossRef]
158. Jespersen, N.Z.; Feizi, A.; Andersen, E.S.; Heywood, S.; Hattel, H.B.; Daugaard, S.; Peijs, L.; Bagi, P.; Feldt-Rasmussen, B.; Schultz, H.S.; et al. Heterogeneity in the perirenal region of humans suggests presence of dormant brown adipose tissue that contains brown fat precursor cells. *Mol. Metab.* **2019**, *24*, 30–43. [CrossRef]
159. Chawla, A.; Lazar, M.A. Peroxisome proliferator and retinoid signaling pathways co-regulate preadipocyte phenotype and survival. *Proc. Natl. Acad. Sci. USA* **1994**, *91*, 1786–1790. [CrossRef]
160. Jiang, Y.; Berry, D.C.; Tang, W.; Graff, J.M. Independent stem cell lineages regulate adipose organogenesis and adipose homeostasis. *Cell Rep.* **2014**, *9*, 1007–1022. [CrossRef]
161. Rodriguez, A.; Catalan, V.; Ramirez, B.; Unamuno, X.; Portincasa, P.; Gomez-Ambrosi, J.; Fruhbeck, G.; Becerril, S. Impact of adipokines and myokines on fat browning. *J. Physiol. Biochem.* **2020**, *76*, 227–240. [CrossRef]
162. Berry, D.C.; Jiang, Y.; Graff, J.M. Mouse strains to study cold-inducible beige progenitors and beige adipocyte formation and function. *Nat. Commun.* **2016**, *7*, 10184. [CrossRef]
163. Berry, D.C.; Jiang, Y.; Arpke, R.W.; Close, E.L.; Uchida, A.; Reading, D.; Berglund, E.D.; Kyba, M.; Graff, J.M. Cellular Aging Contributes to Failure of Cold-Induced Beige Adipocyte Formation in Old Mice and Humans. *Cell Metab.* **2017**, *25*, 166–181. [CrossRef]
164. Min, S.Y.; Desai, A.; Yang, Z.; Sharma, A.; DeSouza, T.; Genga, R.M.J.; Kucukural, A.; Lifshitz, L.M.; Nielsen, S.; Scheele, C.; et al. Diverse repertoire of human adipocyte subtypes develops from transcriptionally distinct mesenchymal progenitor cells. *Proc. Natl. Acad. Sci. USA* **2019**, *116*, 17970–17979. [CrossRef] [PubMed]
165. Bertholet, A.M.; Kazak, L.; Chouchani, E.T.; Bogaczynska, M.G.; Paranjpe, I.; Wainwright, G.L.; Betourne, A.; Kajimura, S.; Spiegelman, B.M.; Kirichok, Y. Mitochondrial Patch Clamp of Beige Adipocytes Reveals UCP1-Positive and UCP1-Negative Cells Both Exhibiting Futile Creatine Cycling. *Cell Metab.* **2017**, *25*, 811–822.e814. [CrossRef]
166. Chance, B.; Lees, H.; Postgate, J.R. The meaning of "reversed electron flow" and "high energy electron" in biochemistry. *Nature* **1972**, *238*, 330–331. [CrossRef] [PubMed]
167. Nedergaard, J.; Cannon, B.; Lindberg, O. Microcalorimetry of isolated mammalian cells. *Nature* **1977**, *267*, 518–520. [CrossRef]
168. Nicholls, D.G. The hunt for the molecular mechanism of brown fat thermogenesis. *Biochimie* **2017**, *134*, 9–18. [CrossRef]
169. Bertholet, A.M.; Kirichok, Y. UCP1: A transporter for H(+) and fatty acid anions. *Biochimie* **2017**, *134*, 28–34. [CrossRef]
170. Fedorenko, A.; Lishko, P.V.; Kirichok, Y. Mechanism of fatty-acid-dependent UCP1 uncoupling in brown fat mitochondria. *Cell* **2012**, *151*, 400–413. [CrossRef] [PubMed]
171. Jastroch, M.; Divakaruni, A.S.; Mookerjee, S.; Treberg, J.R.; Brand, M.D. Mitochondrial proton and electron leaks. *Essays Biochem.* **2010**, *47*, 53–67. [CrossRef]

172. Yao, L.; Cui, X.; Chen, Q.; Yang, X.; Fang, F.; Zhang, J.; Liu, G.; Jin, W.; Chang, Y. Cold-Inducible SIRT6 Regulates Thermogenesis of Brown and Beige Fat. *Cell Rep.* **2017**, *20*, 641–654. [CrossRef]
173. Murano, I.; Barbatelli, G.; Giordano, A.; Cinti, S. Noradrenergic parenchymal nerve fiber branching after cold acclimatisation correlates with brown adipocyte density in mouse adipose organ. *J. Anat.* **2009**, *214*, 171–178. [CrossRef]
174. Li, Y.; Fromme, T.; Schweizer, S.; Schottl, T.; Klingenspor, M. Taking control over intracellular fatty acid levels is essential for the analysis of thermogenic function in cultured primary brown and brite/beige adipocytes. *EMBO Rep.* **2014**, *15*, 1069–1076. [CrossRef] [PubMed]
175. Shin, H.; Shi, H.; Xue, B.; Yu, L. What activates thermogenesis when lipid droplet lipolysis is absent in brown adipocytes? *Adipocyte* **2018**, *7*, 1–5. [CrossRef]
176. Kazak, L.; Rahbani, J.F.; Samborska, B.; Lu, G.Z.; Jedrychowski, M.P.; Lajoie, M.; Zhang, S.; Ramsay, L.C.; Dou, F.Y.; Tenen, D.; et al. Ablation of adipocyte creatine transport impairs thermogenesis and causes diet-induced obesity. *Nat. Metab.* **2019**, *1*, 360–370. [CrossRef]
177. Chouchani, E.T.; Kazak, L.; Jedrychowski, M.P.; Lu, G.Z.; Erickson, B.K.; Szpyt, J.; Pierce, K.A.; Laznik-Bogoslavski, D.; Vetrivelan, R.; Clish, C.B.; et al. Mitochondrial ROS regulate thermogenic energy expenditure and sulfenylation of UCP1. *Nature* **2016**, *532*, 112–116. [CrossRef]
178. Mills, E.L.; Pierce, K.A.; Jedrychowski, M.P.; Garrity, R.; Winther, S.; Vidoni, S.; Yoneshiro, T.; Spinelli, J.B.; Lu, G.Z.; Kazak, L.; et al. Accumulation of succinate controls activation of adipose tissue thermogenesis. *Nature* **2018**, *560*, 102–106. [CrossRef]
179. Enerback, S.; Jacobsson, A.; Simpson, E.M.; Guerra, C.; Yamashita, H.; Harper, M.E.; Kozak, L.P. Mice lacking mitochondrial uncoupling protein are cold-sensitive but not obese. *Nature* **1997**, *387*, 90–94. [CrossRef]
180. Jedrychowski, M.P.; Lu, G.Z.; Szpyt, J.; Mariotti, M.; Garrity, R.; Paulo, J.A.; Schweppe, D.K.; Laznik-Bogoslavski, D.; Kazak, L.; Murphy, M.P.; et al. Facultative protein selenation regulates redox sensitivity, adipose tissue thermogenesis, and obesity. *Proc. Natl. Acad. Sci. USA* **2020**, *117*, 10789–10796. [CrossRef]
181. Ikeda, K.; Kang, Q.; Yoneshiro, T.; Camporez, J.P.; Maki, H.; Homma, M.; Shinoda, K.; Chen, Y.; Lu, X.; Maretich, P.; et al. UCP1-independent signaling involving SERCA2b-mediated calcium cycling regulates beige fat thermogenesis and systemic glucose homeostasis. *Nat. Med.* **2017**, *23*, 1454–1465. [CrossRef]
182. Meyer, C.W.; Willershauser, M.; Jastroch, M.; Rourke, B.C.; Fromme, T.; Oelkrug, R.; Heldmaier, G.; Klingenspor, M. Adaptive thermogenesis and thermal conductance in wild-type and UCP1-KO mice. *Am. J. Physiol. Regul. Integr. Comp. Physiol.* **2010**, *299*, R1396–R1406. [CrossRef]
183. Kazak, L.; Chouchani, E.T.; Jedrychowski, M.P.; Erickson, B.K.; Shinoda, K.; Cohen, P.; Vetrivelan, R.; Lu, G.Z.; Laznik-Bogoslavski, D.; Hasenfuss, S.C.; et al. A creatine-driven substrate cycle enhances energy expenditure and thermogenesis in beige fat. *Cell* **2015**, *163*, 643–655. [CrossRef]
184. Kazak, L.; Chouchani, E.T.; Lu, G.Z.; Jedrychowski, M.P.; Bare, C.J.; Mina, A.I.; Kumari, M.; Zhang, S.; Vuckovic, I.; Laznik-Bogoslavski, D.; et al. Genetic Depletion of Adipocyte Creatine Metabolism Inhibits Diet-Induced Thermogenesis and Drives Obesity. *Cell Metab.* **2017**, *26*, 693. [CrossRef]
185. Gerngross, C.; Schretter, J.; Klingenspor, M.; Schwaiger, M.; Fromme, T. Active Brown Fat During (18)F-FDG PET/CT Imaging Defines a Patient Group with Characteristic Traits and an Increased Probability of Brown Fat Redetection. *J. Nucl. Med.* **2017**, *58*, 1104–1110. [CrossRef]
186. Roesler, A.; Kazak, L. UCP1-independent thermogenesis. *Biochem. J.* **2020**, *477*, 709–725. [CrossRef] [PubMed]
187. Gray, S.G.; Iglesias, A.H.; Lizcano, F.; Villanueva, R.; Camelo, S.; Jingu, H.; Teh, B.T.; Koibuchi, N.; Chin, W.W.; Kokkotou, E.; et al. Functional characterization of JMJD2A, a histone deacetylase- and retinoblastoma-binding protein. *J. Biol. Chem.* **2005**, *280*, 28507–28518. [CrossRef]
188. Xiao, H.; Kang, S. The role of DNA methylation in thermogenic adipose biology. *Epigenetics* **2019**, *14*, 837–843. [CrossRef] [PubMed]
189. Zhao, X.Y.; Li, S.; DelProposto, J.L.; Liu, T.; Mi, L.; Porsche, C.; Peng, X.; Lumeng, C.N.; Lin, J.D. The long noncoding RNA Blnc1 orchestrates homeostatic adipose tissue remodeling to preserve metabolic health. *Mol. Metab.* **2018**, *14*, 60–70. [CrossRef]
190. Amri, E.Z.; Scheideler, M. Small non coding RNAs in adipocyte biology and obesity. *Mol. Cell Endocrinol.* **2017**, *456*, 87–94. [CrossRef]
191. Wu, Z.; Rosen, E.D.; Brun, R.; Hauser, S.; Adelmant, G.; Troy, A.E.; McKeon, C.; Darlington, G.J.; Spiegelman, B.M. Cross-regulation of C/EBP alpha and PPAR gamma controls the transcriptional pathway of adipogenesis and insulin sensitivity. *Mol. Cell* **1999**, *3*, 151–158. [CrossRef]

192. Rosen, E.D.; Sarraf, P.; Troy, A.E.; Bradwin, G.; Moore, K.; Milstone, D.S.; Spiegelman, B.M.; Mortensen, R.M. PPAR gamma is required for the differentiation of adipose tissue in vivo and in vitro. *Mol. Cell* **1999**, *4*, 611–617. [CrossRef]
193. Seale, P.; Kajimura, S.; Yang, W.; Chin, S.; Rohas, L.M.; Uldry, M.; Tavernier, G.; Langin, D.; Spiegelman, B.M. Transcriptional control of brown fat determination by PRDM16. *Cell Metab.* **2007**, *6*, 38–54. [CrossRef] [PubMed]
194. Bostrom, P.; Wu, J.; Jedrychowski, M.P.; Korde, A.; Ye, L.; Lo, J.C.; Rasbach, K.A.; Bostrom, E.A.; Choi, J.H.; Long, J.Z.; et al. A PGC1-alpha-dependent myokine that drives brown-fat-like development of white fat and thermogenesis. *Nature* **2012**, *481*, 463–468. [CrossRef]
195. Harms, M.J.; Lim, H.W.; Ho, Y.; Shapira, S.N.; Ishibashi, J.; Rajakumari, S.; Steger, D.J.; Lazar, M.A.; Won, K.J.; Seale, P. PRDM16 binds MED1 and controls chromatin architecture to determine a brown fat transcriptional program. *Genes Dev.* **2015**, *29*, 298–307. [CrossRef]
196. Angueira, A.R.; Shapira, S.N.; Ishibashi, J.; Sampat, S.; Sostre-Colon, J.; Emmett, M.J.; Titchenell, P.M.; Lazar, M.A.; Lim, H.W.; Seale, P. Early B Cell Factor Activity Controls Developmental and Adaptive Thermogenic Gene Programming in Adipocytes. *Cell Rep.* **2020**, *30*, 2869–2878 e2864. [CrossRef]
197. Stine, R.R.; Shapira, S.N.; Lim, H.W.; Ishibashi, J.; Harms, M.; Won, K.J.; Seale, P. EBF2 promotes the recruitment of beige adipocytes in white adipose tissue. *Mol. Metab.* **2016**, *5*, 57–65. [CrossRef]
198. Scime, A.; Grenier, G.; Huh, M.S.; Gillespie, M.A.; Bevilacqua, L.; Harper, M.E.; Rudnicki, M.A. Rb and p107 regulate preadipocyte differentiation into white versus brown fat through repression of PGC-1alpha. *Cell Metab.* **2005**, *2*, 283–295. [CrossRef] [PubMed]
199. Calo, E.; Quintero-Estades, J.A.; Danielian, P.S.; Nedelcu, S.; Berman, S.D.; Lees, J.A. Rb regulates fate choice and lineage commitment in vivo. *Nature* **2010**, *466*, 1110–1114. [CrossRef]
200. Sato, T.; Vargas, D.; Miyazaki, K.; Uchida, K.; Ariyani, W.; Miyazaki, M.; Okada, J.; Lizcano, F.; Koibuchi, N.; Shimokawa, N. EID1 suppresses lipid accumulation by inhibiting the expression of GPDH in 3T3-L1 preadipocytes. *J. Cell Physiol.* **2020**, *235*, 6725–6735. [CrossRef]
201. Puigserver, P.; Wu, Z.; Park, C.W.; Graves, R.; Wright, M.; Spiegelman, B.M. A cold-inducible coactivator of nuclear receptors linked to adaptive thermogenesis. *Cell* **1998**, *92*, 829–839. [CrossRef]
202. Puigserver, P.; Adelmant, G.; Wu, Z.; Fan, M.; Xu, J.; O'Malley, B.; Spiegelman, B.M. Activation of PPARgamma coactivator-1 through transcription factor docking. *Science* **1999**, *286*, 1368–1371. [CrossRef]
203. Lin, J.; Handschin, C.; Spiegelman, B.M. Metabolic control through the PGC-1 family of transcription coactivators. *Cell Metab.* **2005**, *1*, 361–370. [CrossRef]
204. Wu, Z.; Puigserver, P.; Andersson, U.; Zhang, C.; Adelmant, G.; Mootha, V.; Troy, A.; Cinti, S.; Lowell, B.; Scarpulla, R.C.; et al. Mechanisms controlling mitochondrial biogenesis and respiration through the thermogenic coactivator PGC-1. *Cell* **1999**, *98*, 115–124. [CrossRef]
205. Lehman, J.J.; Barger, P.M.; Kovacs, A.; Saffitz, J.E.; Medeiros, D.M.; Kelly, D.P. Peroxisome proliferator-activated receptor gamma coactivator-1 promotes cardiac mitochondrial biogenesis. *J. Clin. Investig.* **2000**, *106*, 847–856. [CrossRef]
206. Lin, J.; Wu, P.H.; Tarr, P.T.; Lindenberg, K.S.; St-Pierre, J.; Zhang, C.Y.; Mootha, V.K.; Jager, S.; Vianna, C.R.; Reznick, R.M.; et al. Defects in adaptive energy metabolism with CNS-linked hyperactivity in PGC-1alpha null mice. *Cell* **2004**, *119*, 121–135. [CrossRef]
207. Uldry, M.; Yang, W.; St-Pierre, J.; Lin, J.; Seale, P.; Spiegelman, B.M. Complementary action of the PGC-1 coactivators in mitochondrial biogenesis and brown fat differentiation. *Cell Metab.* **2006**, *3*, 333–341. [CrossRef]
208. Kleiner, S.; Mepani, R.J.; Laznik, D.; Ye, L.; Jurczak, M.J.; Jornayvaz, F.R.; Estall, J.L.; Chatterjee Bhowmick, D.; Shulman, G.I.; Spiegelman, B.M. Development of insulin resistance in mice lacking PGC-1alpha in adipose tissues. *Proc. Natl. Acad. Sci. USA* **2012**, *109*, 9635–9640. [CrossRef]
209. Lizcano, F. The Beige Adipocyte as a Therapy for Metabolic Diseases. *Int. J. Mol. Sci.* **2019**, *20*, 5058. [CrossRef]
210. Tchang, B.G.; Abel, B.; Zecca, C.; Saunders, K.H.; Shukla, A.P. An up-to-date evaluation of lorcaserin hydrochloride for the treatment of obesity. *Expert Opin. Pharm.* **2020**, *21*, 21–28. [CrossRef]
211. Folgueira, C.; Beiroa, D.; Porteiro, B.; Duquenne, M.; Puighermanal, E.; Fondevila, M.F.; Barja-Fernandez, S.; Gallego, R.; Hernandez-Bautista, R.; Castelao, C.; et al. Hypothalamic dopamine signaling regulates brown fat thermogenesis. *Nat. Metab.* **2019**, *1*, 811–829. [CrossRef] [PubMed]
212. Silvester, A.J.; Aseer, K.R.; Yun, J.W. Dietary polyphenols and their roles in fat browning. *J. Nutr. Biochem.* **2019**, *64*, 1–12. [CrossRef]

213. Ohyama, K.; Nogusa, Y.; Shinoda, K.; Suzuki, K.; Bannai, M.; Kajimura, S. A Synergistic Antiobesity Effect by a Combination of Capsinoids and Cold Temperature through Promoting Beige Adipocyte Biogenesis. *Diabetes* **2016**, *65*, 1410–1423. [CrossRef]
214. Baskaran, P.; Krishnan, V.; Ren, J.; Thyagarajan, B. Capsaicin induces browning of white adipose tissue and counters obesity by activating TRPV1 channel-dependent mechanisms. *Br. J. Pharm.* **2016**, *173*, 2369–2389. [CrossRef]
215. Perkins, M.N.; Rothwell, N.J.; Stock, M.J.; Stone, T.W. Activation of brown adipose tissue thermogenesis by the ventromedial hypothalamus. *Nature* **1981**, *289*, 401–402. [CrossRef]
216. Yoneshiro, T.; Aita, S.; Matsushita, M.; Kayahara, T.; Kameya, T.; Kawai, Y.; Iwanaga, T.; Saito, M. Recruited brown adipose tissue as an antiobesity agent in humans. *J. Clin. Investig.* **2013**, *123*, 3404–3408. [CrossRef] [PubMed]
217. Lagouge, M.; Argmann, C.; Gerhart-Hines, Z.; Meziane, H.; Lerin, C.; Daussin, F.; Messadeq, N.; Milne, J.; Lambert, P.; Elliott, P.; et al. Resveratrol improves mitochondrial function and protects against metabolic disease by activating SIRT1 and PGC-1alpha. *Cell* **2006**, *127*, 1109–1122. [CrossRef]
218. Baur, J.A.; Pearson, K.J.; Price, N.L.; Jamieson, H.A.; Lerin, C.; Kalra, A.; Prabhu, V.V.; Allard, J.S.; Lopez-Lluch, G.; Lewis, K.; et al. Resveratrol improves health and survival of mice on a high-calorie diet. *Nature* **2006**, *444*, 337–342. [CrossRef]
219. Hui, S.; Liu, Y.; Huang, L.; Zheng, L.; Zhou, M.; Lang, H.; Wang, X.; Yi, L.; Mi, M. Resveratrol enhances brown adipose tissue activity and white adipose tissue browning in part by regulating bile acid metabolism via gut microbiota remodeling. *Int. J. Obes.* **2020**, *44*, 1678–1690. [CrossRef]
220. Andrade, J.M.; Frade, A.C.; Guimaraes, J.B.; Freitas, K.M.; Lopes, M.T.; Guimaraes, A.L.; de Paula, A.M.; Coimbra, C.C.; Santos, S.H. Resveratrol increases brown adipose tissue thermogenesis markers by increasing SIRT1 and energy expenditure and decreasing fat accumulation in adipose tissue of mice fed a standard diet. *Eur. J. Nutr.* **2014**, *53*, 1503–1510. [CrossRef]
221. Zietak, M.; Kovatcheva-Datchary, P.; Markiewicz, L.H.; Stahlman, M.; Kozak, L.P.; Backhed, F. Altered Microbiota Contributes to Reduced Diet-Induced Obesity upon Cold Exposure. *Cell Metab.* **2016**, *23*, 1216–1223. [CrossRef] [PubMed]
222. Moreno-Navarrete, J.M.; Serino, M.; Blasco-Baque, V.; Azalbert, V.; Barton, R.H.; Cardellini, M.; Latorre, J.; Ortega, F.; Sabater-Masdeu, M.; Burcelin, R.; et al. Gut Microbiota Interacts with Markers of Adipose Tissue Browning, Insulin Action and Plasma Acetate in Morbid Obesity. *Mol. Nutr. Food Res.* **2018**, *62*, 1700721. [CrossRef] [PubMed]
223. Li, G.; Xie, C.; Lu, S.; Nichols, R.G.; Tian, Y.; Li, L.; Patel, D.; Ma, Y.; Brocker, C.N.; Yan, T.; et al. Intermittent Fasting Promotes White Adipose Browning and Decreases Obesity by Shaping the Gut Microbiota. *Cell Metab.* **2017**, *26*, 801. [CrossRef]
224. Hanatani, S.; Motoshima, H.; Takaki, Y.; Kawasaki, S.; Igata, M.; Matsumura, T.; Kondo, T.; Senokuchi, T.; Ishii, N.; Kawashima, J.; et al. Acetate alters expression of genes involved in beige adipogenesis in 3T3-L1 cells and obese KK-Ay mice. *J. Clin. Biochem. Nutr.* **2016**, *59*, 207–214. [CrossRef]
225. Moreno-Navarrete, J.M.; Fernandez-Real, J.M. The gut microbiota modulates both browning of white adipose tissue and the activity of brown adipose tissue. *Rev. Endocr. Metab. DISORD* **2019**, *20*, 387–397. [CrossRef]
226. Kinoshita, K.; Ozaki, N.; Takagi, Y.; Murata, Y.; Oshida, Y.; Hayashi, Y. Glucagon is essential for adaptive thermogenesis in brown adipose tissue. *Endocrinology* **2014**, *155*, 3484–3492. [CrossRef]
227. Beiroa, D.; Imbernon, M.; Gallego, R.; Senra, A.; Herranz, D.; Villarroya, F.; Serrano, M.; Ferno, J.; Salvador, J.; Escalada, J.; et al. GLP-1 agonism stimulates brown adipose tissue thermogenesis and browning through hypothalamic AMPK. *Diabetes* **2014**, *63*, 3346–3358. [CrossRef]
228. White, J.D.; Dewal, R.S.; Stanford, K.I. The beneficial effects of brown adipose tissue transplantation. *Mol. Asp. Med.* **2019**, *68*, 74–81. [CrossRef]
229. Shankar, K.; Kumar, D.; Gupta, S.; Varshney, S.; Rajan, S.; Srivastava, A.; Gupta, A.; Gupta, A.P.; Vishwakarma, A.L.; Gayen, J.R.; et al. Role of brown adipose tissue in modulating adipose tissue inflammation and insulin resistance in high-fat diet fed mice. *Eur. J. Pharm.* **2019**, *854*, 354–364. [CrossRef]
230. Arroyave, F.; Montano, D.; Lizcano, F. Adipose tissue browning for the treatment of obesity and metabolic diseases. *CellR4 (Repair Replace Regen Reprogram)* **2020**, *8*, e2877. [CrossRef]
231. Finlin, B.S.; Memetimin, H.; Zhu, B.; Confides, A.L.; Vekaria, H.J.; El Khouli, R.H.; Johnson, Z.R.; Westgate, P.M.; Chen, J.; Morris, A.J.; et al. The beta3-adrenergic receptor agonist mirabegron improves glucose homeostasis in obese humans. *J. Clin. Investig.* **2020**, *130*, 2319–2331. [CrossRef] [PubMed]

232. Cypess, A.M.; Weiner, L.S.; Roberts-Toler, C.; Franquet Elia, E.; Kessler, S.H.; Kahn, P.A.; English, J.; Chatman, K.; Trauger, S.A.; Doria, A.; et al. Activation of human brown adipose tissue by a beta3-adrenergic receptor agonist. *Cell Metab.* **2015**, *21*, 33–38. [CrossRef]
233. Finlin, B.S.; Memetimin, H.; Confides, A.L.; Kasza, I.; Zhu, B.; Vekaria, H.J.; Harfmann, B.; Jones, K.A.; Johnson, Z.R.; Westgate, P.M.; et al. Human adipose beiging in response to cold and mirabegron. *JCI Insight* **2018**, *3*. [CrossRef] [PubMed]
234. Baskin, A.S.; Linderman, J.D.; Brychta, R.J.; McGehee, S.; Anflick-Chames, E.; Cero, C.; Johnson, J.W.; O'Mara, A.E.; Fletcher, L.A.; Leitner, B.P.; et al. Regulation of Human Adipose Tissue Activation, Gallbladder Size, and Bile Acid Metabolism by a beta3-Adrenergic Receptor Agonist. *Diabetes* **2018**, *67*, 2113–2125. [CrossRef]
235. Tseng, Y.H.; Kokkotou, E.; Schulz, T.J.; Huang, T.L.; Winnay, J.N.; Taniguchi, C.M.; Tran, T.T.; Suzuki, R.; Espinoza, D.O.; Yamamoto, Y.; et al. New role of bone morphogenetic protein 7 in brown adipogenesis and energy expenditure. *Nature* **2008**, *454*, 1000–1004. [CrossRef]
236. Fisher, F.M.; Kleiner, S.; Douris, N.; Fox, E.C.; Mepani, R.J.; Verdeguer, F.; Wu, J.; Kharitonenkov, A.; Flier, J.S.; Maratos-Flier, E.; et al. FGF21 regulates PGC-1alpha and browning of white adipose tissues in adaptive thermogenesis. *Genes Dev.* **2012**, *26*, 271–281. [CrossRef] [PubMed]
237. Qian, S.W.; Tang, Y.; Li, X.; Liu, Y.; Zhang, Y.Y.; Huang, H.Y.; Xue, R.D.; Yu, H.Y.; Guo, L.; Gao, H.D.; et al. BMP4-mediated brown fat-like changes in white adipose tissue alter glucose and energy homeostasis. *Proc. Natl. Acad. Sci. USA* **2013**, *110*, E798–E807. [CrossRef]
238. Charles, E.D.; Neuschwander-Tetri, B.A.; Pablo Frias, J.; Kundu, S.; Luo, Y.; Tirucherai, G.S.; Christian, R. Pegbelfermin (BMS-986036), PEGylated FGF21, in Patients with Obesity and Type 2 Diabetes: Results from a Randomized Phase 2 Study. *Obesity* **2019**, *27*, 41–49. [CrossRef] [PubMed]
239. Sanyal, A.; Charles, E.D.; Neuschwander-Tetri, B.A.; Loomba, R.; Harrison, S.A.; Abdelmalek, M.F.; Lawitz, E.J.; Halegoua-DeMarzio, D.; Kundu, S.; Noviello, S.; et al. Pegbelfermin (BMS-986036), a PEGylated fibroblast growth factor 21 analogue, in patients with non-alcoholic steatohepatitis: A randomised, double-blind, placebo-controlled, phase 2a trial. *Lancet* **2019**, *392*, 2705–2717. [CrossRef]
240. Sumida, Y.; Yoneda, M. Current and future pharmacological therapies for NAFLD/NASH. *J. Gastroenterol.* **2018**, *53*, 362–376. [CrossRef]
241. Singh, A.K.; Aryal, B.; Chaube, B.; Rotllan, N.; Varela, L.; Horvath, T.L.; Suarez, Y.; Fernandez-Hernando, C. Brown adipose tissue derived ANGPTL4 controls glucose and lipid metabolism and regulates thermogenesis. *Mol. Metab.* **2018**, *11*, 59–69. [CrossRef]
242. McQueen, A.E.; Kanamaluru, D.; Yan, K.; Gray, N.E.; Wu, L.; Li, M.L.; Chang, A.; Hasan, A.; Stifler, D.; Koliwad, S.K.; et al. The C-terminal fibrinogen-like domain of angiopoietin-like 4 stimulates adipose tissue lipolysis and promotes energy expenditure. *J. Biol. Chem.* **2017**, *292*, 16122–16134. [CrossRef]
243. Agudelo, L.Z.; Ferreira, D.M.S.; Cervenka, I.; Bryzgalova, G.; Dadvar, S.; Jannig, P.R.; Pettersson-Klein, A.T.; Lakshmikanth, T.; Sustarsic, E.G.; Porsmyr-Palmertz, M.; et al. Kynurenic Acid and Gpr35 Regulate Adipose Tissue Energy Homeostasis and Inflammation. *Cell Metab.* **2018**, *27*, 378–392 e375. [CrossRef]
244. Chondronikola, M.; Volpi, E.; Borsheim, E.; Porter, C.; Annamalai, P.; Enerback, S.; Lidell, M.E.; Saraf, M.K.; Labbe, S.M.; Hurren, N.M.; et al. Brown adipose tissue improves whole-body glucose homeostasis and insulin sensitivity in humans. *Diabetes* **2014**, *63*, 4089–4099. [CrossRef]
245. Hanssen, M.J.; van der Lans, A.A.; Brans, B.; Hoeks, J.; Jardon, K.M.; Schaart, G.; Mottaghy, F.M.; Schrauwen, P.; van Marken Lichtenbelt, W.D. Short-term Cold Acclimation Recruits Brown Adipose Tissue in Obese Humans. *Diabetes* **2016**, *65*, 1179–1189. [CrossRef] [PubMed]
246. Schrauwen-Hinderling, V.B.; Kooi, M.E.; Schrauwen, P. Mitochondrial Function and Diabetes: Consequences for Skeletal and Cardiac Muscle Metabolism. *Antioxid. Redox Signal.* **2016**, *24*, 39–51. [CrossRef] [PubMed]
247. Hanssen, M.J.; Hoeks, J.; Brans, B.; van der Lans, A.A.; Schaart, G.; van den Driessche, J.J.; Jorgensen, J.A.; Boekschoten, M.V.; Hesselink, M.K.; Havekes, B.; et al. Short-term cold acclimation improves insulin sensitivity in patients with type 2 diabetes mellitus. *Nat. Med.* **2015**, *21*, 863–865. [CrossRef]
248. Giordano, A.; Frontini, A.; Cinti, S. Convertible visceral fat as a therapeutic target to curb obesity. *Nat. Rev. Drug Discov.* **2016**, *15*, 405–424. [CrossRef]
249. Armani, A.; Infante, M.; Fabbri, A.; Caprio, M. Comment on "mineralocorticoid antagonism enhances brown adipose tissue function in humans: A randomized placebo-controlled cross-over study". *Diabetes Obes. Metab.* **2019**, *21*, 2024–2026. [CrossRef] [PubMed]

250. Infante, M.; Armani, A.; Marzolla, V.; Fabbri, A.; Caprio, M. Adipocyte Mineralocorticoid Receptor. *Vitam. Horm.* **2019**, *109*, 189–209. [CrossRef]
251. Infante, M.; Armani, A.; Mammi, C.; Fabbri, A.; Caprio, M. Impact of Adrenal Steroids on Regulation of Adipose Tissue. *Compr. Physiol.* **2017**, *7*, 1425–1447. [CrossRef]
252. Marzolla, V.; Armani, A.; Feraco, A.; De Martino, M.U.; Fabbri, A.; Rosano, G.; Caprio, M. Mineralocorticoid receptor in adipocytes and macrophages: A promising target to fight metabolic syndrome. *Steroids* **2014**, *91*, 46–53. [CrossRef]
253. Feraco, A.; Marzolla, V.; Scuteri, A.; Armani, A.; Caprio, M. Mineralocorticoid Receptors in Metabolic Syndrome: From Physiology to Disease. *Trends Endocrinol. Metab.* **2020**, *31*, 205–217. [CrossRef] [PubMed]
254. Peng, X.R.; Gennemark, P.; O'Mahony, G.; Bartesaghi, S. Unlock the Thermogenic Potential of Adipose Tissue: Pharmacological Modulation and Implications for Treatment of Diabetes and Obesity. *Front. Endocrinol.* **2015**, *6*, 174. [CrossRef]
255. Claussnitzer, M.; Dankel, S.N.; Kim, K.H.; Quon, G.; Meuleman, W.; Haugen, C.; Glunk, V.; Sousa, I.S.; Beaudry, J.L.; Puviindran, V.; et al. FTO Obesity Variant Circuitry and Adipocyte Browning in Humans. *N. Engl. J. Med.* **2015**, *373*, 895–907. [CrossRef]
256. Symonds, M.E.; Farhat, G.; Aldiss, P.; Pope, M.; Budge, H. Brown adipose tissue and glucose homeostasis—The link between climate change and the global rise in obesity and diabetes. *Adipocyte* **2019**, *8*, 46–50. [CrossRef]
257. Singh, A.M.; Dalton, S. What Can 'Brown-ing' Do for You? *Trends Endocrinol. Metab.* **2018**, *29*, 349–359. [CrossRef]

Article

Impact of Long-Term HFD Intake on the Peripheral and Central IGF System in Male and Female Mice

Santiago Guerra-Cantera [1,2,3], Laura M. Frago [1,2,3], María Jiménez-Hernaiz [1,3], Purificación Ros [2,4], Alejandra Freire-Regatillo [1,2,3], Vicente Barrios [1,3], Jesús Argente [1,2,3,5,*] and Julie A. Chowen [1,3,5,*]

1 Department of Endocrinology, Hospital Infantil Universitario Niño Jesús, Instituto de Investigación La Princesa, E-28009 Madrid, Spain; santiago.guerra@estudiante.uam.es (S.G.-C.); laura.frago@uam.es (L.M.F.); maria_jhernaiz@hotmail.es (M.J.-H.); alefreirefr@gmail.com (A.F.-R.); vicente.barriossa@salud.madrid.org (V.B.)

2 Department of Pediatrics, Faculty of Medicine, Universidad Autónoma de Madrid, E-28029 Madrid, Spain; prosmon@hotmail.com

3 Centro de Investigación Biomédica en Red de Fisiopatología de la Obesidad y Nutrición (CIBEROBN), Instituto de Salud Carlos III, E-28029 Madrid, Spain

4 Department of Pediatrics, Hospital Universitario Puerta de Hierro-Majadahonda, E-28222 Madrid, Spain

5 IMDEA Food Institute, CEI UAM + CSIC, Carretera de Cantoblanco 8, E-28049 Madrid, Spain

* Correspondence: jesus.argente@uam.es (J.A.); julieann.chowen@salud.madrid.org (J.A.C.)

Received: 22 October 2020; Accepted: 11 November 2020; Published: 13 November 2020

Abstract: The insulin-like growth factor (IGF) system is responsible for growth, but also affects metabolism and brain function throughout life. New IGF family members (i.e., pappalysins and stanniocalcins) control the availability/activity of IGFs and are implicated in growth. However, how diet and obesity modify this system has been poorly studied. We explored how intake of a high-fat diet (HFD) or commercial control diet (CCD) affects the IGF system in the circulation, visceral adipose tissue (VAT) and hypothalamus. Male and female C57/BL6J mice received HFD (60% fat, 5.1 kcal/g), CCD (10% fat, 3.7 kcal/g) or chow (3.1 % fat, 3.4 kcal/g) for 8 weeks. After 7 weeks of HFD intake, males had decreased glucose tolerance ($p < 0.01$) and at sacrifice increased plasma insulin ($p < 0.05$) and leptin ($p < 0.01$). Circulating free IGF1 ($p < 0.001$), total IGF1 ($p < 0.001$), IGF2 ($p < 0.05$) and IGFBP3 ($p < 0.01$) were higher after HFD in both sexes, with CCD increasing IGFBP2 in males ($p < 0.001$). In VAT, HFD reduced mRNA levels of IGF2 ($p < 0.05$), PAPP-A ($p < 0.001$) and stanniocalcin (STC)-1 ($p < 0.001$) in males. HFD increased hypothalamic IGF1 ($p < 0.01$), IGF2 ($p < 0.05$) and IGFBP5 ($p < 0.01$) mRNA levels, with these changes more apparent in females. Our results show that diet-induced changes in the IGF system are tissue-, sex- and diet-dependent.

Keywords: IGF1; IGF2; IGFBP2; high-fat diet; obesity; sex differences; neuropeptides

1. Introduction

The insulin-like growth factor (IGF) system has been widely studied in both pre- and postnatal growth and development, but this system is involved in a myriad of functions throughout life with much less known regarding many of these diverse roles. Members of this family include the ligands IGF1 and IGF2, which are crucial for longitudinal bone growth [1] and both also have additional important functions throughout life. Studies of IGF2 have focused mainly on its fundamental role during fetal life, as its expression decreases in multiple organs during aging in mice [2] as well as in the circulation in humans [3], but less is known about its postnatal actions. IGF1 and IGF2 bind to six different IGF binding proteins (IGFBPs) to increase their half-life and for transport to target tissues and importantly, to reduce their potential hypoglycemic effect. IGFBP3 is the most abundant binding protein in circulation, carrying from 70% to 90% of circulating IGF1 and 2 [4]. When IGF1 or IGF2 are bound to IGFBP3 or IGFBP5, it also binds to an acid-labile subunit (ALS) forming a trimolecular 150 KDa complex [5]. It is estimated that around 80% of the IGFs are bound in these trimolecular complexes, approximately 20% are bound to other IGFBPs and less than 1% are in the free form [6]. IGFBP2 is the second most abundant binding protein in serum, at least in some conditions [7,8], with a slight preference for IGF2 over IGF1 [9]. However, expression levels of the IGFBPs are tissue-specific, with, for example, IGFBP2 being the most abundant IGFBP in the postnatal central nervous system (CNS) [10], as well as in white preadipocytes during adipogenesis [11]; while IGFBP4 is the most abundant in cultured adult human adipose tissue [12].

When released from their binding proteins, the IGFs can activate the IGF1 receptor (IGF1R), which is structurally similar to the insulin receptor and when phosphorylated activates the PI3K/AKT pathway. Activation of this pathway underlies the anabolic effects of the IGFs [13]. In contrast, IGF2R is classified as a scavenger receptor for removing excess IGF2 from the circulation linked to lysosomal enzymes [14], with IGF2 having a 100-fold higher binding affinity for the IGF2R than IGF1 does [14].

New members of this system identified in recent years include pregnancy-associated plasma protein-A (PAPP-A) and PAPP-A2. These pappalysins modulate the biological activity of IGFs by cleaving the ligand from the binding protein and thus allowing them to bind to their receptors [15]. In addition, stanniocalcin (STC)-1 and 2 (STC-2) inhibit the proteolytic activity of pappalysins, reducing the release of IGF1 and 2 from the IGFBPs, hence their binding receptors [16,17].

The IGF system is important in anabolism, causing muscle hypertrophy [18], and is crucial in adipocyte maturation and differentiation [19]. Moreover, it can be altered by nutritional status [20]. In the brain it is important during development, but also throughout life. Indeed, in addition to peripheral IGF1 and IGF2 crossing the blood–brain barrier to reach the brain [21,22], IGFs are locally produced by brain cells, especially microglia and astrocytes [23,24]. These centrally produced IGFs can act as neuroprotective factors during events that result in gliosis and neuroinflammation. In the CNS, IGF1 is involved in postnatal synaptogenesis and neurogenesis [25], amyloid clearance [26], protection against neuroinflammation [24] and in neuroprotection [27], as well as with regulation of brain glucose metabolism [28]. The functions of IGF2 in the postnatal brain have been studied less, despite the fact that, for example, it is highly expressed in the hippocampus and has been associated with memory consolidation [29].

Gliosis, involving both astrocytes and microglia, is initially a protective response [30,31] but when prolonged can become damaging [32]. In response to high-fat diet (HFD) consumption, astrocytes are thought to participate in both the early protective response [33] and the detrimental effects of prolonged HFD intake and obesity [34]. The neuroprotective actions of astrocytes against oxidative stress [27], apoptosis and inflammation [35] are executed at least partially through the release of IGF1 [36,37] and consequently the downstream activation of the pro-survival pathway PI3K [38]. Thus, it is possible that the IGF system is involved in neuroprotection against the noxious effects of poor nutrition.

We have previously reported that short-term dietary changes modify the IGF system, with both a HFD and commercial control diet (CCD) inducing sex-specific changes peripherally and centrally compared to a normal rodent chow diet [39]. Here, our objective was to evaluate the long-term effects of HFD and CCD consumption on the central and peripheral IGF systems in male and female mice.

2. Results

2.1. Body Composition

Body weight changed with time throughout the study ($F_{(8,53)} = 325.6$, $p < 0.001$; Figure 1A),and was also influenced by sex ($F_{(1,53)} = 235.6$, $p < 0.001$) and diet ($F_{(2,53)} = 19.4$, $p < 0.001$). At study onset, there were differences between the sexes ($F_{(1,53)} = 27.4$, $p < 0.001$) in body weight (Figure 1A), with males weighing more than females and these sex differences were maintained throughout the study.

HFD increased body weight in males from the first week ($F_{(2,26)} = 5.2$, $p < 0.05$) until the last week of the study ($F_{(2,26)} = 62.2$, $p < 0.001$). However, in females HFD intake induced a significant increase in weight only at the last week of the study ($F_{(2,26)} = 4.0$, $p < 0.05$).

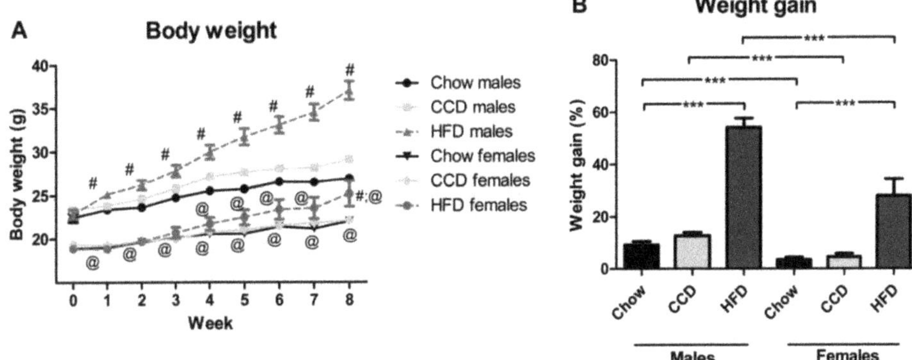

Figure 1. Body weight progression (**A**) and accumulated weight gain in percentage (**B**) in male and female mice exposed to a high-fat diet (HFD), commercial control diet (CCD) or a chow diet for 8 weeks. *** $p < 0.001$; #: different from chow in the same sex, @: different between sexes on the same diet. $n = 6$.

Weight gain (Figure 1B) was influenced by both sex ($F_{(1,53)} = 27.4$, $p < 0.001$), with males gaining more weight than females, and diet ($F_{(2,53)} = 77.2$, $p < 0.001$), with an interaction between these factors ($F_{(2,53)} = 6.4$, $p < 0.01$). HFD induced weight gain in both males ($F_{(2,26)} = 116.2$, $p < 0.001$) and females ($F_{(2,26)} = 13.5$, $p < 0.001$).

Body weight at sacrifice (Table 1) was determined by sex ($F_{(1,53)} = 134.5$, $p < 0.001$), with males weighing more than females, and by diet ($F_{(2,53)} = 55.4$, $p < 0.001$), with an interaction between these two factors ($F_{(2,53)} = 8.5$, $p < 0.01$). Body weight was higher after HFD consumption in both males ($F_{(2,26)} = 72.2$, $p < 0.001$) and females ($F_{(2,26)} = 8.2$, $p < 0.01$), with no effect of CCD intake on final body weight.

The amount of visceral adipose tissue (VAT; Table 1) was affected by both sex ($F_{(1,53)} = 24.8$, $p < 0.001$), with males having higher levels than females, and diet ($F_{(2,53)} = 60.7$, $p < 0.001$), with an interaction between these factors ($F_{(2,53)} = 9.1$, $p < 0.001$). HFD intake increased the percentage of VAT in both males ($F_{(2,26)} = 94.9$, $p < 0.001$) and females ($F_{(2,26)} = 8.6$, $p < 0.01$). Subcutaneous adipose tissue content (Table 1) was influenced by diet ($F_{(2,53)} = 6.3$, $p < 0.001$), with an interaction between sex and diet ($F_{(2,53)} = 4.9$, $p < 0.05$). The percentage of subcutaneous adipose tissue was increased by HFD intake in both males ($F_{(2,26)} = 58.4$, $p < 0.001$) and females ($F_{(2,26)} = 10.5$, $p < 0.01$).

The number of kcal ingested per day was influenced by sex ($F_{(1,17)} = 62.5$, $p < 0.001$; Table 1) and diet ($F_{(2,17)} = 267.9$, $p < 0.001$). There was also an interaction between sex and diet ($F_{(2,17)} = 135.9$, $p < 0.001$), with HFD only increasing energy intake per mouse in females ($F_{(2,8)} = 331.0$, $p < 0.001$). When adjusted for body weight, females consumed more energy than males ($F_{(1,17)} = 232.4$, $p < 0.001$), with this also being affected by diet ($F_{(2,17)} = 255.5$, $p < 0.001$). There was an interaction between sex and diet ($F_{(2,17)} = 232.4$, $p < 0.001$) as energy intake/g body weight was only increased by HFD in females ($F_{(2,8)} = 331.0$, $p < 0.001$).

Energy efficiency, determined as the amount of weight gained per kilocalories consumed, was higher in males compared to females ($F_{(1,17)} = 156.4$, $p < 0.001$; Table 1). Energy efficiency was also influenced by diet ($F_{(2,17)} = 37.3$, $p < 0.001$), with an interaction between sex and diet ($F_{(2,17)} = 51.9$, $p < 0.001$). This parameter was higher in males fed a HFD ($F_{(2,8)} = 63.2$, $p < 0.001$) compared to those on chow.

Circulating leptin levels were affected by diet ($F_{(2,35)} = 19.8$, $p < 0.01$; Table 1) with an interaction between sex and diet ($F_{(2,35)} = 4.3$, $p < 0.05$). HFD increased plasma leptin levels in both males ($F_{(2,18)} = 16.3$, $p < 0.001$) and females ($F_{(2,16)} = 5.5$, $p < 0.05$).

Table 1. Body weight, fat mass, glycemia, energy intake and circulating leptin levels in male and female mice exposed for 8 weeks to a high-fat diet (HFD), commercial control diet (CCD) or a chow diet. a: effect of sex, b: effect of diet, c: interaction between sex and diet, #: different from chow on the same sex, @: differences between sexes on the same diet, $n = 6$, except energy intake was $n = 3$ (number of cages/group).

	Chow Males	CCD Males	HFD Males	Chow Females	CCD Females	HFD Females	Significance
Final body weight (g)	24.6 ± 0.5	26.4 ± 0.3	35.2 ± 1.0 #	19.6 ± 0.3 @	20.3 ± 0.3 @	24.2 ± 1.5 #,@	a, $p < 0.001$ b, $p < 0.001$ c, $p < 0.01$
Visceral adipose tissue (%)	1.16 ± 0.07	1.22 ± 0.14	4.31 ± 0.28 #	0.62 ± 0.07 @	0.99 ± 0.12	2.19 ± 0.46 #,@	a, $p < 0.001$ b, $p < 0.001$ c, $p < 0.001$
Subcutaneous adipose tissue (%)	0.52 ± 0.04	0.57 ± 0.05	1.86 ± 0.16 #	0.67 ± 0.04 @	0.96 ± 0.09 @	1.52 ± 0.21 #,@	b, $p < 0.001$ c, $p < 0.001$
Glycemia (mg/dl)	67.8 ± 2.0	71.8 ± 3.7	80.6 ± 3.2	63.6 ± 3.9	68.7 ± 4.3	79.6 ± 4.9	b, $p < 0.01$
Kcal/mouse/ day	11.7 ± 0.1	10.9 ± 0.8	13.8 ± 0.5	10.0 ± 0.2 @	9.9 ± 0.2	24.9 ± 0.4 #,@	a, $p < 0.001$ b, $p < 0.001$ c, $p < 0.001$
Kcal/mouse/ day/100 g	46.2 ± 0.7	42.3 ± 2.0	45.8 ± 0.7	48.7 ± 1.2	47.6 ± 0.9	114.5 ± 3.3 #,@	a, $p < 0.001$ b, $p < 0.001$ c, $p < 0.001$
Energy efficiency (%)	0.68 ± 0.02	0.96 ± 0.11	1.84 ± 0.06 #	0.56 ± 0.02 @	0.50 ± 0.05 @	0.45 ± 0.07 @	a, $p < 0.001$ b, $p < 0.001$ c, $p < 0.001$
Leptin (ng/mL)	0.77 ± 0.28	0.65 ± 0.21	10.05 ± 2.10 #	1.06 ± 0.34	1.22 ± 0.49	4.52 ± 1.43 #	b, $p < 0.01$ c, $p < 0.05$

2.2. Glucose Tolerance Test and Insulin Levels

Glycemia at sacrifice was affected by diet ($F_{(2,53)} = 7.6$, $p < 0.01$), with an increase after HFD in both sexes (Table 1).

In the glucose tolerance test (Figure 2A), diet had an overall effect ($F_{(2,34)} = 7.9$, $p < 0.01$) as did sex ($F_{(1,34)} = 7.8$, $p < 0.01$), with an interaction between these factors ($F_{(2,34)} = 3.3$, $p < 0.05$) and a change over time ($F_{(4,34)} = 79.3$, $p < 0.001$). Basal glycemia levels showed differences according to sex ($F_{(1,34)} = 18.9$, $p < 0.001$) and diet ($F_{(2,34)} = 5.3$, $p < 0.05$), with an interaction between these factors ($F_{(2,34)} = 3.2$, $p = 0.05$). At baseline, there were sex differences when CCD was consumed ($F_{(1,11)} = 6.6$, $p < 0.05$) as well as with HFD ($F_{(1,11)} = 30.1$, $p < 0.001$), with males presenting a higher glycemia than females in both cases.

In males, HFD increased glycemia ($F_{(2,17)} = 7.0$, $p < 0.01$). At 30 min ($F_{(2,34)} = 4.9$, $p < 0.05$), 60 min ($F_{(2,34)} = 4.5$, $p < 0.05$) and 90 min ($F_{(2,34)} = 6.6$, $p < 0.01$) there was an effect of diet on the response to a glucose bolus with an increase in glycemia after HFD consumption. There was an effect of sex at 60 ($F_{(1,34)} = 4.5$, $p < 0.05$) and 90 min ($F_{(1,34)} = 12.8$, $p < 0.01$), with overall higher levels in males than females.

At 120 min, there was an effect of sex ($F_{(1,34)} = 9.5$, $p < 0.01$) and diet ($F_{(2,34)} = 5.0$, $p < 0.05$), and an interaction between these two factors ($F_{(2,34)} = 5.0$, $p < 0.05$). There were differences between males and females on chow ($F_{(1,10)} = 5.3$, $p < 0.05$) and HFD ($F_{(1,11)} = 12.1$, $p < 0.01$), with males having increased glycemia in both cases. Males on a HFD had a higher glycemia compared to when consuming chow ($F_{(2,17)} = 6.3$, $p < 0.05$).

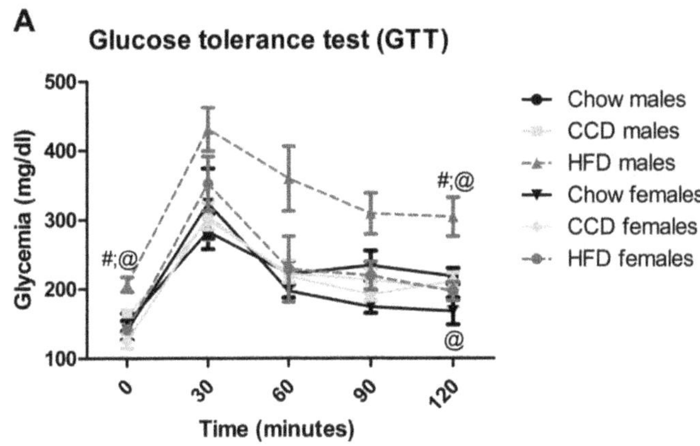

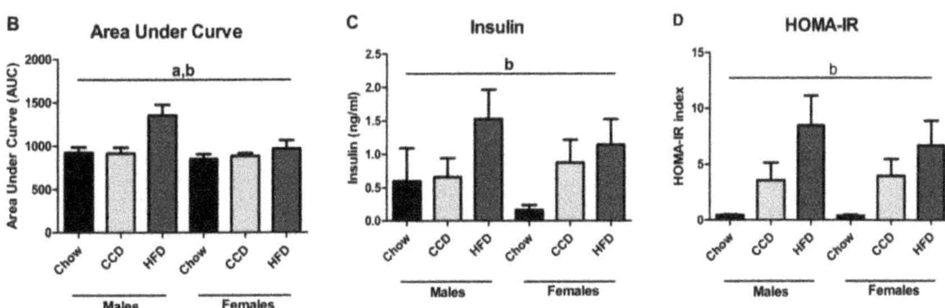

Figure 2. Glucose tolerance test and insulin levels. Glucose levels over time (**A**), area under curve (**B**), plasma insulin levels (**C**) and Homeostatic Model Assessment for Insulin Resistance (HOMA-IR; **D**) in male and female mice on a high-fat diet (HFD), commercial control diet (CCD) or a chow diet for 8 weeks. a: overall effect of sex, b: overall effect of diet, #: different from chow on the same sex, @: differences between sexes on the same diet, $n = 6$.

The area under the curve (AUC; Figure 2B) was affected by sex ($F_{(1,34)} = 6.2$, $p < 0.05$) and diet ($F_{(2,34)} = 7.6$, $p < 0.01$), with the increase after HFD consumption being more apparent in males.

Plasma insulin levels ($F_{(2,33)} = 3.7$, $p < 0.05$: Figure 2C) and the Homeostatic Model Assessment for Insulin Resistance (HOMA-IR) ($F_{(2,31)} = 7.6$, $p < 0.01$; Figure 2D) were affected by diet, with both HFD and CCD inducing an overall increase in these parameters, with the CCD effect on insulin being more apparent in females.

2.3. Circulating Levels of the IGF System

Free IGF1 levels were modified by diet ($F_{(2,29)} = 18.6$, $p < 0.001$; Figure 3A), with an increase after HFD consumption in both sexes. Total IGF1 was affected by sex ($F_{(1,34)} = 19.0$, $p < 0.001$; Figure 3B), with males having higher levels than females, and by diet ($F_{(2,34)} = 8.2$, $p < 0.01$) with an increase after HFD intake in both sexes.

Circulating IGF2 levels were affected by sex ($F_{(1,35)} = 7.6$, $p < 0.05$; Figure 3C), with females having overall higher levels than males, and diet ($F_{(2,35)} = 8.2$, $p < 0.01$), with HFD inducing an overall increase in IGF2 levels. There was an effect of sex on plasma IGFBP2 levels ($F_{(1,35)} = 15.9$, $p < 0.001$; Figure 3D), with males having higher levels of IGFBP2 than females. There was an interaction between sex and diet ($F_{(2,35)} = 5.1$, $p < 0.05$) with sex differences in IGFBP2 levels after CCD consumption ($F_{(1,11)} = 19.6$, $p < 0.01$). In males, there was an increase in IGFBP2 after CCD consumption ($F_{(2,17)} = 5.5$, $p = 0.05$).

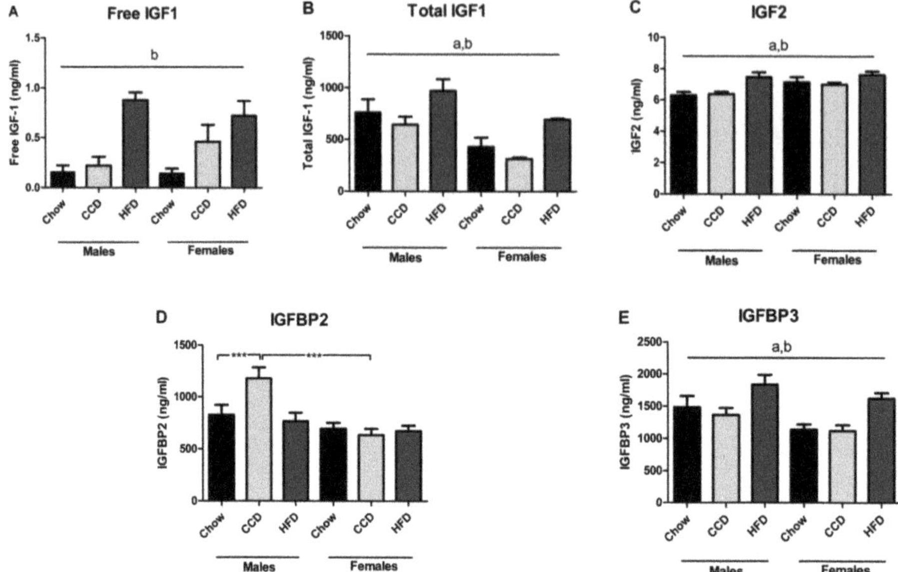

Figure 3. Circulating levels of free IGF-1 (**A**), total IGF-1 (**B**), IGF2 (**C**), IGFBP2 (**D**) and IGFBP3 (**E**) in mice on a high-fat diet (HFD), commercial control diet (CCD) or chow diet for 8 weeks. *** $p < 0.001$, a: effect of sex, b: effect of diet, $n = 6$.

There was also an effect of sex ($F_{(1,35)} = 7.3$, $p < 0.05$), with males having higher levels than females, and diet ($F_{(2,35)} = 9.5$, $p < 0.01$) on plasma IGFBP3 levels (Figure 3E), with HFD increasing the levels of this binding protein in both sexes.

2.4. The IGF System in Visceral Adipose Tissue (VAT)

Relative IGF1 mRNA levels were modified by diet ($F_{(2,33)} = 4.5$, $p < 0.05$; Figure 4A), with a decrease after HFD in males. There was an interaction between sex and diet regarding IGF2 mRNA levels ($F_{(2,34)} = 3.2$, $p = 0.05$; Figure 4B). In males, relative IGF2 mRNA levels were decreased after HFD intake ($F_{(2,16)} = 4.9$, $p < 0.05$), resulting in males having lower levels than females when on a HFD ($F_{(1,10)} = 11.9$, $p < 0.01$). There was no effect of sex or diet on relative mRNA levels of IGF1R (Figure 4C), IGF2R (Figure 4D) or IGFBP2 (Figure 4E).

Relative PAPP-A mRNA levels were affected by both diet ($F_{(2,34)} = 5.3$, $p < 0.05$; Figure 4F) and sex ($F_{(1,34)} = 12.1$, $p < 0.01$), with an interaction between these factors ($F_{(2,34)} = 8.8$, $p < 0.01$). In males, PAPP-A mRNA levels were reduced after HFD intake ($F_{(2,16)} = 7.7$, $p < 0.01$), with no HFD effect in females resulting in males having lower levels when on a HFD ($F_{(1,10)} = 47.3$, $p < 0.001$). Relative PAPP-A2 mRNA levels were only affected by sex ($F_{(1,33)} = 28.2$, $p < 0.001$; Figure 4G), being higher in males than females.

Relative STC-1 mRNA levels were modified by diet ($F_{(2,32)} = 18.2$, $p < 0.001$; Figure 4H) and sex ($F_{(1,32)} = 58.3$, $p < 0.001$), with an interaction between these factors ($F_{(2,32)} = 33.8$, $p < 0.001$). STC-1 mRNA levels were higher in males than females when on chow ($F_{(1,10)} = 52.8$, $p < 0.001$) or a CCD ($F_{(1,10)} = 16.8$, $p < 0.01$). HFD consumption led to decreased levels in males ($F_{(2,15)} = 25.8$, $p < 0.001$) but increased in females ($F_{(2,16)} = 26.2$, $p < 0.001$). In males, STC-1 mRNA levels were also decreased after CCD intake ($F_{(2,15)} = 25.8$, $p < 0.001$). No significant differences in STC-2 mRNA levels were found (Figure 4I).

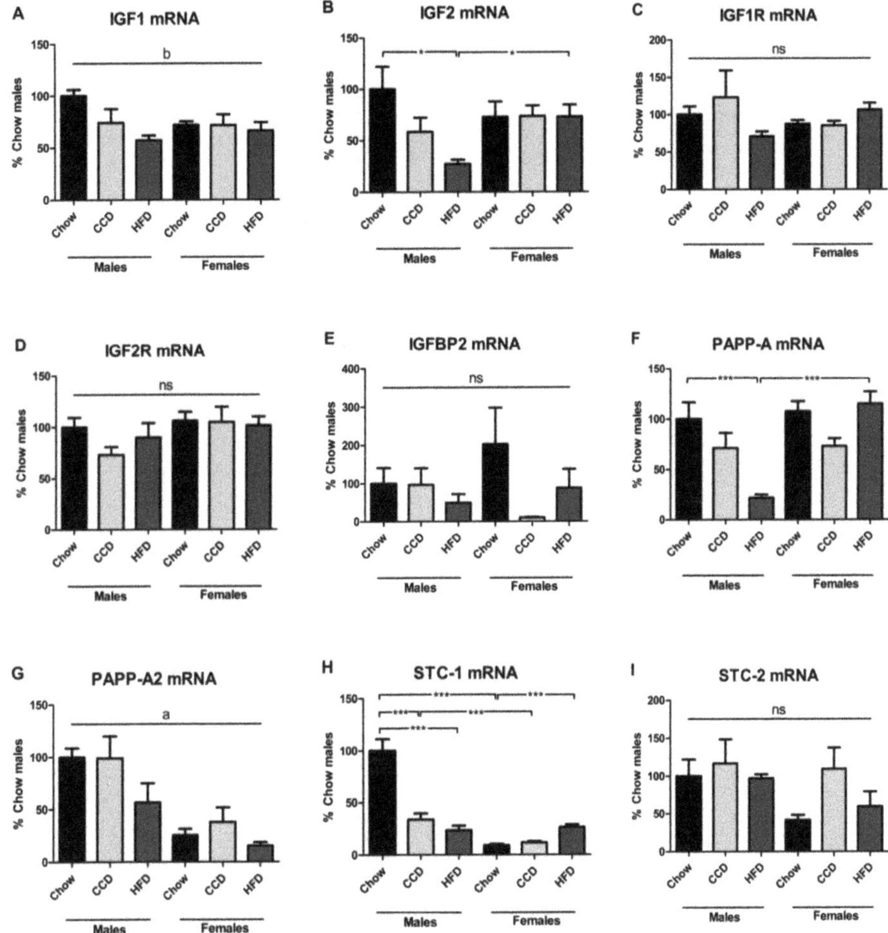

Figure 4. Relative gene expression of the insulin-like growth factor (IGF) system in visceral adipose tissue: IGF1 (**A**), IGF2 (**B**), IGF1R (**C**), IGF2R (**D**), IGFBP2 (**E**), PAPP-A (**F**), PAPP-A2 (**G**), stanniocalcin (STC)-1 (**H**) and STC-2 (**I**) in mice on a high-fat diet (HFD), commercial control diet (CCD) or a chow diet for 8 weeks. * $p < 0.05$, *** $p < 0.001$, a: effect of sex, b: effect of diet, ns = non-significant, $n = 6$.

2.5. Hypothalamic Response to Dietary Change

2.5.1. The IGF System in the Hypothalamus

The hypothalamic levels of IGF1 mRNA were affected by diet ($F_{(2,43)} = 6.8$, $p < 0.01$; Figure 5A), being increased after HFD consumption in both sexes. There was also an effect of diet on the hypothalamic mRNA levels of IGF2 ($F_{(2,42)} = 4.7$, $p < 0.05$; Figure 5B), with an overall increase in response to HFD, with this being more apparent in females. Relative IGF1R mRNA levels were altered by diet ($F_{(2,35)} = 3.3$, $p = 0.05$; Figure 5C) with an overall increase in response to the CCD. However, there were no effects of either sex or diet on relative IGF2R mRNA levels (Figure 5D).

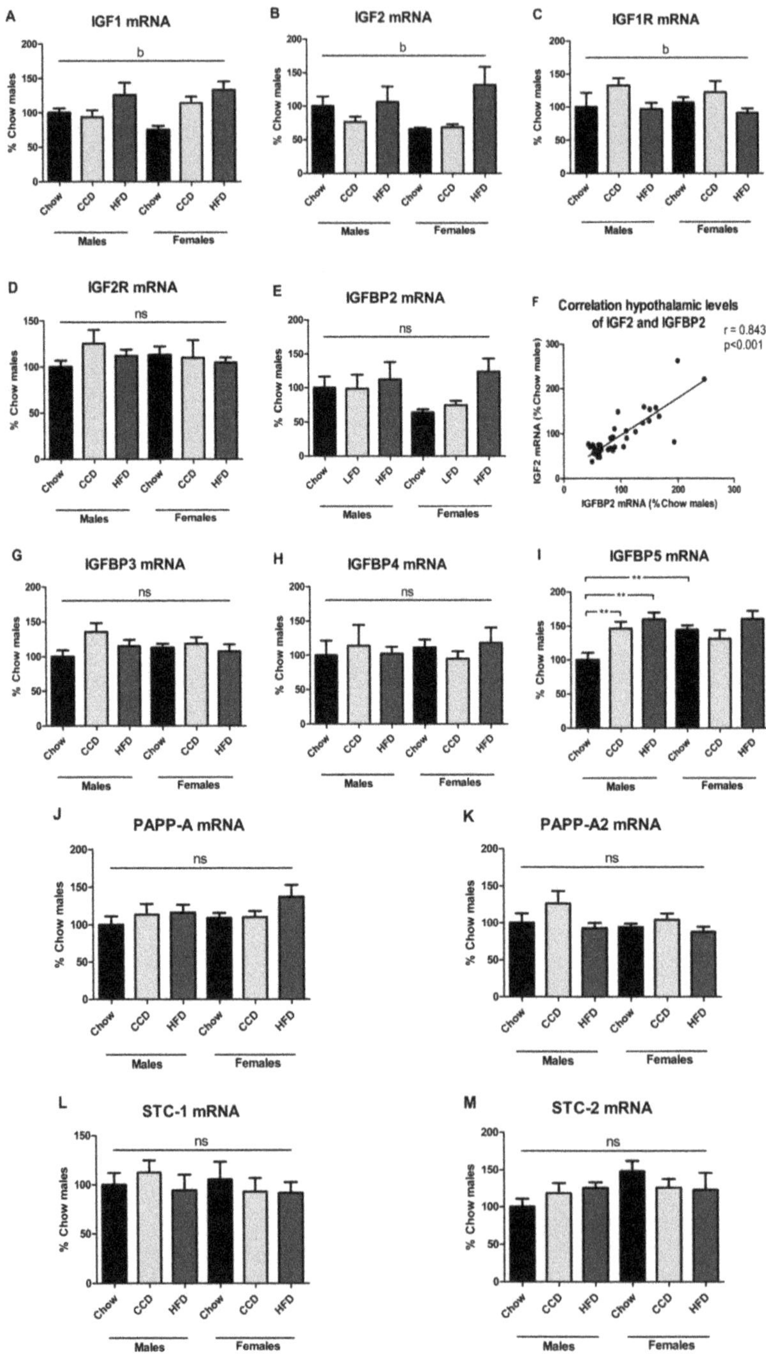

Figure 5. Relative hypothalamic mRNA levels of the IGF system: IGF1 (**A**), IGF2 (**B**), IGF1R (**C**), IGF2R (**D**), IGFBP2 (**E**), IGFBP3 (**G**), IGFBP4 (**H**), IGFBP5 (**I**), PAPP-A (**J**), PAPP-A2 (**K**), STC-1 (**L**) and STC-2 (**M**). Correlation of relative hypothalamic mRNA levels of IGF2 and IGFBP2 (**F**). ** $p < 0.01$, b: effect of diet, ns = non-significant, HFD = high-fat diet, CCD = commercial control diet, $n = 6$.

Regarding IGFBP2 mRNA expression, despite the apparent increase after HFD consumption in females, there was no overall effect of sex or diet (Figure 5E). As previously reported [39], the relative mRNA levels of IGF2 and IGFBP2 showed a similar profile, with a significant positive correlation between these two factors ($r = 0.843$, $p < 0.001$; Figure 5F).

We observed no differences in the relative mRNA levels of IGFBP3 (Figure 5G) or IGFBP4 (Figure 5H). However, IGFBP5 mRNA levels were affected by diet ($F_{(2,35)} = 6.8$, $p < 0.01$; Figure 5I) with an interaction between sex and diet ($F_{(2,35)} = 4.5$, $p < 0.05$). When on a chow diet, females were found to have higher levels than males ($F_{(1,11)} = 13.1$, $p < 0.01$). When separated by sex, CCD and HFD consumption were found to increase the mRNA levels of IGFBP5 in the hypothalamus in males ($F_{(2,17)} = 10.0$, $p < 0.01$), with no significant effect of diet in females.

Relative hypothalamic mRNA levels of PAPP-A (Figure 5J), PAPP-A2 (Figure 5K), STC-1 (Figure 5L) and STC-2 (Figure 5M) were not affected by sex or diet.

IGFBP2 is proposed to be an antidiabetic factor, protecting against obesity onset [40,41]. In addition, taking into consideration that IGFBP2 has preference for IGF2 over IGF1 [9], we determined the statistical correlations between these factors in the circulation, VAT, and the hypothalamus with body weight and glycemia (Table 2). Plasma IGF2 was positively and IGFBP2 negatively correlated with body weight, but exclusively in males. No significant correlations were found with glycemia in either sex. In VAT, the only correlation observed was that relative IGF2 mRNA levels were negatively correlated with body weight in males. In the hypothalamus there was a positive correlation between IGF2 mRNA levels and both body weight and glycemia, but only in females. Hypothalamic IGFBP2 mRNA relative levels also correlated positively with both body weight and glycemia and again only in females.

Table 2. Linear correlation between circulating levels of IGF2 and IGFBP2, relative mRNA levels of IGF2 and IGFBP2 in visceral adipose tissue (VAT) and in the hypothalamus with body weight and glycemia in both sexes. * $p < 0.05$, ** $p < 0.01$, $n = 6$.

	Plasma IGF2	Plasma IGFBP2	VAT IGF2 mRNA	VAT IGFBP2 mRNA	HPT IGF2 mRNA	HPT IGFBP2 mRNA
Body weight (males)	0.661 **	−0.470 *	−0.523 *	−0.306	0.088	0.090
Body weight (females)	0.286	−0.225	−0.107	−0.132	0.579 **	0.598 **
Glycemia (males)	0.159	−0.102	−0.218	0.075	−0.092	0.138
Glycemia (females)	0.268	−0.016	−0.157	−0.293	0.570 **	0.623 **

2.5.2. Hypothalamic Neuropeptides

The relative mRNA levels of neuropeptide Y (NPY) ($F_{(2,35)} = 5.8$, $p < 0.01$; Figure 6A) and Agouti-related protein (AgRP) ($F_{(2,34)} = 11.9$, $p < 0.001$; Figure 6B) in the hypothalamus were affected by diet, with the mRNA levels of both orexigenic neuropeptides decreasing after HFD intake.

Proopiomelanocortin (POMC) mRNA levels (Figure 6C) were not altered by sex or diet, whereas relative cocaine and amphetamine regulated transcript (CART) expression was affected by diet ($F_{(2,35)} = 3.6$, $p < 0.05$; Figure 6D), with HFD inducing an overall increase.

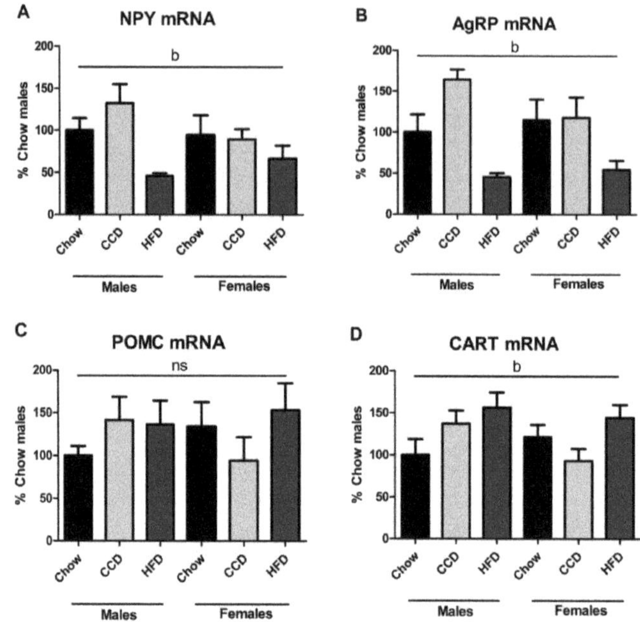

Figure 6. Relative mRNA levels of orexigenic and anorexigenic neuropeptides in the hypothalamus: Neuropeptide Y (NPY; **A**), Agouti-related protein (AgRP; **B**), proopiomelanocortin (POMC; **C**) and cocaine and amphetamine regulated transcript (CART; **D**). b: effect of diet, ns = non-significant. HFD = high-fat diet; CCD = commercial control diet, $n = 6$.

2.5.3. Gliosis and Hypothalamic Stress

We analyzed markers of gliosis and endoplasmic reticulum (ER) stress in the hypothalamus (Table 3). Gliosis (GFAP and Iba1) and ER stress markers (pJNK) were not altered in response to dietary intake (Table 3).

Table 3. Effects of 8 weeks on a high-fat diet (HFD), commercial control diet (CCD) or chow diet on glial structural protein (GFAPs: Glial fibrillary acidic protein, Iba1: Ionized calcium binding adaptor molecule 1) and endoplasmic reticulum (ER)-stress markers (pJNK: phosphorylated c-Jun N-terminal kinases). ns = non-significant, $n = 6$.

	Chow Males	CCD Males	HFD Males	Chow Females	CCD Females	HFD Females	Significance
GFAP	100.0 ± 2.1	115.6 ± 6.8	107.4 ± 9.5	103.9 ± 7.6	97.9 ± 3.0	103.3 ± 4.5	ns
Iba1	100.0 ± 3.2	117.1 ± 3.6	110.4 ± 12.5	114.8 ± 7.4	101.5 ± 4.5	97.7 ± 6.9	ns
pJNK	100.0 ± 10.8	84.0 ± 11.5	71.6 ± 5.2	107.5 ± 18.3	86.7 ± 13.9	95.3 ± 14.1	ns

3. Discussion

As previously reported, long-term HFD consumption increased body weight, fat mass and circulating leptin levels in both sexes [42–45]. However, female mice needed more time before a significant weight gain was observed which is in concordance with previous observations in rodents [44,46], at least in young animals. Energy intake was also increased, particularly in females, with females on the HFD consuming more energy than males. Although both sexes gained significant weight, glucose tolerance was only affected in males, as previously shown after 16 weeks of HFD intake in C57/BL mice [45]. However, basal insulin levels and the HOMA index were increased in both sexes after HFD in agreement with previous studies [43].

Sex differences in the metabolic response to long-term dietary challenges have been previously reported [46–48]. Likewise, we and others have shown that the circulating IGF system also differs between male and female rodents, both at baseline [49,50] and in response to a short-term dietary change [39]. Male mice were found to have higher circulating levels of total IGF1, IGFBP2 and IGFBP3, whereas females have higher IGF2 levels, with no sex differences in free IGF1 levels, as previously reported in rats [50]. We previously reported higher levels of free and total IGF1, IGF2 and IGFBP3, but reduced IGFBP2 levels at baseline, in male compared to female rats [39]. The circulating IGF system has also been shown to differ between the sexes in humans [51,52]. Together, the sex differences in the growth hormone (GH)/IGF system in conjunction with sex steroids [53] underlie the differences in longitudinal growth between males and females.

In both male and female mice, the circulating IGF system was modified by HFD-induced obesity, similar to that observed in patients with obesity. In both sexes, circulating levels of free IGF1 [54], total IGF1, IGF2 [55–57] and IGFBP3 [58] are reported to be increased in humans and rodents with obesity, although some studies found no changes in total IGF1 in patients with obesity [54]. This general increase in the IGF system that is associated with obesity occurs despite GH hyposecretion, which contributes to adiposity [59,60]. Increased hepatic GH sensitivity most likely contributes to the maintenance of circulating IGF1 levels despite low GH secretion [61].

Rapid changes can be seen in the IGF system in response to a HFD even before weight gain is observed [39], suggesting that they may be caused by the diet per se. In contrast, the changes found here after long-term HFD consumption could be due to the modifications in overall nutritional status in addition to possible direct effects of the diet, whereas, in response to a short-term dietary change, the modifications observed in the IGF system were more abundant in response to CCD [39]—here, long-term effects were associated with HFD intake. This supports the hypothesis that changes associated with HFD intake are largely due to body weight gain.

Circulating IGFBP2, one of the most abundant IGFBPs in serum [62], is suggested to play a role in protection against HFD-induced obesity and diabetes onset [40,41]. A negative correlation between circulating IGFBP2 and body mass index (BMI) has been reported in obese patients [63] and, in children with obesity, IGFBP2 levels are reduced, whereas in anorexic adolescents they are increased [64]. We observed a negative correlation between IGFBP2 and body weight in male mice independently of the diet consumed, although the increase in IGFBP2 in male mice on a CCD was not associated with weight change. It is possible that changes in this binding protein are time-dependent and occur differently between males and females. Indeed, after one week of CCD IGFBP2 was decreased in female rats [39], while after a more long-term dietary change, this binding protein was not different from those on a chow diet. The augmented IGFBP2 in males after more long-term CCD could be an attempt to protect against the higher sucrose intake. The susceptibility to diet-induced obesity and its metabolic complications differs between the sexes [46,65] and this could be related to the differential changes in IGFBP2. However, more studies are clearly needed for a better comprehension of the role of IGFBP2 in metabolism.

IGF1 is important for adipocyte metabolism, maturation and differentiation [19,66,67]. In male C57 mice, IGF1 mRNA levels were reported to be decreased in adipocytes purified from perigonadal adipose tissue of obese compared to lean mice, despite the lack of change in IGF1 expression in lysates of this same tissue [68]. This could suggest that although IGF1 mRNA levels may be reduced in adipocytes, other cell types in adipose tissue, such as macrophages, could increase their expression of *Igf1* such that global IGF1 mRNA levels in VAT do not change. Here, IGF1 mRNA levels were reduced in VAT after HFD in males, with the possibility that adipocytes in VAT decrease their IGF1 mRNA levels after HFD intake with the VAT macrophages or circulating IGF1 providing the source of IGF1 required to maintain adipocyte hypertrophy and hyperplasia [68]. This system is not altered in females, which are metabolically less affected by HFD consumption despite the increase in the VAT percentage observed in response to a HFD intake. If a longer exposition to this diet is required to alter it in females, or if this system is sex-dependent, deserves further study. As seen here, IGF2 mRNA levels in VAT were previously reported to be reduced in C57 mice in response to HFD-induced obesity; however in PWK mice, which are resistant to HFD-induced obesity, they are not [69]. These authors suggest that IGF2 has an anti-inflammatory effect on TNF-α-induced MCP-1 expression in adipocytes [69]. They also reported increased expression of IGF2R in VAT in HFD-exposed C57 mice [69], which was not found here possibly due to the shorter exposure to the HFD in our study.

During adipogenesis, IGFBP2 is the principal IGFBP produced by visceral white adipocytes, inhibiting both adipogenesis and lipogenesis [70], with levels reported to be reduced in obese *ob/ob*, *db/db* and high-fat-fed mice [71]. This reduction was accompanied by decreased IGFBP2 in circulation, suggesting that the decreased levels in VAT after HFD intake may lead to a decrease in circulating IGFBP2 levels [71]. Here we found no effect of HFD on IGFBP2 mRNA levels in VAT or circulating IGFBP2; again, it is possible that a longer exposition to a HFD may be needed.

In PAPP-A knockout mice, adipocyte size and lipid uptake are reduced in mesenteric fat, and visceral fat did not expand in PAPP-A KO mice exposed to a HFD, suggesting that the local cleavage of IGFBP4 is mandatory for adipocyte expansion [72]. Local free IGF1 levels are suggested to increase in adipose tissue in response to PAPP-A secretion in pregnant women [73], supporting the idea that IGF1 may contribute to the adipocyte hypertrophy associated with HFD intake through PAPP-A activity, and the reduction in these factors could be a negative feedback effect in attempt to block excess expansion. However, we found no effect of diet on PAPP-A2 expression, although males had significantly higher levels than females. Whether this protease is involved in sex differences in VAT function remains to be determined.

Both the CCD and HFD reduced STC-1 mRNA levels in male mice. In cultured adipocytes from male rats, STC-1 was shown to increase glucose uptake and the storage of triacylglycerol during the postprandial period [74]. Thus, this reduction in STC-1 mRNA levels may be a response conducted to control the lipid accumulation by VAT and this could possibly be induced by the higher amount of fat content in both the CCD and HFD compared to the chow diet. STC-2 mRNA levels were not altered in VAT, but the possible role of STCs regulating the IGFs availability in VAT, and potentially modifying the adipocyte size, deserves further research.

Hypothalamic IGF1 mRNA levels were increased after HFD consumption in both sexes, in agreement with our previous results in male mice after 7 weeks of HFD [75]. This increase could be a homeostatic mechanism to counteract the central effects of a HFD as increased central expression of IGF1 has been shown to improve glucose tolerance and insulin sensitivity [76]. Hypothalamic IGF2 mRNA levels were also increased after HFD, especially in females. Maternal HFD was shown to increase hypothalamic IGF2 mRNA levels in female Sprague Dawley rat offspring at postnatal day 10, with no changes in males [77], but little is known regarding central IGF2 in metabolism. In the adult brain, IGF2 promotes memory consolidation, neurogenesis and cognitive function [78] and it could be involved in structural changes induced by HFD and/or obesity, although this remains to be determined.

Hypothalamic levels of IGF2 and IGFBP2 were positively correlated, as previously reported in rats [39]. In females, both IGF2 and IGFBP2 were positively correlated with glycemia. A role of both IGFBP2 and IGF2 in glucose metabolism has been reported [41,79], but whether these factors participate in central glucose regulation remains unknown. The fact that females were metabolically less affected by long-term HFD consumption compared to males could suggest a possible metabolic effect of these factors at the level of the hypothalamus.

The role of IGFBP5 in metabolism has been poorly studied, particularly at the central level. Male IGFBP5-deficient mice are reported to have a greater body weight, impaired glucose tolerance at baseline and a larger increase in adipose tissue on a HFD compared to wild type [80]. This suggests that IGFBP5 is protective against obesity and glucose metabolism impairment, but the specific role of hypothalamic IGFPB5 in metabolic control is unknown. As this binding protein is increased after both HFD and CCD consumption in males one might speculate that it could be a protective response against these diets at the central level, and that this could be induced by the higher fat content in both of these diets compared to chow. No other members of the IGF system analyzed in the hypothalamus were affected by either diet or sex.

No evidence of gliosis or ER stress was found in the hypothalamus. Although various authors have reported both phenomena as well as central inflammation after 8 weeks of HFD consumption [81–83] or even earlier [81], other authors report no alterations in hypothalamic inflammation in male mice [84]. Thus, the causal relationship between hypothalamic inflammation and gliosis in weight gain is unclear [85]. In addition, the absence of gliosis observed in our study in response to HFD is accompanied by an increase in the gene expression of IGF1 and 2, which may play a neuroprotective role by reducing cell stress [27].

Diet composition is reported to be more important than caloric intake per se in determining NPY/AgRP neuron activity [86] and NPY and AgRP were reduced in the hypothalamus after HFD consumption in both sexes, as previously reported in male rats [75,87,88]. This is most likely related to the satiating properties of the HFD inducing a response that attempts to reduce the amount of energy consumed. However, in females this response is attenuated, at least regarding NPY, and this could underlie their higher energy intake compared to both HFD males and control females. Insulin decreases both NPY and AgRP expression in the hypothalamic neuronal line mHypoE-46 [89]. Considering the insulin-like actions of IGF1 and 2, it is possible that the rise in these factors with HFD intake contributes to the inhibition of orexigeneic neuropeptide expression. Administration of IGF icv has been shown to increase POMC mRNA levels in chicks [90], which was not observed here. This difference could be due to the difference in species. An overall effect of HFD to increase the anorexic neuropeptide CART was found, but no changes were observed in POMC expression. On the contrary, we previously found POMC mRNA levels to increase after HFD in male mice with no effect on CART relative expression [75], whereas other authors reported no significant changes in the hypothalamic POMC and CART mRNA expression in female mice on a long-term HFD [91], employing a different source of HFD (45% kcal from fat, 4.73 kcal/g). In our previous study [75], the exposition to HFD was for only 50 days instead of the 8 weeks here, which may explain some of these differences.

The CCD used here is what has been traditionally referred to as a control low-fat diet (LFD) as it was developed to be used as the control to the commercial HFD. However, it is now clear that these commercially developed control diets can also have metabolic effects as their nutrient composition differs from a normal chow diet. Here, the CCD did not modify body weight, but basal insulin levels and HOMA index were increased in both sexes. Although the caloric content is similar between the rodent chow and CCD, the amount of sucrose, 33% in CCD and 0.9% in chow, differs considerably. Dry sucrose or polycose consumption does not affect final body weight in rats [92,93], nor does consumption of a 33% sucrose solution, despite the metabolic alterations which are found [94,95]. However, these studies found fat mass and specific metabolic parameters to be modified. Thus, it is possible that the high sucrose composition in the CCD induces these observed metabolic alterations, again indicating that not only caloric intake, but dietary composition, is important.

The response of the IGF system to CCD intake is time- and sex-dependent, whereas, in female rats exposed to 1 week of CCD, IGFBP2 levels were reduced [39] and no changes were found in this sex after 8 weeks. In contrast, males showed no differences at one week of CCD intake [39] but had increased serum IGFBP2 levels after 8 weeks. This sex-specific response may be related to the protective role of IGFBP2 in glucose metabolism [40,41] as glucose metabolism is differently affected in males and females. Short- and a long-term HFD intake also differently affected IGF2 levels in males. Some authors suggest that an initial reduction in serum IGF2 levels may indicate a bad prognosis for weight gain [96], whereas longer HFD exposition leads to an increase in the circulating IGF2 as in obesity [56]. What is clear is that each diet has differential metabolic implications that are time- and sex-dependent.

In conclusion, long-term HFD intake modulates the IGF system in a tissue-dependent manner, which may respond to different tissue requirements, and suggests an active role of this system in the metabolic response to this dietary challenge. Moreover, most of the observed changes are sex-specific and this could be involved in the differential metabolic responses between males and females.

4. Materials and Methods

4.1. Ethical Statement

This study was designed following the European Communities Council Directive (2010/63/UE) and the Spanish Royal Decree 53/2013 concerning the protection of experimental animals. In addition, the study was approved by the Ethical Committee of Animal Experimentation of the Hospital Puerta de Hierro de Madrid and the Animal Welfare Organ of the Comunidad Autónoma de Madrid (07/346225.9/15, 9 April 2015). All care was taken to use the minimum number of animals.

4.2. Animals and Diets

In this study, 54 seven-week-old C57BL/6J mice (27 males and 27 females) were purchased from Charles River Laboratories. Upon arrival, animals were weighed and randomly distributed into cages according to sex (3 mice per cage) and allowed to acclimate for a week with free access to a normal chow diet and tap water. The mice were weighed again and randomly distributed into the different experimental groups. They were fed ad libitum with a high-fat diet (HFD; 62% kcal from fat, 18% kcal from proteins, 20% kcal from carbohydrates, 5.1 kcal/g, LabDiet), a commercial control diet (CCD; 10% kcal from fat, 18% kcal from proteins, 72% kcal from carbohydrates, 3.76 kcal/g, LabDiet) or standard rodent chow (6% kcal from fat, 17% kcal from proteins, 77% from carbohydrates, 3.41 kcal/g, Panlab) for 8 weeks, with free access to tap water throughout the study. This resulted in a total of 6 groups with nine animals per diet and sex ($n = 9$). Body weight and food intake were monitored each week. Animals were maintained at 22 ± 2 °C throughout the study.

4.3. Glucose Tolerance Test (GTT)

A week before sacrifice, six mice per group ($n = 6$) were fasted for 6 h prior to testing. They were weighed and then intraperitoneally injected with 2 mg of D-glucose per gram of body weight. Glycemia was determined in a drop of blood from the tail by using a Freestyle Optimum Neo glucometer (Abbott, Witney, UK). Glycemia was measured just prior to the injection (basal, 0 min) and at 30, 60, 90 and 120 min after the D-glucose injection.

4.4. Tissue Collection

Mice were weighed and fasted 12 h before sacrifice by decapitation, which took place between 9:00 and 11:00 am. A few days before sacrifice, sawdust from male cages was mixed in the females' cage to equalize estrous cycle in females, which was determined by vaginal cytology at sacrifices. In total, 18.5% of the females were on proestrus phase, 77.8% were in estrus, 0% in metaestrus and 3.7% in diestrus. Trunk blood was collected in tubes containing a 0.5 M ethylenediaminetetraacetic acid (EDTA) solution. Samples were then centrifuged at 3000 rpm for 15 min at 4 °C and plasma was collected and aliquoted (to avoid freeze–thaw cycles) and stored at −80 °C until used. Perigonadal visceral adipose tissue (VAT) was dissected, weighted and frozen at −80 °C. The hypothalamus (rostrally limited by the optic chiasm and caudally by the mammillary bodies) was also dissected and kept at −80 °C until processed.

4.5. ELISA Assays

Plasma levels of free IGF1 (AnshLabs, Webster, TX, USA), total IGF1 (Mediagnost, Reutlingen, Germany), IGF2 (R&D Systems, Minneapolis, MN, USA), IGFBP2 (Millipore, Burlington, MA, USA), IGFBP3 (Mediagnost), insulin (Millipore) and leptin (Millipore) were measured following the manufacturers' instructions. Absorbance was read by spectrophotometry (Tecan Infinite M200, Grödig, Austria).

Homeostatic Model Assessment for Insulin Resistance (HOMA-IR) was calculated by using the following equation: HOMA-IR = glycemia (mmol/L) × insulin (mU/L)/22.5

4.6. RNA and Protein Extraction

An RNeasy Lipid Tissue Mini Kit (Qiagen, Hilden, Germany) was used for visceral adipose tissue RNA extraction according to manufacturer's instructions, whereas an RNeasy Plus Mini Kit (Qiagen) was used following the manufacturer's instructions for hypothalamic RNA isolation. Protein was extracted from the eluate after tissue lysis. The eluate was mixed with 4 volumes of cold acetone and stored overnight at −20 °C. The samples were centrifuged 10 min at 3000 rpm and the pellets containing the protein were resuspended in a CHAPS hydrate (Sigma-Aldrich, Saint Louis, MO, USA) solution, containing 7 M urea, 2 M thiourea, 4% CHAPS, 0.5% 1M Tris pH 8.8 in distilled water and stored at −80°C. For protein quantification, Protein Assay Dye Reagent Concentrate (Bio-Rad Laboratories, Hercules, CA, USA) was used to perform the Bradford assay.

4.7. Western Blotting

Depending on the expected signal and molecular weight of each target, 10 to 40 μg of protein were resolved on 8, 10 or 12% sodium dodecyl sulphate-denaturing polyacrylamide gels. After electrophoresis, proteins were transferred to a previously activated polyvinylidine difluoride (PVDF) membrane and then blocked with a Tris-buffered saline buffer containing 0.1% Tween 20 and 5% non-fat dried milk or bovine serum albumin (BSA) when phosphorylated proteins were assayed. Primary antibodies (Table 4) were diluted in the same buffer and incubated O/N with agitation at 4 °C. The next day, the corresponding horseradish peroxidase-conjugated secondary antibody was diluted in the same buffer and the membranes incubated for 1.5 h. Clarity Western ECL Substrate (Bio-Rad Laboratories, Hercules, CA, USA) was employed to visualize the peroxidase activity and the chemiluminescent signal was detected by using ImageQuant Las 4000 Software (GE Healthcare Life Sciences, Barcelona, Spain). To normalize for protein loading, GAPDH was used, as indicated.

Table 4. Antibodies used for Western blotting.

Antibody	Class	Dilution	Host	Commercial Source	Reference
pJNK	Polyclonal	1:3000	Rabbit	Promega	V7932
GAPDH	Polyclonal	1:10,000	Rabbit	Sigma-Aldrich	G9545
GFAP	Polyclonal	1:5000	Guinea pig	Synaptic Systems	173004
Iba1	Polyclonal	1:1000	Rabbit	Synaptic Systems	234003
α-guinea pig HRP-conjugated	Polyclonal	1:2000	Goat	AbD Serotec	AHP861P
α-rabbit HRP-conjugated	Polyclonal	1:20,000	Goat	Invitrogen	31460

4.8. Real-Time qPCR

For RT-qPCR, RNA (0.5–1 μg) was retro-transcribed to copy DNA (cDNA) by using an NZY First-Strand cDNA Synthesis Kit (NZYTech, Lisbon, Portugal). TaqMan probes of target genes (Table 5) were used for qPCR in a QuantStudio 3 Real-Time PCR System (Applied Biosystems, Carlsbad, CA, USA). Mouse glyceraldehyde 3-phosphate dehydrogenase (GAPDH) endogenous control (Applied Biosystems) was employed as the housekeeping gene, except on visceral adipose tissue, for which *Ppia* was chosen (Table 5). For the mathematical analysis, the ΔΔCT method was performed. Relative levels of expression are expressed as percentage of the chow male group.

Table 5. List of TaqMan probes used for qPCR.

Name	Gene	Reference	Commercial Source
Agouti-related protein	*Agrp*	Mm00475829_g1	Applied Biosystems
Cyclophilin A (Peptidylprolyl isomerase A)	*Ppia*	Mm02342430_g1	Applied Biosystems
Cocaine and amphetamine regulated transcript prepropeptide	*Cartpt*	Mm04210469_m1	Applied Biosystems
Insulin-like growth factor 1	*Igf1*	Mm00439560_m1	Applied Biosystems
Insulin-like growth factor 1 receptor	*Igf1r*	Mm00802831_m1	Applied Biosystems
Insulin-like growth factor 2	*Igf2*	Mm00439564_m1	Applied Biosystems
Insulin-like growth factor 2 receptor	*Igf2r*	Mm00439576_m1	Applied Biosystems
Insulin-like growth factor-binding protein 2	*Igfbp2*	Mm00492632_m1	Applied Biosystems
Insulin-like growth factor-binding protein 3	*Igfbp3*	Mm01187817_m1	Applied Biosystems
Insulin-like growth factor-binding protein 4	*Igfbp4*	Mm00494922_m1	Applied Biosystems
Insulin-like growth factor-binding protein 5	*Igfbp5*	Mm00516037_m1	Applied Biosystems
Neuropeptide Y	*Npy*	Mm03048253_m1	Applied Biosystems
Pregnancy-associated plasma protein A	*Pappa*	Mm01259244_m1	Applied Biosystems
Pregnancy-associated plasma protein A-2	*Pappa2*	Mm01284029_m1	Applied Biosystems
Pro-opiomelanocortin	*Pomc*	Mm00435874_m1	Applied Biosystems
Stanniocalcin-1	*Stc1*	Mm01322191_m1	Applied Biosystems
Stanniocalcin-2	*Stc2*	Mm00441560_m1	Applied Biosystems

4.9. Statistical Analysis

Statistics analyses were performed using SPSS 15.0 (SPSS Inc., Chicago, IL, USA) software. A two-way ANOVA was employed using the diet and sex factors. Weight gain and food intake over time were calculated by a repeated measures ANOVA, as well as the weekly data for these variables. For the glucose tolerance test, the area under the curve (AUC) was calculated by using GraphPad Prism 5 software (San Diego, CA, USA). Data are presented as the mean ± standard error of the mean (SEM). Graphs were made by GraphPad Prism 5 software. For the linear correlation between variables, a Pearson correlation coefficient was calculated. $p < 0.05$ was considered significant.

Author Contributions: Conceptualization: J.A.C., L.M.F. and J.A.; methodology: J.A.C., L.M.F. and J.A.; validation: S.G.-C. and L.M.F.; formal analysis: S.G.-C. and J.A.C.; investigation: S.G.-C., M.J.-H., P.R., A.F.-R. and V.B.; resources: J.A.C., L.F.M. and J.A.; data curation: S.G.-C. and J.A.C.; writing—original draft preparation: S.G.-C.; writing—review and editing: J.A.C., L.M.F. and J.A.; visualization: S.G.-C. and J.A.C.; supervision: L.M.F. and J.A.C.; project administration: J.A.C.; funding acquisition: J.A.C., L.M.F. and J.A. All authors have read and agreed to the published version of the manuscript.

Funding: This research was funded by grants from the Spanish Ministry of Science and Innovation (BFU2017-82565-C21-R2 to J.A.C. and L.M.F.), Spanish Ministry of Education, Culture and Sports (university training grant PU13/00909 to A.F.-R.), Fondo de Investigación Sanitaria (PI1900166 to J.A.) and Fondos FEDER. Centro de Investigación Biomédica en Red Fisiopatología de Obesidad y Nutrición (CIBEROBN), Instituto de Salud Carlos III (J.A.).

Acknowledgments: We would like to thank Francisca Díaz and Sandra Canelles for their excellent technical assistance.

Conflicts of Interest: The authors declare no conflict of interest.

References

1. Yakar, S.; Rosen, C.J.; Beamer, W.G.; Ackert-Bicknell, C.L.; Wu, Y.; Liu, J.L.; Ooi, G.T.; Setser, J.; Frystyk, J.; Boisclair, Y.R.; et al. Circulating levels of IGF-1 directly regulate bone growth and density. *J. Clin. Investig.* **2002**, *110*, 771–781. [CrossRef] [PubMed]
2. Lui, J.C.; Finkielstain, G.P.; Barnes, K.M.; Baron, J. An imprinted gene network that controls mammalian somatic growth is down-regulated during postnatal growth deceleration in multiple organs. *Am. J. Physiol. Regul. Integr. Comp. Physiol.* **2008**, *295*, R189–R196. [CrossRef] [PubMed]
3. Bann, D.; Holly, J.M.; Lashen, H.; Hardy, R.; Adams, J.; Kuh, D.; Ong, K.K.; Ben-Shlomo, Y. Changes in insulin-like growth factor-I and -II associated with fat but not lean mass in early old age. *Obesity* **2015**, *23*, 692–698. [CrossRef] [PubMed]

4. Blum, W.F.; Ranke, M.B. Insulin-like growth factor binding proteins (IGFBPs) with special reference to IGFBP-3. *Acta Paediatr. Scand. Suppl.* **1990**, *367*, 55–62. [CrossRef] [PubMed]
5. Baxter, R.C. Insulin-like growth factor (IGF)-binding proteins: Interactions with IGFs and intrinsic bioactivities. *Am. J. Physiol. Endocrinol. Metab.* **2000**, *278*, E967–E976. [CrossRef] [PubMed]
6. Ranke, M.B. Insulin-like growth factor binding-protein-3 (IGFBP-3). *Best Pract. Res. Clin. Endocrinol. Metab.* **2015**, *29*, 701–711. [CrossRef]
7. Probst-Hensch, N.M.; Steiner, J.H.; Schraml, P.; Varga, Z.; Zürrer-Härdi, U.; Storz, M.; Korol, D.; Fehr, M.K.; Fink, D.; Pestalozzi, B.C.; et al. IGFBP2 and IGFBP3 protein expressions in human breast cancer: Association with hormonal factors and obesity. *Clin. Cancer Res.* **2010**, *16*, 1025–1032. [CrossRef]
8. Russo, V.C.; Azar, W.J.; Yau, S.W.; Sabin, M.A.; Werther, G.A. IGFBP-2: The dark horse in metabolism and cancer. *Cytokine Growth Factor Rev.* **2015**, *26*, 329–346. [CrossRef]
9. Bach, L.A. IGF-binding proteins. *J. Mol. Endocrinol.* **2018**, *61*, T11–T28. [CrossRef]
10. Lee, W.H.; Michels, K.M.; Bondy, C.A. Localization of insulin-like growth factor binding protein-2 messenger RNA during postnatal brain development: Correlation with insulin-like growth factors I and II. *Neuroscience* **1993**, *53*, 251–265. [CrossRef]
11. Boney, C.M.; Moats-Staats, B.M.; Stiles, A.D.; D'Ercole, A.J. Expression of insulin-like growth factor-I (IGF-I) and IGF-binding proteins during adipogenesis. *Endocrinology* **1994**, *135*, 1863–1868. [CrossRef] [PubMed]
12. Gude, M.F.; Frystyk, J.; Flyvbjerg, A.; Bruun, J.M.; Richelsen, B.; Pedersen, S.B. The production and regulation of IGF and IGFBPs in human adipose tissue cultures. *Growth Horm. IGF Res.* **2012**, *22*, 200–205. [CrossRef] [PubMed]
13. Velloso, C.P. Regulation of muscle mass by growth hormone and IGF-I. *Br. J. Pharmacol.* **2008**, *154*, 557–568. [CrossRef] [PubMed]
14. Kiess, W.; Yang, Y.; Kessler, U.; Hoeflich, A. Insulin-like growth factor II (IGF-II) and the IGF-II/mannose-6-phosphate receptor: The myth continues. *Horm. Res.* **1994**, *41*, 66–73. [CrossRef]
15. Overgaard, M.T.; Boldt, H.B.; Laursen, L.S.; Sottrup-Jensen, L.; Conover, C.A.; Oxvig, C. Pregnancy-associated plasma protein-A2 (PAPP-A2), a novel insulin-like growth factor-binding protein-5 proteinase. *J. Biol. Chem.* **2001**, *276*, 21849–21853. [CrossRef]
16. Kløverpris, S.; Mikkelsen, J.H.; Pedersen, J.H.; Jepsen, M.R.; Laursen, L.S.; Petersen, S.V.; Oxvig, C. Stanniocalcin-1 Potently Inhibits the Proteolytic Activity of the Metalloproteinase Pregnancy-associated Plasma Protein-A. *J. Biol. Chem.* **2015**, *290*, 21915–21924. [CrossRef]
17. Jepsen, M.R.; Kløverpris, S.; Mikkelsen, J.H.; Pedersen, J.H.; Füchtbauer, E.M.; Laursen, L.S.; Oxvig, C. Stanniocalcin-2 inhibits mammalian growth by proteolytic inhibition of the insulin-like growth factor axis. *J. Biol. Chem.* **2015**, *290*, 3430–3439. [CrossRef]
18. Hennebry, A.; Oldham, J.; Shavlakadze, T.; Grounds, M.D.; Sheard, P.; Fiorotto, M.L.; Falconer, S.; Smith, H.K.; Berry, C.; Jeanplong, F.; et al. IGF1 stimulates greater muscle hypertrophy in the absence of myostatin in male mice. *J. Endocrinol.* **2017**, *234*, 187–200. [CrossRef]
19. Scavo, L.M.; Karas, M.; Murray, M.; Leroith, D. Insulin-like growth factor-I stimulates both cell growth and lipogenesis during differentiation of human mesenchymal stem cells into adipocytes. *J. Clin. Endocrinol. Metab.* **2004**, *89*, 3543–3553. [CrossRef]
20. Argente, J.; Caballo, N.; Barrios, V.; Muñoz, M.T.; Pozo, J.; Chowen, J.A.; Hernández, M. Disturbances in the growth hormone-insulin-like growth factor axis in children and adolescents with different eating disorders. *Horm. Res.* **1997**, *48*, 16–18. [CrossRef]
21. Reinhardt, R.R.; Bondy, C.A. Insulin-like growth factors cross the blood-brain barrier. *Endocrinology* **1994**, *135*, 1753–1761. [CrossRef] [PubMed]
22. Pan, W.; Kastin, A.J. Interactions of IGF-1 with the blood-brain barrier in vivo and in situ. *Neuroendocrinology* **2000**, *72*, 171–178. [CrossRef] [PubMed]
23. Quesada, A.; Romeo, H.E.; Micevych, P. Distribution and localization Patterns of Estrogen Receptor-β and Insulin-Like Growth Factor-1 Receptors in Neurons and Glial Cells of the Female Rat Substantia Nigra. *J. Comp. Neurol.* **2007**, *503*, 198–208. [CrossRef] [PubMed]
24. Labandeira-Garcia, J.L.; Costa-Besada, M.A.; Labandeira, C.M.; Villar-Cheda, B.; Rodríguez-Perez, A.I. Insulin-Like Growth Factor-1 and Neuroinflammation. *Front. Aging Neurosci.* **2017**, *9*, 365. [CrossRef] [PubMed]

25. O'Kusky, J.R.; Ye, P.; D'Ercole, A.J. Insulin-like growth factor-I promotes neurogenesis and synaptogenesis in the hippocampal dentate gyrus during postnatal development. *J. Neurosci.* **2000**, *20*, 8435–8442. [CrossRef]
26. Carro, E.; Torres-Aleman, I. The role of insulin and insulin-like growth factor I in the molecular and cellular mechanisms underlying the pathology of Alzheimer's disease. *Eur. J. Pharmacol.* **2004**, *490*, 127–133. [CrossRef]
27. Genis, L.; Dávila, D.; Fernandez, S.; Pozo-Rodrigálvarez, A.; Martínez-Murillo, R.; Torres-Aleman, I. Astrocytes require insulin-like growth factor I to protect neurons against oxidative injury. *F1000 Res.* **2014**, *3*, 28. [CrossRef]
28. Hernandez-Garzón, E.; Fernandez, A.M.; Perez-Alvarez, A.; Genis, L.; Bascuñana, P.; Fernandez de la Rosa, R.; Delgado, M.; Angel Pozo, M.; Moreno, E.; McCormick, P.J.; et al. The insulin-like growth factor I receptor regulates glucose transport by astrocytes. *Glia* **2016**, *64*, 1962–1971. [CrossRef]
29. Chen, D.Y.; Stern, S.A.; Garcia-Osta, A.; Saunier-Rebori, B.; Pollonini, G.; Bambah-Mukku, D.; Blitzer, R.D.; Alberini, C.M. A critical role for IGF-II in memory consolidation and enhancement. *Nature* **2011**, *469*, 491–497. [CrossRef]
30. Myer, D.J.; Gurkoff, G.G.; Lee, S.M.; Hovda, D.A.; Sofroniew, M.V. Essential protective roles of reactive astrocytes in traumatic brain injury. *Brain* **2006**, *129*, 2761–2772. [CrossRef]
31. Faulkner, J.R.; Herrmann, J.E.; Woo, M.J.; Tansey, K.E.; Doan, N.B.; Sofroniew, M.V. Reactive astrocytes protect tissue and preserve function after spinal cord injury. *J. Neurosci.* **2004**, *24*, 2143–2155. [CrossRef] [PubMed]
32. Liddelow, S.; Guttenplan, K.; Clarke, L.; Bennett, F.C.; Bohlen, C.J.; Schirmer, L.; Bennett, M.L.; Münch, A.E.; Chung, W.S.; Peterson, T.C.; et al. Neurotoxic reactive astrocytes are induced by activated microglia. *Nature* **2017**, *541*, 481–487. [CrossRef] [PubMed]
33. Buckman, L.B.; Thompson, M.M.; Lippert, R.N.; Blackwell, T.S.; Yull, F.E.; Ellacott, K.L. Evidence for a novel functional role of astrocytes in the acute homeostatic response to high-fat diet intake in mice. *Mol. Metab.* **2014**, *4*, 58–63. [CrossRef] [PubMed]
34. Douglass, J.D.; Dorfman, M.D.; Fasnacht, R.; Shaffer, L.D.; Thaler, J.P. Astrocyte IKKβ/NF-κB signaling is required for diet-induced obesity and hypothalamic inflammation. *Mol. Metab.* **2017**, *6*, 366–373. [CrossRef] [PubMed]
35. Iglesias, J.; Morales, L.; Barreto, G.E. Metabolic and Inflammatory Adaptation of Reactive Astrocytes: Role of PPARs. *Mol. Neurobiol.* **2017**, *54*, 2518–2538. [CrossRef] [PubMed]
36. Komoly, S.; Hudson, L.D.; Webster, H.D.; Bondy, C.A. Insulin-like growth factor I gene expression is induced in astrocytes during experimental demyelination. *Proc. Natl. Acad. Sci. USA* **1992**, *89*, 1894–1898. [CrossRef] [PubMed]
37. Fernandez, A.M.; Garcia-Estrada, J.; Garcia-Segura, L.M.; Torres-Aleman, I. Insulin-like growth factor I modulates c-fos induction and astrocytosis in response to neurotoxic insult. *Neuroscience* **1996**, *76*, 117–122. [CrossRef]
38. Ryu, B.R.; Ko, H.W.; Jou, I.; Noh, J.S.; Gwag, B.J. Phosphatidylinositol 3-kinase-mediated regulation of neuronal apoptosis and necrosis by insulin and IGF-I. *J. Neurobiol.* **1999**, *39*, 536–546. [CrossRef]
39. Guerra-Cantera, S.; Frago, L.M.; Díaz, F.; Ros, P.; Jiménez-Hernaiz, M.; Freire-Regatillo, A.; Barrios, V.; Argente, J.; Chowen, J.A. Short-Term Diet Induced Changes in the Central and Circulating IGF Systems Are Sex Specific. *Front. Endocrinol.* **2020**, *11*, 513. [CrossRef]
40. Hedbacker, K.; Birsoy, K.; Wysocki, R.W.; Asilmaz, E.; Ahima, R.S.; Farooqi, I.S.; Friedman, J.M. Antidiabetic effects of IGFBP2, a leptin-regulated gene. *Cell Metab.* **2010**, *11*, 11–22. [CrossRef]
41. Wheatcroft, S.B.; Kearney, M.T.; Shah, A.M.; Ezzat, V.A.; Miell, J.R.; Modo, M.; Williams, S.C.; Cawthorn, W.P.; Medina-Gomez, G.; Vidal-Puig, A.; et al. IGF-binding protein-2 protects against the development of obesity and insulin resistance. *Diabetes* **2007**, *56*, 285–294. [CrossRef] [PubMed]
42. Paruthiyil, S.; Hagiwara, S.I.; Kundassery, K.; Bhargava, A. Sexually dimorphic metabolic responses mediated by CRF2 receptor during nutritional stress in mice. *Biol. Sex Differ.* **2018**, *9*, 49. [CrossRef] [PubMed]
43. de Wilde, J.; Smit, E.; Mohren, R.; Boekschoten, M.V.; de Groot, P.; van den Berg, S.A.; Bijland, S.; Voshol, P.J.; van Dijk, K.W.; de Wit, N.W.; et al. An 8-week high-fat diet induces obesity and insulin resistance with small changes in the muscle transcriptome of C57BL/6J mice. *J. Nutrigenet. Nutr.* **2009**, *2*, 280–291. [CrossRef]

44. Ros, P.; Díaz, F.; Freire-Regatillo, A.; Argente-Arizón, P.; Barrios, V.; Argente, J.; Chowen, J.A. Sex Differences in Long-term Metabolic Effects of Maternal Resveratrol Intake in Adult Rat Offspring. *Endocrinology* **2020**, *161*, bqaa090. [CrossRef]
45. Samuel, P.; Khan, M.A.; Nag, S.; Inagami, T.; Hussain, T. Angiotensin AT(2) receptor contributes towards gender bias in weight gain. *PLoS ONE* **2013**, *8*, e48425. [CrossRef] [PubMed]
46. Pettersson, U.S.; Waldén, T.B.; Carlsson, P.O.; Jansson, L.; Phillipson, M. Female mice are protected against high-fat diet induced metabolic syndrome and increase the regulatory T cell population in adipose tissue. *PLoS ONE* **2012**, *7*, e46057. [CrossRef] [PubMed]
47. Pinos, H.; Carrillo, B.; Díaz, F.; Chowen, J.A.; Collado, P. Differential vulnerability to adverse nutritional conditions in male and female rats: Modulatory role of estradiol during development. *Front. Neuroendocrinol.* **2018**, *48*, 13–22. [CrossRef]
48. Chowen, J.A.; Freire-Regatillo, A.; Argente, J. Neurobiological characteristics underlying metabolic differences between males and females. *Prog. Neurobiol.* **2019**, *176*, 18–32. [CrossRef]
49. Peshti, V.; Obolensky, A.; Nahum, L.; Kanfi, Y.; Rathaus, M.; Avraham, M.; Tinman, S.; Alt, F.W.; Banin, E.; Cohen, H.Y. Characterization of physiological defects in adult SIRT6-/- mice. *PLoS ONE* **2017**, *12*, e0176371. [CrossRef]
50. Frystyk, J.; Grønbaek, H.; Skjaerbaek, C.; Flyvbjerg, A.; Orskov, H.; Baxter, R.C. Developmental changes in serum levels of free and total insulin-like growth factor I (IGF-I), IGF-binding protein-1 and -3, and the acid-labile subunit in rats. *Endocrinology* **1998**, *139*, 4286–4292. [CrossRef]
51. Argente, J.; Barrios, V.; Pozo, J.; Muñoz, M.T.; Hervás, F.; Stene, M.; Hernández, M. Normative data for insulin-like growth factors (IGFs), IGF-binding proteins, and growth hormone-binding protein in a healthy Spanish pediatric population: Age- and sex-related changes. *J. Clin. Endocrinol. Metab.* **1993**, *77*, 1522–1528. [CrossRef] [PubMed]
52. Waters, D.L.; Yau, C.L.; Montoya, G.D.; Baumgartner, R.N. Serum Sex Hormones, IGF-1, and IGFBP3 Exert a Sexually Dimorphic Effect on Lean Body Mass in Aging. *J. Gerontol. A Biol. Sci. Med. Sci.* **2003**, *58*, 648–652. [CrossRef] [PubMed]
53. Borski, R.J.; Tsai, W.; DeMott-Friberg, R.; Barkan, A.L. Regulation of somatic growth and the somatotropic axis by gonadal steroids: Primary effect on insulin-like growth factor I gene expression and secretion. *Endocrinology* **1996**, *137*, 3253–3259. [CrossRef] [PubMed]
54. Nam, S.Y.; Lee, E.J.; Kim, K.R.; Cha, B.S.; Song, Y.D.; Lim, S.K.; Lee, H.C.; Huh, K.B. Effect of obesity on total and free insulin-like growth factor (IGF)-1, and their relationship to IGF-binding protein (BP)-1, IGFBP-2, IGFBP-3, insulin, and growth hormone. *Int. J. Obes. Relat. Metab. Disord.* **1997**, *21*, 355–359. [CrossRef] [PubMed]
55. Doyle, S.L.; Donohoe, C.L.; Finn, S.P.; Howard, J.M.; Lithander, F.E.; Reynolds, J.V.; Pidgeon, G.P.; Lysaght, J. IGF-1 and its receptor in esophageal cancer: Association with adenocarcinoma and visceral obesity. *Am. J. Gastroenterol.* **2012**, *107*, 196–204. [CrossRef] [PubMed]
56. Xuan, L.; Ma, J.; Yu, M.; Yang, Z.; Huang, Y.; Guo, C.; Lu, Y.; Yan, L.; Shi, S. Insulin-like growth factor 2 promotes adipocyte proliferation, differentiation and lipid deposition in obese type 2 diabetes. *J. Transl. Sci.* **2019**, *6*, 1–7. [CrossRef]
57. Buchanan, C.M.; Phillips, A.R.; Cooper, G.J. Preptin derived from proinsulin-like growth factor II (proIGF-II) is secreted from pancreatic islet beta cells and enhances insulin secretion. *Biochem. J.* **2001**, *360*, 431–439. [CrossRef]
58. Argente, J.; Caballo, N.; Barrios, V.; Pozo, J.; Muñoz, M.T.; Chowen, J.A.; Hernández, M. Multiple endocrine abnormalities of the growth hormone and insulin-like growth factor axis in prepubertal children with exogenous obesity: Effect of short- and long-term weight reduction. *J. Clin. Endocrinol. Metab.* **1997**, *82*, 2076–2083. [CrossRef]
59. Vahl, N.; Klausen, I.; Christiansen, J.S.; Jørgensen, J.O. Growth hormone (GH) status is an independent determinant of serum levels of cholesterol and triglycerides in healthy adults. *Clin. Endocrinol.* **1999**, *51*, 309–316. [CrossRef]
60. Berryman, D.E.; Glad, C.A.; List, E.O.; Johannsson, G. The GH/IGF-1 axis in obesity: Pathophysiology and therapeutic considerations. *Nat. Rev. Endocrinol.* **2013**, *9*, 346–356. [CrossRef]
61. Lewitt, M.S.; Dent, M.S.; Hall, K. The Insulin-Like Growth Factor System in Obesity, Insulin Resistance and Type 2 Diabetes Mellitus. *J. Clin. Med.* **2014**, *3*, 1561–1574. [CrossRef] [PubMed]

62. Clemmons, D.R.; Snyder, D.K.; Busby, W.H., Jr. Variables controlling the secretion of insulin-like growth factor binding protein-2 in normal human subjects. *J. Clin. Endocrinol. Metab.* **1991**, *73*, 727–733. [CrossRef] [PubMed]
63. Yau, S.W.; Harcourt, B.E.; Kao, K.T.; Alexander, E.J.; Russo, V.C.; Werther, G.A.; Sabin, M.A. Serum IGFBP-2 levels are associated with reduced insulin sensitivity in obese children. *Clin. Obes.* **2018**, *8*, 184–190. [CrossRef] [PubMed]
64. Barrios, V.; Buño, M.; Pozo, J.; Muñoz, M.T.; Argente, J. Insulin-like growth factor-binding protein-2 levels in pediatric patients with growth hormone deficiency, eating disorders and acute lymphoblastic leukemia. *Horm. Res.* **2000**, *53*, 221–227. [CrossRef] [PubMed]
65. Hwang, L.L.; Wang, C.H.; Li, T.L.; Chang, S.D.; Lin, L.C.; Chen, C.P.; Chen, C.T.; Liang, K.C.; Ho, I.K.; Yang, W.S.; et al. Sex differences in high-fat diet-induced obesity, metabolic alterations and learning, and synaptic plasticity deficits in mice. *Obesity* **2010**, *18*, 463–469. [CrossRef]
66. Wabitsch, M.; Hauner, H.; Heinze, E.; Teller, W.M. The role of growth hormone/insulin-like growth factors in adipocyte differentiation. *Metabolism* **1995**, *44*, 45–49. [CrossRef]
67. Smith, P.J.; Wise, L.S.; Berkowitz, R.; Wan, C.; Rubin, C.S. Insulin-like growth factor-I is an essential regulator of the differentiation of 3T3-L1 adipocytes. *J. Biol. Chem.* **1988**, *263*, 9402–9408.
68. Chang, H.R.; Kim, H.J.; Xu, X.; Ferrante, A.W., Jr. Macrophage and adipocyte IGF1 maintain adipose tissue homeostasis during metabolic stresses. *Obesity* **2016**, *24*, 172–183. [CrossRef]
69. Morita, S.; Horii, T.; Kimura, M.; Arai, Y.; Kamei, Y.; Ogawa, Y.; Hatada, I. Paternal allele influences high fat diet-induced obesity. *PLoS ONE* **2014**, *9*, e85477. [CrossRef]
70. Yau, S.W.; Russo, V.C.; Clarke, I.J.; Dunshea, F.R.; Werther, G.A.; Sabin, M.A. IGFBP-2 inhibits adipogenesis and lipogenesis in human visceral, but not subcutaneous, adipocytes. *Int. J. Obes.* **2015**, *39*, 770–781. [CrossRef]
71. Li, Z.; Picard, F. Modulation of IGFBP2 mRNA expression in white adipose tissue upon aging and obesity. *Horm. Metab. Res.* **2010**, *42*, 787–791. [CrossRef] [PubMed]
72. Conover, C.A.; Harstad, S.L.; Tchkonia, T.; Kirkland, J.L. Preferential impact of pregnancy-associated plasma protein-A deficiency on visceral fat in mice on high-fat diet. *Am. J. Physiol. Endocrinol. Metab.* **2013**, *305*, E1145–E1153. [CrossRef] [PubMed]
73. Rojas-Rodriguez, R.; Lifshitz, L.M.; Bellve, K.D.; Min, S.Y.; Pires, J.; Leung, K.; Boeras, C.; Sert, A.; Draper, J.T.; Corvera, S.; et al. Human adipose tissue expansion in pregnancy is impaired in gestational diabetes mellitus. *Diabetologia* **2015**, *58*, 2106–2114. [CrossRef] [PubMed]
74. Sarapio, E.; De Souza, S.K.; Model, J.F.A.; Trapp, M.; Da Silva, R.S.M. Stanniocalcin-1 and -2 effects on glucose and lipid metabolism in white adipose tissue from fed and fasted rats. *Can. J. Physiol. Pharmacol.* **2019**, *97*, 916–923. [CrossRef]
75. Baquedano, E.; Ruiz-Lopez, A.M.; Sustarsic, E.G.; Herpy, J.; List, E.O.; Chowen, J.A.; Frago, L.M.; Kopchick, J.J.; Argente, J. The absence of GH signaling affects the susceptibility to high-fat diet-induced hypothalamic inflammation in male mice. *Endocrinology* **2014**, *155*, 4856–4867. [CrossRef]
76. Hong, H.; Cui, Z.Z.; Zhu, L.; Fu, S.P.; Rossi, M.; Cui, Y.H.; Zhu, B.M. Central IGF1 improves glucose tolerance and insulin sensitivity in mice. *Nutr. Diabetes* **2017**, *7*, 2. [CrossRef]
77. Barrand, S.; Crowley, T.M.; Wood-Bradley, R.J.; De Jong, K.A.; Armitage, J.A. Impact of maternal high fat diet on hypothalamic transcriptome in neonatal Sprague Dawley rats. *PLoS ONE* **2017**, *12*, e0189492. [CrossRef]
78. Iwamoto, T.; Ouchi, Y. Emerging evidence of insulin-like growth factor 2 as a memory enhancer: A unique animal model of cognitive dysfunction with impaired adult neurogenesis. *Rev. Neurosci.* **2014**, *25*, 559–574. [CrossRef]
79. Uchimura, T.; Hollander, J.M.; Nakamura, D.S.; Liu, Z.; Rosen, C.J.; Georgakoudi, I.; Zeng, L. An essential role for IGF2 in cartilage development and glucose metabolism during postnatal long bone growth. *Development* **2017**, *144*, 3533–3546. [CrossRef]
80. Gleason, C.E.; Ning, Y.; Cominski, T.P.; Gupta, R.; Kaestner, K.H.; Pintar, J.E.; Birnbaum, M.J. Role of insulin-like growth factor-binding protein 5 (IGFBP5) in organismal and pancreatic beta-cell growth. *Mol. Endocrinol.* **2010**, *24*, 178–192. [CrossRef]
81. Thaler, J.P.; Yi, C.X.; Schur, E.A.; Guyenet, S.J.; Hwang, B.H.; Dietrich, M.O.; Zhao, X.; Sarruf, D.A.; Izgur, V.; Maravilla, K.R.; et al. Obesity is associated with hypothalamic injury in rodents and humans. *J. Clin. Investig.* **2012**, *122*, 153–162. [CrossRef] [PubMed]

82. Valdearcos, M.; Robblee, M.M.; Benjamin, D.I.; Nomura, D.K.; Xu, A.W.; Koliwad, S.K. Microglia dictate the impact of saturated fat consumption on hypothalamic inflammation and neuronal function. *Cell Rep.* **2014**, *9*, 2124–2138. [CrossRef] [PubMed]
83. Lemus, M.B.; Bayliss, J.A.; Lockie, S.H.; Santos, V.V.; Reichenbach, A.; Stark, R.; Andrews, Z.B. A stereological analysis of NPY, POMC, Orexin, GFAP astrocyte, and Iba1 microglia cell number and volume in diet-induced obese male mice. *Endocrinology* **2015**, *156*, 1701–1713. [CrossRef] [PubMed]
84. Baufeld, C.; Osterloh, A.; Prokop, S.; Miller, K.R.; Heppner, F.L. High-fat diet-induced brain region-specific phenotypic spectrum of CNS resident microglia. *Acta Neuropathol.* **2016**, *132*, 361–375. [CrossRef] [PubMed]
85. Dorfman, M.D.; Thaler, J.P. Hypothalamic inflammation and gliosis in obesity. *Curr. Opin. Endocrinol. Diabetes Obes.* **2015**, *22*, 325–330. [CrossRef]
86. Wei, W.; Pham, K.; Gammons, J.W.; Sutherland, D.; Liu, Y.; Smith, A.; Kaczorowski, C.C.; O'Connell, K.M. Diet composition, not calorie intake, rapidly alters intrinsic excitability of hypothalamic AgRP/NPY neurons in mice. *Sci. Rep.* **2015**, *5*, 16810. [CrossRef]
87. Hassan, A.M.; Mancano, G.; Kashofer, K.; Fröhlich, E.E.; Matak, A.; Mayerhofer, R.; Reichmann, F.; Olivares, M.; Neyrinck, A.M.; Delzenne, N.M.; et al. High-fat diet induces depression-like behaviour in mice associated with changes in microbiome, neuropeptide Y, and brain metabolome. *Nutr. Neurosci.* **2019**, *22*, 877–893. [CrossRef]
88. de Araujo, T.M.; Razolli, D.S.; Correa-da-Silva, F.; de Lima-Junior, J.C.; Gaspar, R.S.; Sidarta-Oliveira, D.; Victorio, S.C.; Donato, J., Jr.; Kim, Y.B.; Velloso, L.A. The partial inhibition of hypothalamic IRX3 exacerbates obesity. *EBioMedicine* **2019**, *39*, 448–460. [CrossRef]
89. Mayer, C.M.; Belsham, D.D. Insulin directly regulates NPY and AgRP gene expression via the MAPK MEK/ERK signal transduction pathway in mHypoE-46 hypothalamic neurons. *Mol. Cell Endocrinol.* **2009**, *307*, 99–108. [CrossRef]
90. Fujita, S.; Honda, K.; Yamaguchi, M.; Fukuzo, S.; Saneyasu, T.; Kamisoyama, H. Role of Insulin-like Growth Factor-1 in the Central Regulation of Feeding Behavior in Chicks. *J. Poult. Sci.* **2019**, *56*, 270–276. [CrossRef]
91. Roepke, T.A.; Yasrebi, A.; Villalobos, A.; Krumm, E.A.; Yang, J.A.; Mamounis, K.J. Loss of ERα partially reverses the effects of maternal high-fat diet on energy homeostasis in female mice. *Sci. Rep.* **2017**, *7*, 6381. [CrossRef] [PubMed]
92. Sclafani, A. Carbohydrate-induced hyperphagia and obesity in the rat: Effects of saccharide type, form, and taste. *Neurosci. Biobehav. Rev.* **1987**, *11*, 155–162. [CrossRef]
93. Sclafani, A.; Xenakis, S. Influence of diet form on the hyperphagia-promoting effect of polysaccharide in rats. *Life Sci.* **1984**, *34*, 1253–1259. [CrossRef]
94. Fuente-Martín, E.; García-Cáceres, C.; Granado, M.; Sánchez-Garrido, M.A.; Tena-Sempere, M.; Frago, L.M.; Argente, J.; Chowen, J.A. Early postnatal overnutrition increases adipose tissue accrual in response to a sucrose-enriched diet. *Am. J. Physiol. Endocrinol. Metab.* **2012**, *302*, E1586–E1598. [CrossRef] [PubMed]
95. Fuente-Martín, E.; Granado, M.; García-Cáceres, C.; Sanchez-Garrido, M.A.; Frago, L.M.; Tena-Sempere, M.; Argente, J.; Chowen, J.A. Early nutritional changes induce sexually dimorphic long-term effects on body weight gain and the response to sucrose intake in adult rats. *Metabolism* **2012**, *61*, 812–822. [CrossRef]
96. Sandhu, M.S.; Gibson, J.M.; Heald, A.H.; Dunger, D.B.; Wareham, N.J. Low circulating IGF-II concentrations predict weight gain and obesity in humans. *Diabetes* **2003**, *52*, 1403–1408. [CrossRef]

Article

Distinct Changes in Gut Microbiota Are Associated with Estradiol-Mediated Protection from Diet-Induced Obesity in Female Mice

Kalpana D. Acharya [1], Hye L. Noh [2], Madeline E. Graham [1], Sujin Suk [2], Randall H. Friedline [2], Cesiah C. Gomez [1], Abigail E. R. Parakoyi [1], Jun Chen [3], Jason K. Kim [2] and Marc J. Tetel [1,*]

1 Neuroscience Department, Wellesley College, Wellesley, MA 02481, USA; kacharya@wellesley.edu (K.D.A.); mgraham2@wellesley.edu (M.E.G.); cgomez3@wellesley.edu (C.C.G.); aparakoy@wellesley.edu (A.E.R.P.)
2 Program in Molecular Medicine, Division of Endocrinology, Metabolism, and Diabetes, Department of Medicine, University of Massachusetts Medical School, Worcester, MA 01605, USA; HyeLim.Noh@crl.com (H.L.N.); scarlet108@gmail.com (S.S.); Randall.Friedline@umassmed.edu (R.H.F.); Jason.Kim@umassmed.edu (J.K.K.)
3 Department of Health Sciences Research & Center for Individualized Medicine, Mayo Clinic, Rochester, MN 55905, USA; Chen.Jun2@mayo.edu
* Correspondence: mtetel@wellesley.edu

Abstract: A decrease in ovarian estrogens in postmenopausal women increases the risk of weight gain, cardiovascular disease, type 2 diabetes, and chronic inflammation. While it is known that gut microbiota regulates energy homeostasis, it is unclear if gut microbiota is associated with estradiol regulation of metabolism. In this study, we tested if estradiol-mediated protection from high-fat diet (HFD)-induced obesity and metabolic changes are associated with longitudinal alterations in gut microbiota in female mice. Ovariectomized adult mice with vehicle or estradiol (E2) implants were fed chow for two weeks and HFD for four weeks. As reported previously, E2 increased energy expenditure, physical activity, insulin sensitivity, and whole-body glucose turnover. Interestingly, E2 decreased the tight junction protein occludin, suggesting E2 affects gut epithelial integrity. Moreover, E2 increased *Akkermansia* and decreased Erysipleotrichaceae and Streptococcaceae. Furthermore, *Coprobacillus* and *Lactococcus* were positively correlated, while *Akkermansia* was negatively correlated, with body weight and fat mass. These results suggest that changes in gut epithelial barrier and specific gut microbiota contribute to E2-mediated protection against diet-induced obesity and metabolic dysregulation. These findings provide support for the gut microbiota as a therapeutic target for treating estrogen-dependent metabolic disorders in women.

Keywords: diabetes; estrogens; gut permeability/integrity; insulin sensitivity; *Akkermansia*; gut microbiome

Citation: Acharya, K.D.; Noh, H.L.; Graham, M.E.; Suk, S.; Friedline, R.H.; Gomez, C.C.; Parakoyi, A.E.R.; Chen, J.; Kim, J.K.; Tetel, M.J. Distinct Changes in Gut Microbiota Are Associated with Estradiol-Mediated Protection from Diet-Induced Obesity in Female Mice. *Metabolites* **2021**, *11*, 499. https://doi.org/10.3390/metabo11080499

Academic Editors: Giancarlo Panzica, Stefano Gotti and Paloma Collado Guirao

Received: 10 June 2021
Accepted: 27 July 2021
Published: 30 July 2021

1. Introduction

More than 40% of the US population is obese (CDC, 2018), which is a leading cause of morbidity and mortality worldwide [1]. The latest example of the increasing impact of obesity on human health is the strong association of obesity with the number of hospitalized COVID-19 positive patients [2]. Obesity is more prevalent in women [3], in particular during menopause, and is positively associated with a steep decline in ovarian hormones. Increased fat weight gain in postmenopausal women elevates their risk of hyperglycemia, insulin resistance, hyperlipidemia, low-grade inflammation, osteoporosis, cognitive decline, breast cancer and colorectal cancer [4–9]. Estrogen replacement therapy decreases the postmenopausal adiposity and protects women from diabetes, coronary heart disease, and increases overall lifespan [8,10]. Ovariectomized rodents provide excellent models for studying estrogen-dependent effects on energy homeostasis. Ovariectomy causes diet-induced obesity, hyperglycemia and insulin resistance in rodents, which can be rescued by estradiol (E2) treatment [11–15]. In further support of a protective role for estrogens, mice lacking estrogen receptors or the estrogen synthesizing enzyme, aromatase, develop obesity [16–18].

Another key regulator of energy homeostasis is the gut microbiota, a community of bacteria, fungi, viruses, and archaea that reside on the gastrointestinal epithelium [19]. The gut microbiota influences host physiology through nutrient harvest, synthesis of vitamins, hormones, and neurotransmitters, and imprinting and strengthening of the immune system [20,21]. Gut microbes produce energy from fermentation of non-digestible carbohydrates, in particular, short-chain fatty acids, that are linked to improved insulin sensitivity and health in humans [22]. Notably, germ-free mice and rats have a profound reduction in energy harvest capacity, a compromised immune system, and abnormal intestinal features compared to conventionally raised animals, indicating an important function of gut microbiota on host health. [23,24].

Diet strongly modulates gut microbial composition and activity in humans and rodents. For example, high-fat diet (HFD) profoundly decreases microbial diversity [21,25–29]. HFD promotes the endotoxin-producing gram-negative communities, inducing abnormal immune responses and inflammation, characteristic pathologies of obesity and diabetes [30,31]. HFD also increases intestinal permeability, allowing microbiota-induced toxins into the circulation and alters expression of multiple genes in the intestinal epithelium of male and female mice [31–33]. Given that HFD increases the risk of developing obesity and metabolic syndromes in postmenopausal women [34], it is important to gain a better understanding of the functions of gut microbiota in metabolic health in females.

Recent evidence suggests there is cross-talk between estrogens and gut microbiota. Urinary estrogens in postmenopausal women positively correlate with gut microbiota taxa diversity [35]. In further support of estrogens' influence on gut microbiota in women, phytoestrogens increase *Lactobacillus*, *Enterococcus* and *Bifidobacterium* [36,37]. In mice, ovariectomy alters gut microbial diversity, in particular, by shifting abundances of the two major bacterial phyla, Bacteroidetes and Firmicutes and by increasing *Bifidobacterium* [38–40]. In mice fed a high-sucrose, high-fat containing western diet, chronic E2 administration via drinking water decreased lipopolysaccharide-producing microbes, such as *Escherichia* and *Shigella*, and increased *Bifidobacterium* and *Akkermansia* [41]. In addition, we recently found that estrogens alter gut microbiota in leptin-deficient (*ob/ob*) obese female mice. E2 decreased gut microbial evenness in both lean and obese (*ob/ob*) mice and increased S24-7 abundance [42]. Taken together, these studies suggest that estrogens can influence gut microbiota in females.

While diet and gut microbes profoundly affect energy metabolism, it is not known if the estrogen-mediated protective effects in females are linked to changes in gut microbiota. In the present study, we tested the hypothesis that the E2-mediated protection against HFD-induced obesity and metabolic disorders is associated with changes in gut microbiota and intestinal morphology in female mice. This study investigated the effects of E2 and diet in female mice on their metabolic profiles, associated longitudinal changes in gut microbiota, and gut epithelial integrity. These findings enhance our understanding of how estrogens function in women's metabolic health and help identify potential gut microbial modulators in estrogen-dependent protection from metabolic syndromes.

2. Results

2.1. Estradiol Attenuates Body Weight and Fat Mass Gain in Female Mice on HFD

Ten-week-old female C57BL/6J mice were bilaterally ovariectomized (OVX) and received implants containing 17β-estradiol (E2) or vehicle (Veh, n = 6/group) [11,42]. Metabolic and gut microbiota data were collected at different points through the study (Figure 1A). Analysis of longitudinal data, including both STND and HFD feeding, showed a main effect of E2 on body weight. During the two weeks on STND, Veh and E2 mice did not differ in body weight. After switching to HFD, Veh mice gained weight, whereas E2 mice were protected from the weight gain. Veh mice weighed more than E2 mice from D21 till the end of the study (Figure 1B), due to increased fat mass (Figure 1C). The effect of E2 on fat mass was profound during HFD, although no effect was seen during STND. Lean mass was not affected by E2 during either diet (Figure 1D).

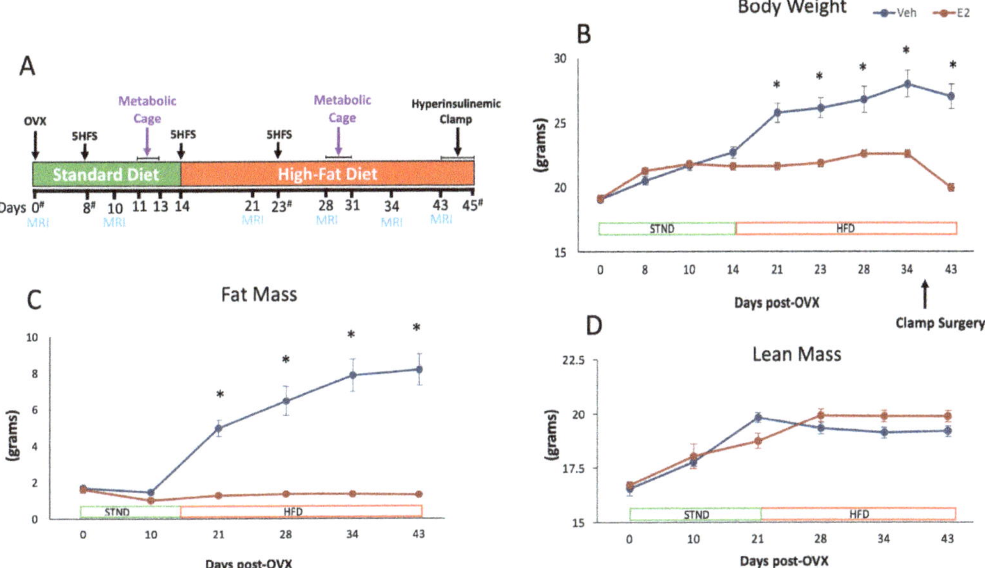

Figure 1. Estradiol attenuates weight gain in female mice on a high-fat diet (HFD). Experimental timeline (**A**). Ten-week old mice were ovariectomized and implanted with capsules containing E2 (50 µg) or Veh implants (n = 6/group). Animals were placed in metabolic cages for 3 days, once each during STND and HFD. Body weight (**B**) Fat mass (**C**) Lean mass (**D**). Mice were switched from standard diet (STND) to HFD on day 14. Surgery for the hyperinsulinemic- euglycemic clamp on days 37–39 (4 mice/day) resulted in weight change in both groups. Error bars are shown as ± SEM. * denotes $p < 0.001$, using repeated measures ANOVA followed by t-test. OVX: ovariectomy; 5HFS: 5-h fasting blood glucose; MRI: body composition measurement using proton magnetic resonance spectroscopy (1H-MRS); Clamp: hyperinsulinemic-euglycemic clamp; # indicates fecal sample collection days.

2.2. Estradiol Reduces Food Intake and Energy Expenditure in Female Mice on STND

Analysis of food intake (calories), including both STND and HFD feeding, showed a main effect of E2 treatment during night. During STND, E2 reduced food intake during 24 h and day (Figure 2A). An interaction between E2 and diet was also observed on food intake during 24 h, day and night. E2 altered water intake during STND such that E2-mice consumed less water during day, but more during night, while no effect was detected in cumulative 24 h data (Figure S1A).

Locomotor activity was measured using metabolic cages (TSE Systems, Germany) for a 72-h period on D11–D13 and D29–31 during STND and HFD, respectively. A main effect of E2 treatment was present on locomotor activity during 24 h and night. E2 mice on STND were less active during the light phase compared to Veh mice (Figure 2B).

VO_2 consumption and VCO_2 production were also measured using metabolic cages. E2 altered VO_2 consumption during day and night, and VCO_2 production during day in longitudinal data. During STND, E2 decreased VO_2 consumption during 24 h and day (Figure 2C) and VCO_2 production at 24 h, day, and night (Figure 2D). Respiratory exchange rate (RER, VO_2/VCO_2), a predictor of relative contribution of carbohydrate (value > 0.85) vs lipid (value < 0.8) on energy production, was decreased in E2 group during day, but not during 24 h or night (Figure S2A). E2 affected resting energy expenditure (EE) during day and night, with no effect on 24 h data. During STND, EE was attenuated in E2-treated mice during 24 h and day (Figure 2E).

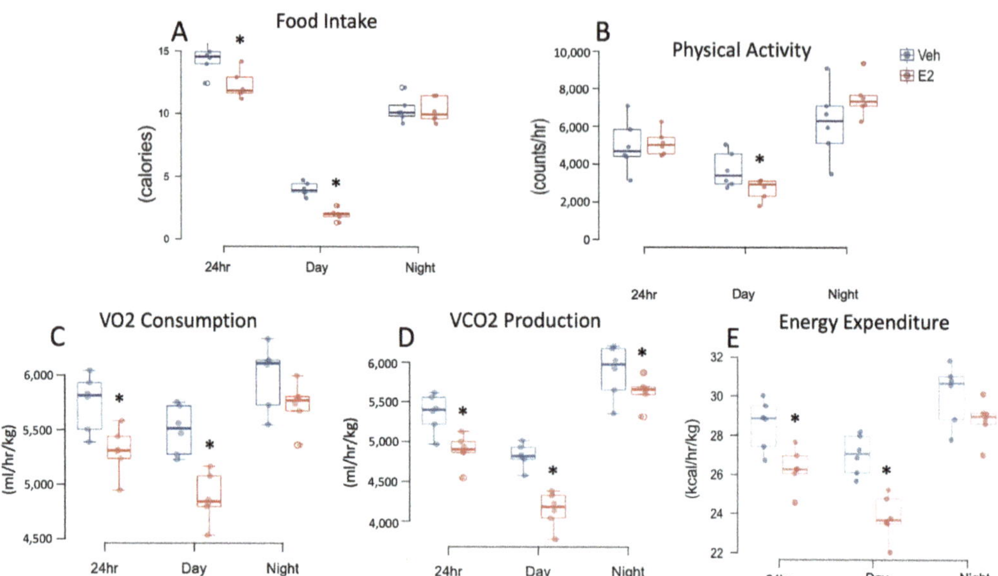

Figure 2. Estradiol decreases food intake and energy expenditure in female mice fed a STND. Data for Food intake (**A**) Physical activity (**B**) VO_2 consumption (**C**) VCO_2 production (**D**) Resting energy expenditure (**E**) were collected from mice in metabolic cages on days 11–13. The average 24-h data were obtained from 72-h data and used for statistical analysis. * indicates differences between E2 and Veh mice ($n = 6$/group) ($p < 0.05$; t-test).

2.3. Estradiol Increases Food Intake and Energy Expenditure in Female Mice during HFD

E2 increased HFD consumption during 24 h and night (Figure 3A). Similarly, water intake was increased in E2 mice during 24 h and night (Figure S1B). E2 profoundly increased locomotor activity during 24 h and night (Figure 3B). Increased metabolic capacity in E2 mice during HFD was further confirmed by increases in VO_2 consumption during 24 h and night and VCO_2 production, during 24 h and night (Figure 3C,D). Similarly, EE was increased in E2 mice during 24 h and night (Figure 3E). RER was not affected by E2 during HFD feeding (Figure S2B).

2.4. HFD Increases Body Weight and Fat Mass Gain in Female Mice

To determine the effects of E2 on energy metabolism and gut microbiota under different diet conditions, mice were fed a chow diet (STND) for the first 14 days after OVX and then fed HFD for days 14–45 (Figure 1A). HFD had a profound effect on body weight and fat mass. An interaction between diet and E2 treatment was also present on body weight and fat mass (Figure 1B,C). For the lean mass, while an effect of diet was present on longitudinal data, there was no effect on separate data during STND or HFD (Figure 1D).

2.5. HFD Alters Food Intake and Energy Expenditure in Female Mice

A comparison of food intake, in calories, between STND and HFD feeding revealed a main effect of diet during 24 h and day, but not at night. However, there was an interaction of diet and hormone treatment during 24 h, day, and night. Interestingly, Veh mice on HFD ate less calories during 24 h compared to Veh mice on STND. In contrast, E2 mice on HFD ate more calories during 24 h, than during STND (Figures 2 and 3). Both Veh and E2-mice had a greater water intake during STND than HFD, during 24 h and night (Figure S1A,B).

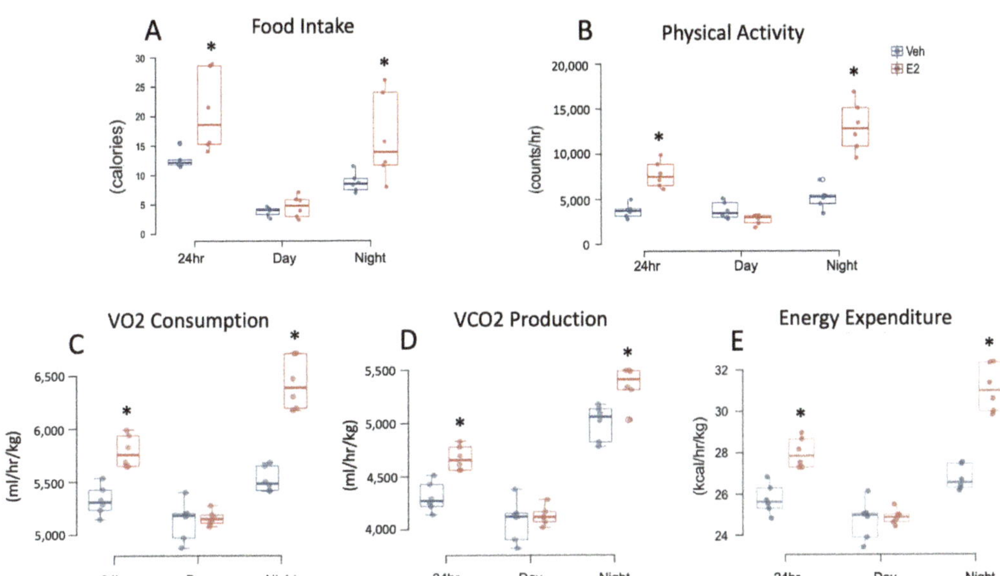

Figure 3. Estradiol increases food intake, water intake, and energy expenditure in female mice fed HFD. Food intake (**A**) Physical activity (**B**) VO_2 consumption (**C**) VCO_2 production (**D**) Resting energy expenditure (**E**) were measured in mice in metabolic cages on days 15–17 of HFD feeding. The average 24-h data were obtained from 72-h data and used for statistical analysis. * indicates differences between E2 and Veh mice ($n = 6$/group) ($p < 0.05$; t-test).

Diet affected locomotor activity during 24 h, day, and night. Within E2 group, mice were more active at 24 h and night following the switch to HFD compared to STND, suggesting that one mechanism by which estrogens prevent HFD-induced obesity is by increasing activity. In contrast, Veh mice were less active during HFD feeding, than on STND, during day. Diet also affected VCO_2 production (during 24 h, day, and night). Veh mice had reduced VCO_2 production during HFD than on STND (on 24 h, day, and night data). E2 mice also had reduced VCO_2 production during HFD compared to STND, but only during night. Moreover, an interaction of E2 and diet was present on VCO_2 production (on 24 h, day and night data). Similar to VCO_2 production, Veh mice on HFD had decreased VO_2 consumption (during 24 h, day, and night), compared to STND. In contrast, VO_2 consumption was increased in E2-mice during HFD (during 24 h, day, and night) (Figures 2 and 3). An interaction of E2 and diet was also present on VO_2 consumption (during 24 h, day, and night).

A main effect of diet was also observed on RER, with a lower RER during HFD (during 24 h, day and night) compared to STND. As expected, RER was decreased in both E2 mice (during 24 h data, day, and night) and Veh mice (also during 24 h, day, and night), during HFD feeding due to lipid oxidation. In contrast, EE did not show a main effect of diet, but an interaction of treatment and diet was detected (during 24 h, day, and night). E2-treated mice had an increased EE during HFD (on 24 h, day, and night data) whereas Veh mice during HFD had a decreased EE (during 24 h, day, and night) (Figures 2 and 3).

2.6. Estradiol Attenuates Fasting Glucose Levels and Plasma Adipokines in Female Mice

Five-hour fasting blood glucose was measured at different times during STND and HFD, which was lower in E2-treated mice than Veh mice on D8 and D14 (during STND), and D23 (during HFD) (Figure 4A). As a response to changes in plasma glucose and lipids, adipokines are produced, many of which are regulated by E2 [13,43,44]. We investigated if the adipokines, leptin and resistin, are altered by E2 during STND or HFD feeding. Leptin was increased in Veh mice on STND as early as D8 ($p = 0.029$) and on D23 during

HFD ($p < 0.001$) (Figure 4B). Compared to E2-treated mice, plasma resistin levels increased in Veh mice during both STND and HFD (Figure 4C). E2 did not alter plasma levels of the pro-inflammatory cytokines, IL-6 and TNF-α (Figure S3A,B). Diet had no effect on plasma glucose and adipokines on the days examined (Figure 4 and Figure S3A,B). Plasma estradiol was measured on D23 of the implant to confirm its release into the circulation, which was significantly higher in E2 group compared to controls (Figure 4D). The intestinal hormones ghrelin and GLP-1 in plasma were undetectable.

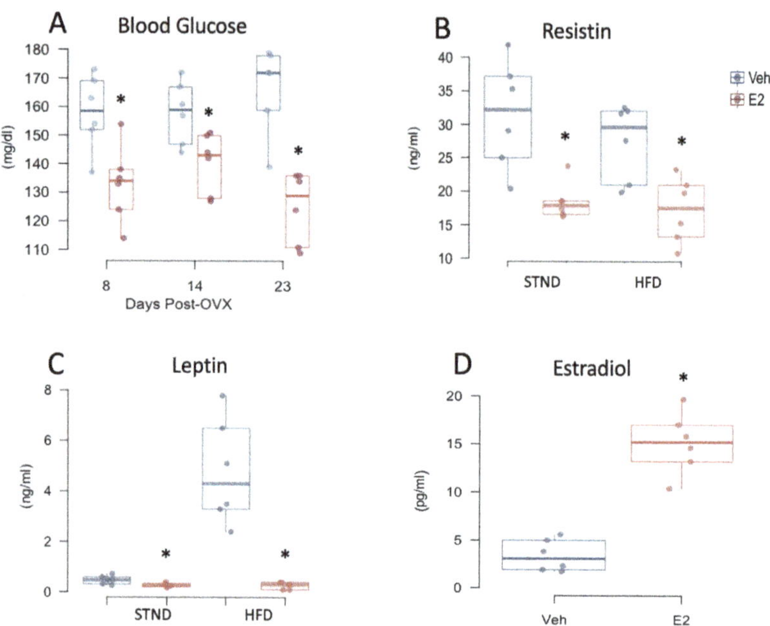

Figure 4. Estradiol decreases plasma glucose and adipokines in female mice independent of diet. 5 h-fasting blood glucose (**A**) on days 8 and 14, both during STND and on day 23, during HFD. Resistin (**B**) and leptin (**C**) were measured on D8 during STND and on D23 during HFD. Plasma estradiol levels were measured on D23 to confirm physiological levels in the treatment group (**D**). * indicates differences between E2 and Veh mice ($n = 6$/group) ($p < 0.05$, repeated measures ANOVA followed by *t*-test for A, B, and C, and *t*-test for D).

2.7. Estradiol Improves Insulin Sensitivity in Female Mice on HFD

We performed hyperinsulinemic-euglycemic clamp to measure insulin sensitivity and glucose metabolism in awake mice. E2 mice had an increased glucose infusion rate and increased whole-body glucose turnover compared to Veh controls during clamp (Figure 5A,B). Whole-body glycogen synthesis was increased in E2-treated mice compared to Veh mice (Figure 5C). Insulin-stimulated glucose uptake in skeletal muscle (gastrocnemius) did not differ between E2 and Veh mice, although a trend towards a decrease ($p = 0.08$) was observed in E2 mice, suggesting that skeletal muscle is not primarily responsible for the insulin-stimulated energy utilization in females as an effect of E2 (Figure 5D). E2 did not alter basal or clamp plasma glucose levels, although a trend towards a decrease in basal glucose ($p = 0.08$) was observed in E2 mice (Figure S4A). Consistently, a trend towards an increase in basal hepatic glucose production (HGP; $p = 0.076$) was observed in E2 groups, whereas clamp HGP was not affected (Figure S4B). Whole-body glycolysis, hepatic insulin action, or liver triglyceride levels were not affected by E2 (Figure S4C–E).

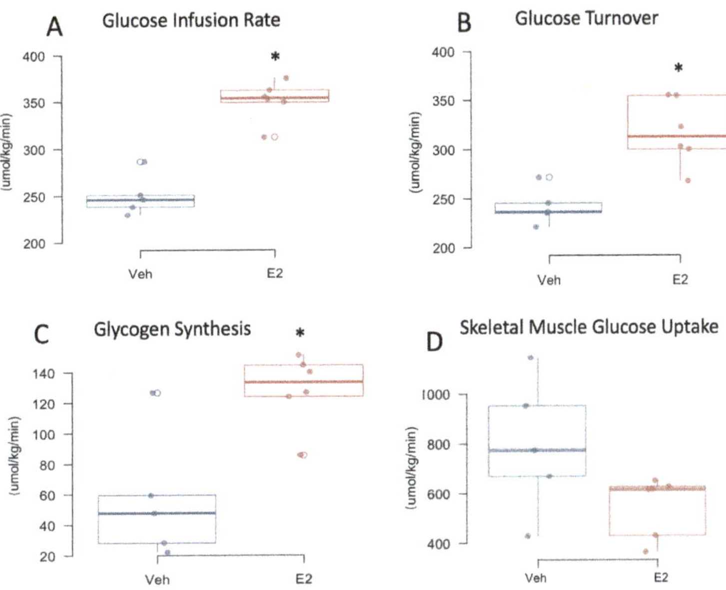

Figure 5. Estradiol increases insulin sensitivity and glucose utilization in female mice on HFD. Mice underwent hyperinsulinemic-euglycemic clamp on days 43–45, a week following jugular vein surgery. Glucose infusion rate (**A**) Glucose turnover (**B**) Glycogen synthesis (**C**) Skeletal muscle glucose uptake (**D**). * indicates differences between E2 ($n = 6$) and Veh ($n = 5$) mice * ($p < 0.05$, t-test).

2.8. Estradiol Decreases Occludin Expression in Colon in Female Mice Fed HFD

Tight junction proteins provide an indirect measure of intestinal epithelial integrity. Thus, to investigate the role of E2 on healthy epithelial barrier, the tight junction proteins occludin and ZO-1 were measured in female mice after 2 weeks on HFD. Interestingly, E2 treatment reduced the area and intensity of occludin immunoreactivity in the mid-colon compared to Veh mice (Figure 6B,C). There was no effect of E2 on occludin in the proximal and distal colon, or on ZO-1 expression throughout the colon (Figure S5A,B).

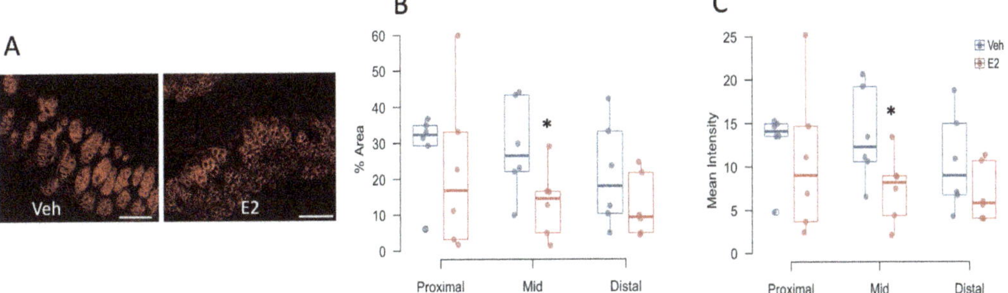

Figure 6. Estradiol decreases occludin immunoreactivity in colonic epithelium in female mice fed HFD. Colon tissues were collected on day 22 (15 days after start of HFD). Occludin immunolabeling in representative mid-colon sections from a Veh (**left**) and an E2-treated (**right**) mouse (**A**), Percent area (**B**) and mean intensity (**C**) of occludin immunoreactivity in the three subdivisions of the colon. Scale bar = 100 μm. * indicates differences between E2 and Veh mice ($n = 6$/group) ($p < 0.05$, t-test).

2.9. *Estradiol Alters Gut Microbial Diversity in Female Mice*

To identify the effects of E2 on gut microbial diversity during STND or HFD feeding, fresh fecal samples from D1 and D8 (during STND), and from D23 and D42 (during HFD), were analyzed. α-diversity, a measure of within-sample diversity as measured by richness and evenness of species within a population, was not significantly associated with E2 during STND or HFD (Figure S6A,B). β-diversity, a measure of dissimilarity between microbial communities, revealed a distinct clustering of the microbiota communities (Bray-Curtis distance) due to E2 during HFD on D23 ($p = 0.003$) and D42 ($p = 0.007$) (Figure 7B,C, respectively). There were no significant effects of E2 during STND on the aggregate data from D0 and D8 (Figure 7D). These data suggest a profound effect of E2 on gut microbiota diversity in mice fed HFD.

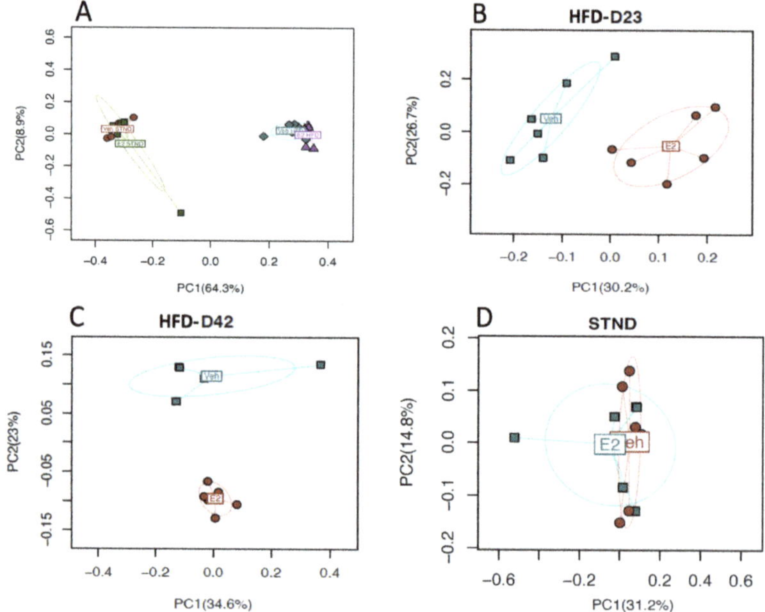

Figure 7. Estradiol and HFD alter gut microbiota β-diversity in female mice. Bray-Curtis distance between aggregate microbiota communities shows a distinct clustering between STND and HFD (**A**) and between E2 and Veh-treated animals during HFD, on Day 23 (**B**) and Day 42 (**C**), but not during STND (**D**). For A, (n = 12/group) and for B-D, (n = 6/group). $p < 0.05$ (PERMANOVA) considered significant.

2.10. *Estradiol Alters Relative Abundances of Gut Microbiota in Female Mice*

The generalized mixed effects models with FDR control was used to identify differentially abundant taxa (q-value < 0.05). A total of 14 taxa differed between E2 and Veh groups (Figure 8A,B). Of these 14 taxa, Verrucomicrobia (phylum) and all its lower taxa levels, including Verrucomicrobiae (order), Verrucomicrobiales (class), Verrucomicrobiaceae (family), and the genus *Akkermansia*, were increased in E2 mice, with the most pronounced differences during HFD feeding (Figure 8B). *Dorea* spp. were also increased in E2 mice compared to Veh mice (Figure 8A,B).

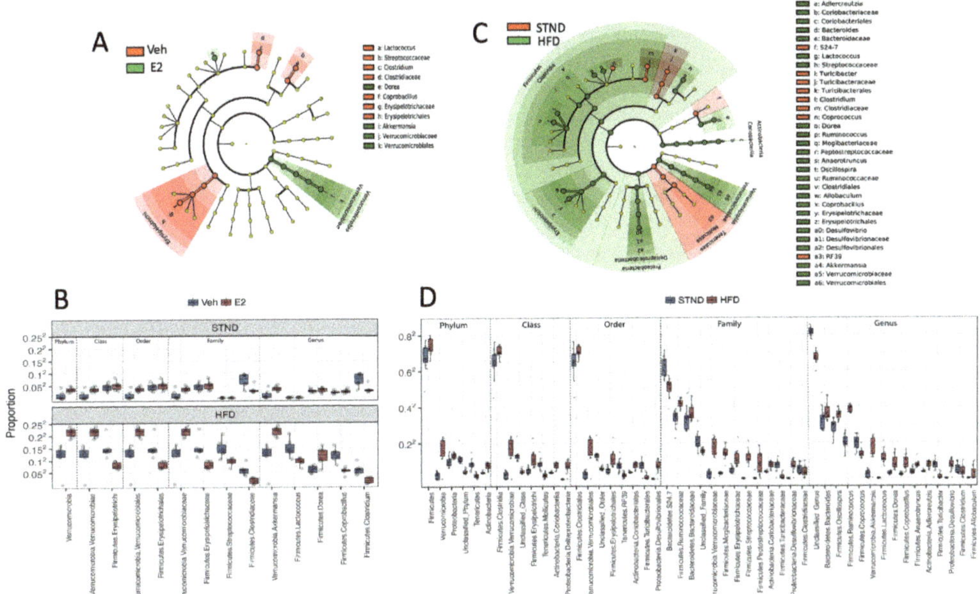

Figure 8. Estradiol and HFD alter gut microbiota taxa diversity in female mice. Cladogram (**A**) and relative proportions (**B**) of taxa associated with E2 or Veh, subgrouped as phylum, class, order, family and genus. Samples during STND (D0 and D8) and HFD feeding (D23 and D45) were analyzed. Cladogram (**C**) and relative proportions (**D**) of taxa associated with STND (D0 and D8) or HFD (D23 and D45). The innermost nodes in the cladogram represent phyla and the connecting outer nodes represent lower taxa within the phylum. q-value < 0.05 (generalized mixed effects model) considered significant. $n = 24$/group for (**A**,**C**), and $n = 12$/group for (**B**,**D**).

Other taxa, including Erysipelotrichi (order) and its genus *Coprobacillus,* and Streptococceae (family) and its genus *Lactococcus* were decreased in E2 mice compared to Veh controls during HFD feeding. In addition, the family Clostridiaceae and its genus *Clostridium,* were decreased in E2 mice both during STND and HFD feeding (Figure 8B).

2.11. HFD Alters Gut Microbiota Diversity and Relative Abundances in Female Mice

Similar to previous reports mostly in males [31,32,45–48], HFD profoundly affected gut microbiota composition in female mice. HFD decreased microbiota richness (Chao1; $p = 0.002$; Figure S5A) and increased evenness, the measure of homogeneity of species distribution in a population (Pielou's; $p < 0.001$; Figure S5B).

Diet also profoundly altered microbiota community structures. The microbial communities distinctly clustered between STND and HFD feeding ($p < 0.001$, PERMANOVA), as depicted by the PC1 (64.3%; Figure 7A). The effect of E2 was strong during HFD (Figure 7B), while no effects of E2 were detected during STND (Figure 7C). Moreover, a total of 49 taxa were differentially associated with STND vs HFD, of which 39 were positively associated with HFD, while only 10 were positively associated with STND, further supporting a profound effect of HFD on gut microbiota (Figure 8C,D).

HFD increased 39 taxa including Firmicutes and its lower taxa Clostridia (order), Clostridiales (class), the families Mogibacteriaceae and Peptostreptococcaceae, and the genera *Dorea, Ruminococcus, Anaerotruncus,* and *Oscillospira.* HFD increased additional taxa within the Firmicutes, including Erysipelotrichi (order) and its lower taxa Erysipelotrichales (class), Erysipelotrichaceae (family) and *Allobaculum* and *Coprobacillus.* Two other families, Bacteroidaceae and Streptococcaceae, and their genera *Bacteroides* and *Lactococcus,* respectively, were also positively associated with HFD. Furthermore, HFD increased Proteobacteria and its lower taxa, including Deltaproteobacteria (order), Desulfovibrionales (class), Desulfovibrionaceae (family), and *Desulfovibrio.* In addition, the phylum Verrucomicrobia and all of its lower taxa levels, including Verrucomicrobiae (order), Verrucomicrobiales (class), Verrucomicrobiaceae (family), and the genus *Akkermansia,* were increased during HFD feeding compared to STND. Actinobacteria (phylum) and its lower taxa at all levels, including Coriobacteriia (order), Coriobacteriales (class), Coriobacteriaceae (family) and its genus *Adlercreutzia,* were also increased as a result of HFD feeding.

Ten taxa were increased during STND compared to HFD, including Turibacteriales (class), its family Turibacteriaceae and the genus *Turibacter*, as reported previously [49]. The relative abundances of the family Clostridiaceae and its genus *Clostridium* and *Coprococcus* were also increased during STND compared to HFD. Similarly, Tenericutes (phylum), its lower taxa Mollicutes (order), and the genus *RF39* were increased during STND.

2.12. *Gut Microbiota Associates with Metabolic Status in Female Mice*

To investigate if metabolic changes are associated with changes in the gut microbiota community, correlation analysis was performed between the measures. PERMANOVA test based on Bray-Curtis distance followed by FDR correction (q-value < 0.1) revealed significant correlations of body weight (q = 0.015), plasma glucose (q = 0.025) and physical activity (q = 0.09) with microbial community distances. To identify specific taxa that are linked to E2-dependent metabolic effects, correlation analysis was done between the microbial taxa that were altered by E2 treatment and the major metabolic profiles (Figure 9). Verrucomicrobia, along with its lower taxa levels, including *Akkermansia,* negatively correlated with body weight, fat mass, and leptin, suggesting *Akkermansia* as a microbial mediator of E2-dependent protection against obesity. Verrucomicrobiae and *Dorea*, both increased in E2 mice, were negatively associated with blood glucose levels. In addition, *Dorea* was positively associated with physical activity. In contrast, some taxa that were increased in Veh mice, including Streptococcaceae and its genus *Lactococcus,* were positively associated with body weight, fat mass, and leptin, suggesting these taxa are predictors of obesity. Similarly, Erysipelotrichi, its family Erysipelotrichaceae and genus *Coprobacillus,* were positively associated with body weight. Interestingly, *Coprobacillus* was positively correlated with fat mass, but negatively correlated with physical activity and basal energy expenditure, suggesting a negative impact of this microbe on metabolic health in female mice.

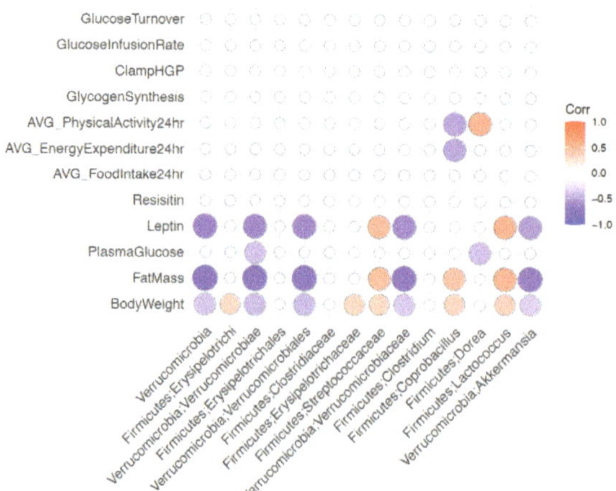

Figure 9. Metabolic changes correlate with multiple gut microbiota taxa in female mice. Spearman correlation tests were done on the taxa associated with E2 treatment. Red circles represent positive correlation whereas purple circles represent negative correlation, with increasing size of circles indicating stronger associations. q-value < 0.1 considered significant.

3. Discussion

In the current study, we investigated the comprehensive mechanisms by which estrogens protect females against diet-induced obesity and insulin resistance. E2 treatment prevented HFD-induced weight gain and adiposity in ovariectomized adult mice, consistent with earlier work from our group and others [11,13,14,42,44]. The E2-dependent protection against HFD-induced obesity was most strongly associated with increased physical activity and basal energy expenditure. E2 prevented hyperglycemia during both STND and HFD intake. Consistent with previous studies, E2 decreased the plasma adipokines leptin and resistin [13,44]. Furthermore, hyperinsulinemic-euglycemic clamp results showed that E2 improved systemic insulin sensitivity and glucose turnover in HFD-fed mice. However, skeletal muscle glucose uptake, hepatic glucose production, and hepatic triglycerides were not altered by E2 in HFD-fed mice. These data demonstrate tissue-specific effects of E2 in providing the protective mechanisms against HFD-induced obesity and insulin resistance.

Estrogen receptors (ER) exist in two forms, ERα and ERβ, which are transcribed from different genes [50,51]. These subtypes differ in their abilities to bind different ligands, are expressed differently in specific tissues and mediate different functions in behavior and physiology [51,52]. Intestinal epithelium predominantly expresses ERβ [53]. To identify any effects of estrogens in this key metabolic passageway, we analyzed changes in the intestinal epithelium in mice with or without E2. The tight junction protein, occludin, was decreased in the colon of E2-treated mice fed HFD, suggesting that HFD-induced increase in gut permeability, due to the depletion of tight junction proteins and mucus layer thickness [54,55], is modulated by E2. In future work, it will be important to study additional tight junction proteins combined with in vivo gut permeability assays to further explore the effects of E2 and diet on gut integrity and barrier function.

Host metabolic status can be predicted by its gut microbiota community and composition. Metabolic syndrome, characterized by adiposity, hyperlipidemia and hyperglycemia, is linked to dysbiosis of the gut microbial ecosystem [56–58]. However, we currently lack a full understanding of the parallel assessment of gut microbiota and metabolic changes within the same animals as an effect of E2, which limits the knowledge of any direct interactions between the host metabolic status and microbial factors. Thus, in the present study, we assessed if changes in gut microbiota are linked to the protective effects of estrogens against obesity, hyperglycemia, and insulin resistance. E2 altered microbial communities and taxa, with a profound effect during HFD feeding. Notably, the relative abundances of the phylum Verrucomicrobia, including its major constituent genus *Akkermansia, and Dorea* (phylum Firmicutes), were significantly increased by E2 during HFD. An increase in *Akkermansia* abundance is also associated with E2-mediated protection against western diet-induced obesity and metabolic syndrome in ovariectomized mice [41]. In the current study, we further identified an association between *Akkermansia* and multiple metabolic measures. *Akkermansia* negatively correlated with body weight and fat mass, suggesting it functions in the protective effects of E2 on metabolic health. Similarly, *Dorea* was positively associated with physical activity. In support of these findings, ovariectomy increases weight gain in both STND- and HFD-fed rats and is associated with changes in gut microbiota [59]. Similarly, in a PCOS mouse model, FMT from androgen-treated mice disrupts metabolic and endocrine health in germ-free recipients, whereas gut microbiota from control donors protects against metabolic dysregulation [60–62]. In a different study, diet-independent, ovariectomy-induced weight gain was not rescued by cohousing with intact mice, with the goal of transferring of gut microbiota [49]. It is possible that a more complete transfer of microbiota is needed to rescue this ovariectomy-induced weight gain, such as fecal microbiota transfer (FMT) by gavage or co-housing combined with FMT. Nevertheless, the present findings, taken together with previous ones, suggest that gut microbiota functions in metabolic dysregulation caused by diet or sex hormones.

The relative abundance of *Akkermansia,* the only intestinal resident genus of the phylum Verrucomicrobia, was significantly increased in E2-treated mice. *Akkermansia,* a mucin-degrader and a producer of short chain fatty acids [63–66], is decreased in obese humans, including obese pregnant women [65,67–69]. In postmenopausal women, *Akkermansia* is negatively correlated with insulin resistance and dyslipidemia [70]. Similarly, in ovariectomized mice fed a western diet, *Akkermansia* was increased following E2 treatment [41]. Administration of heat-killed *Akkermansia muciniphila* decreased body weight, fat mass, and hip circumference in obese women, highlighting its beneficial role in women's metabolic health [71]. In further support, *A. muciniphila* supplementation in male mice attenuated HFD-induced obesity and inflammation and improved insulin signaling [72–75]. In the present study, the relative abundance of *Dorea* was also increased by E2 treatment in HFD-fed females. Similar to *Akkermansia*, *Dorea* is a mucin degrader, suggesting these two microbes are co-altered in response to changes in diet or hormones, likely due to the similar nutrient environment and/or quorum sensing [76]. Taken together, these findings suggest that *Akkermansia* and *Dorea* contribute to the E2-mediated compensatory protection against HFD-induced metabolic changes.

Coprobacillus, *Lactococcus*, and *Clostridium*, including their families Erysipelotrichaceae, Streptococceae and Clostridiaceae, respectively, were increased in Veh mice. Among these microbes, *Coprobacillus* and *Lactococcus* were positively correlated with body weight and fat mass, suggesting they contribute to obesity in E2-deficient female mice. An inverse correlation of *Lactococcus* and *Coprobacillus* with E2 has been previously demonstrated in female mice on standard diet [77] and female *ob/ob* mice on HFD [42]. *Lactococcus* are efficient energy harvesters through the conversion of glucose to pyruvate [78]. *Coprobacillus* produce β-galactosidases, enzymes necessary for the breakdown of galactosides, such as lactose in food [79]. These and other gut microbes can also affect intestinal endocrine cells through metabolite production [80,81], which can impact the development of type 2 diabetes and obesity [82]. In future studies, it will be important to determine if selective depletion of these microbes mitigates the metabolic insult caused by the loss of estrogens in females.

Intake of a high-calorie diet during menopause, a period characterized by a slowed metabolic rate, further increases the risk of obesity and metabolic disorders in women [4,7,9]. In the current study, the protective effect of E2 treatment on metabolic status was profound during HFD intake in female mice, which is consistent with previous reports [11,13,14,32,42,57,83–85]. The increases in body weight and fat mass, and a decrease in basal energy expenditure, due to HFD feeding were attenuated by E2. E2 corrected HFD-induced positive energy balance primarily by increasing basal energy expenditure and locomotor activity, extending previous findings [14]. Moreover, E2 increased energy utilization in HFD-mice by increasing systemic insulin sensitivity and whole-body glucose turnover. Since these effects were not associated with increased muscle glucose metabolism, other estrogen-sensitive organs might be responsible for increased glucose utilization in E2-treated mice. E2-mediated improvements in some measures of insulin sensitivity have also been demonstrated previously [14,44]. Given that mice lacking ERα are insulin resistant [86], these effects of E2 on metabolic pathways discussed above are most likely mediated by ERα.

The present and previous studies have found that levels of the adipokines, leptin and resistin, were decreased by E2 in female mice [13,44]. The present study reveals that this decrease in leptin levels was associated with gut microbiota. In particular, leptin was negatively associated with *Akkermansia*, a positive microbial predictor of metabolic health, whereas was positively associated with *Lactococcus* [42]. Leptin decreases food intake and increases energy expenditure [87,88]. However, increased circulating leptin is positively linked to metabolic syndrome in women [89,90]. In addition, resistin deficiency is associated with increased insulin sensitivity, particularly through a reduction in hepatic glucose production [91,92]. Therefore, the early changes in adipokines observed in the present and previous studies [13,44] may serve as early markers of diet-induced obesity and insulin resistance, as well as measures of the E2 response against various metabolic insults. Moreover, the E2-dependent downregulation of leptin and its interaction with gut microbiota may provide an essential braking mechanism against the development of diet-induced obesity.

4. Materials and Methods

The following animal experiments were performed at the University of Massachusetts Medical School (PROTO202000104, 11/01/20). All procedures were approved by the Institutional Animal Care and Use Committees of UMass Medical School and Wellesley College and performed in accordance with National Institutes of Health Animal Care and Use Guidelines.

4.1. Diet and Ovariectomy

Female C57BL/6J mice were purchased from The Jackson Laboratory and housed in the animal facility at UMass Medical School. Ten-week-old female C57BL/6J mice were bilaterally ovariectomized (OVX) and silastic capsules filled with 17β-estradiol (E2, 50 μg in 25 μL of 5% ETOH/sesame oil, n = 6) or vehicle (Veh, n = 6) were implanted subcutaneously [11,42]. Mice were singly housed and fed a chow diet (STND; 13.5% calories from fat, #5001, Purina, LabDiet, Fort Worth, TX, USA) for the first 14 days after OVX. To test the effects of E2 on metabolism and gut microbiota under HFD, mice were put on a HFD containing 60% kcal fat (#D12492, Research Diets, New Brunswick, NJ, USA) for the remainder of the study (days 14–45; Figure 1A).

4.2. In Vivo Assessment of Energy Balance Using Metabolic Cages

We performed a 3-day measurement of energy balance (i.e., food intake, VO_2 consumption and VCO_2 production, energy expenditure, respiratory exchange ratios, and physical activity) using metabolic Cages (TSE Systems, Germany) in mice (n = 6 per treatment group) on D11-13 during STND and D29-31 during HFD as described previously [93,94]. The O_2 consumption and CO_2 production were used to calculate the respiratory exchange ratio (RER). The horizontal and vertical movement (XYZ-axis) were measured in the cages as an index of locomotor activity. Body composition (fat/lean mass) was assessed by proton magnetic resonance spectroscopy (1-H MRS; EchoMRI, Houston, TX, USA) once each week (Figure 1A).

4.3. Measurement of Glucose Metabolism Using Hyperinsulinemic-Euglycemic Clamp

On days 37–39, anesthetized mice underwent a survival surgery to establish an in-dwelling catheter in the jugular vein. One week after the surgery, following overnight fasting, a 2-h hyperinsulinemic-euglycemic clamp was conducted in awake mice with a primed (150 mU/kg body weight) and continuous infusion of human insulin at a rate of 2.5 mU/kg/min to raise plasma insulin within a physiological range [95]. D-[3-3H] glucose was intravenously infused using microdialysis pumps during the experiments to assess the whole-body glucose turnover [96]. Blood samples were collected at 10–20 min intervals for the immediate measurement of plasma glucose, and 20% glucose was infused at variable rates to maintain euglycemia. To estimate insulin-stimulated glucose uptake in individual organs, 2-[1-14C] deoxy-D-glucose (2-[14C] DG) was administered as a bolus (10 μCi) at 75 min after the start of clamp. Blood samples were taken for the measurement of plasma [3H] glucose, 3H2O, and 2-[14C] DG concentrations. At the end of the clamp, mice were anesthetized and tissue samples were taken for biochemical and molecular analyses.

4.4. Calculation of In Vivo Glucose Metabolism

Basal whole-body glucose turnover was determined as the ratio of the [3H] glucose infusion rate to the specific activity of plasma glucose at the end of basal period, as previously described [96]. Insulin-stimulated whole-body glucose uptake was determined as the ratio of the [3H] glucose infusion rate to the specific activity of plasma glucose during the final 30 min of clamps. Hepatic glucose production during insulin-stimulated state (clamp) was determined by subtracting the glucose infusion rate from the whole-body glucose uptake. Whole-body glycolysis was calculated from the rate of increase in plasma 3H2O concentration from 90–120 min of clamp. Whole-body glycogen plus lipid synthesis was estimated by subtracting whole-body glycolysis from whole-body glucose uptake. Since 2-DG is a non-metabolizable glucose analog, insulin-stimulated glucose uptake in skeletal muscle were estimated by determining muscle-specific content of 2-[14C] DG-6-P. Skeletal muscle glucose were calculated from plasma 2-[14C] DG decay profile and intracellular 2-[14C] DG-6-P content.

4.5. Biochemical Assays

Blood samples were collected after 5-h fasting on day (D) D8 (during STND) and on D23 (9 days on the start of HFD) by tail vein puncture (Analytic Core, MMPC). Plasma E2 on D23 was measured using Mouse/Rat Estradiol ELISA kit (#ES180S, Calbiotech [12,97]. The cytokines Il-6 and TNF-α, the adipokines leptin and resistin, and intestinal hormones, ghrelin and GLP-1 were measured using an ELISA with a Luminex 200 Multiplex system (Millipore, Darmstadt, Germany).

Glucose concentrations during clamps were analyzed using clinical glucose analyzer, and insulin levels were measured using an ELISA kit. Plasma [3H] glucose, 2-[14C] DG, and $3H_2O$ concentrations were determined following deproteinization of samples using liquid scintillation counter on dual channels for separation of 3H and 14C. The radioactivity of 3H in tissue glycogen was determined by precipitating glycogen with ethanol. Organ-specific 2-[14C] DG-6-phosphate concentrations were determined using ion-exchange column as previously described [98]. Hepatic intracellular triglyceride level was measured using spectrophotometry using triglyceride assay kit after digesting tissue samples in chloroform-methanol.

The following animal experiments were performed at Wellesley College, and all procedures were approved by the Institutional Animal Care and Use Committees of Wellesley College (#2101, 02/05/21) and performed in accordance with National Institutes of Health Animal Care and Use Guidelines.

4.6. Fecal DNA Extraction and Sequencing

DNA was extracted from fresh frozen fecal samples on D0 and D8, during STND and D23 and D42, during HFD, using MO BIO PowerSoil DNA Isolation Kit (Valencia, CA) with minor adjustments to the manufacturer's protocol, as described previously [42]. The DNA quality and quantity were assessed using Nanodrop spectrophotometer (Thermo Scientific, Waltham, MA, USA). 16S rDNA was amplified at the V3-V4 region using the universal 16S rDNA primers: for- ward 341F (5′-CCTACGGGAGGCAGCAG-3′) and reverse 806R (5′-GGACTACHVGGGTWTCTAAT-3′) with sequence adapters on both primers and sample-specific Golay barcodes on the reverse primer [99]. The amplicons were quantified by PicoGreen (Invitrogen, Carlsbad, CA, USA) and pooled in equal concentrations. The pooled amplicons were cleaned using UltraClean PCR Clean-Up Kit (MO BIO, Carlsbad, CA, USA) followed by quantification using the Qubit (Invitrogen, Carlsbad, CA, USA).

Samples were multiplexed and paired-end sequenced using 16S rDNA primers on an Illumina MiSeq (Illumina, San Diego, CA, USA) at the Microbiome Core (Mayo Clinic, Rochester, Minnesota). Paired R1 and R2 sequence reads were processed via the hybrid-denovo bioinformatics pipeline, which clustered a mixture of good-quality paired-end and single-end reads into operational taxonomic units (OTUs) at 97% similarity level [100]. OTUs were assigned taxonomy using the RDP classifier trained on the GreenGenes database (v13.5) [101,102]. A phylogenetic tree based on FastTree algorithm was constructed based on the OTU representative sequences [103]. The total number of reads ranged from 44,869 to 697,323 with a median of 122,445 reads per sample.

4.7. Intestinal Tissue Processing for Histology

Mice used for the intestinal histology analysis were housed in the animal facility at Wellesley College. Mice were ovariectomized and implanted with E2 (n = 6) or Veh (n = 6), as described above. Animals were fed STND for 7 days and switched to HFD containing 60% kcal fat (#D12492, Research Diets) on day 8 (D8). On D22 of OVX (after 2 wks on HFD), mice were euthanized and colons were collected.

Mice treated with E2 or Veh (n = 6/group) were euthanized after 2 wks on HFD to investigate the effects of E2 on intestinal epithelium. Colon was prepared as previously described, with some modifications [104]. In brief, colon tissue was longitudinally cut open and vigorously washed 3 times in 1X PBS (pH 7.2). The colon tissue was then dipped in modified Bouin fixative (50% EtOH in 5% Acetic acid in 1X PBS) for 5 min. The tissue was rolled on a toothpick, transferred to a tissue cassette, and fixed overnight in 10% formalin at room temperature. Prior to paraffin embedding, tissue was processed using a tissue processor (CITADEL 2000, Thermo Fisher) at DERC Morphology Core, UMass Medical School. Briefly, the tissue was incubated in 70% EtOH for 1x, 95% EtOH for 1x, and 100% EtOH and 3x, for 1 h each on an orbital shaker. The tissue was then incubated in xylene (#X5SK, Thermo Fisher) 3x for 1 h each and kept in a tissue mold and incubated in paraffin (Histoplast IM, Cat# 8331, Thermo Fisher, Waltham, MA, USA) at 58 °C 2x for 2 h. The tissue roll was sectioned at 5 µm thickness on a cryostat.

4.8. Triple-Label Immunohistochemistry for Tight Junction Proteins

Triple-label immunohistochemistry was done on colon sections to quantify tight junction proteins, occludin and zona occludens 1 (ZO-1), and a nucleic acid stain (DAPI) in colonic epithelium including crypts as described previously [105]. In brief, the colon sections were deparaffinized in xylene and dehydrated in 100% ethanol, followed by rehydrating in graded ethanol concentrations of 95%, 70% and 50%. Antigen retrieval was done to increase the antigen accessibility by incubating the slides for 30 min in Tris-EDTA (pH 9.0) in boiling water. Slides were allowed to cool and washed in 0.5% sodium borohydride (*w/v*) in TBS for 20 min to remove excess fixative. Following additional washes in TBS, the sections were incubated in blocking buffer (10% normal donkey serum, 0.3% Triton, 1% BSA) for 30 min. The sections were then incubated at 4 °C overnight in rabbit polyclonal antibody directed against human occludin (1:100, #ab168986, Abcam) [106] and goat polyclonal antibody directed against C-terminal of human ZO-1 (1:100; #ab190085, Abcam, Cambridge, MA, USA) [107]. The specificities of occludin and ZO-1 on mouse intestinal tissue have been verified previously [74,108]. The following day, sections were washed then incubated for 1 h in dark at room temperature with a nucleic acid stain, DAPI, at a concentration of 30 uM and fluorescently labeled donkey-anti-rabbit (1:100; Alexa Fluor 647, Invitrogen) [109] and donkey-anti-goat (1:100; Alexa Fluor 488, Invitrogen, Waltham, MA, USA) [110] secondary antibodies for the detection of occludin and ZO-1, respectively. Slides were washed, coverslipped with Fluorogel (Electron Microscopy Sciences, Hatfield, PA, USA), stored overnight in dark, and imaged within two days.

4.9. Imaging by Confocal Microscopy and Analysis

The proximal, middle and distal regions of the colon were imaged using a Leica laser scanning confocal microscope (TCS SP5 II), equipped with 405 Diode, Argon, HeNe 594, and HeNe 633 lasers and with Leica software (LAS version 2.7.3.9) [111]. All images were taken under 200x magnification with the PL APO dry-objective (numerical aperture, 0.7). The gain and offset values for each laser were optimized for each channel and kept constant for all animals. Sections of 1 µm thickness were optically imaged and analyzed using the NIH ImageJ software (version 1.52) [112]. A representative section per animal per subregion was analyzed using uniform regions of interest (ROI), which were kept constant for each subregion across all animals. For the quantification of immunolabeling, threshold for each laser channel was set separately based on a scale of 0–255, to minimize the background. Using three random images per treatment, the threshold that displayed the immunolabeling signal closest to the unprocessed original image was chosen and applied across all animals. Any value below the threshold was considered to be background. The % area with labeling above threshold and the mean pixel intensity were collected within the selected ROI for each subregion.

4.10. Statistical Analysis

4.10.1. Metabolic Data

The effects of diet and E2 treatment were analyzed on longitudinal data using mixed model repeated measures ("lme" in R "nlme" package) [113] using mouse as a random subject. Once the main effects were observed, the separate effects during STND and HFD were measured using repeated measures ANOVA (spss v.24) [114]. Data from metabolic cages, including food and water intake, locomotor activity, and respiration (O_2 consumption and CO_2 production) and its derivative measures, respiratory exchange ratio and resting energy expenditure, were recorded over 72 h and averaged to produce 24-h data, and analyzed using the *t*-test. Fasting blood glucose, plasma hormones and cytokines, and end point measures (including clamp data) were analyzed using the *t*-test. $p < 0.05$ was considered statistically significant.

4.10.2. 16S rRNA Sequence Data

The data were rarefied to the minimum depth of 44,869 prior to the α-diversity and β-diversity analyses [115]. For α-diversity analysis (Chao1 richness and Pielou's evenness indices), linear mixed effects models were fitted to the alpha diversity measures with a random intercept for each mouse ("lme" in R "nlme" package). Wald test was used for assessing the significance. For β-diversity analysis (Bray-Curtis distance), PERMANOVA test (R adonis, 1000 permutations) was used to test whether overall microbiota composition is associated with E2 or diet. For testing the E2 effects, the mouse (not individual sample) was the permutation unit; for testing the diet effects, the mouse was the permutation stratum (i.e., permutation only occurred within the same mice) [116]. The R^2 was given as the effect size.

Differential abundance analysis of treatment (E2 vs Veh) and diet (HFD vs STND) effect was performed on the phylum, class, order, family and genus level. Only taxa with prevalence >10% and maximum proportion >0.2% were tested. Generalized linear mixed effects model (R "glmmPQL" function, over-dispersed Poisson regression, random intercept) was fitted to the aggregated counts accounting for within-mouse correlation [117]. The library size was estimated using GMPR method [118]. The log library size was included as an offset in the regression model. Treatment and diet variables were included as covariates. Potential treatment and diet interaction (GxD) was also investigated by including the interaction term in the regression model. Wald test was used to test the significance of the association. Data were winsorized at 95% quantile (i.e., we replace outlier counts with 95% quantile) to reduce the influence of potential outliers. False discovery rate (FDR) control (BH procedure, R p.adjust function) was used for multiple testing correction and performed on each taxonomic level from phylum down to genus. The taxa with an FDR-adjusted *p* value (or q value) < 0.05 were considered as statistically significant.

4.10.3. Correlation Analysis of Microbiome and Metabolic Data

To identify if any changes in E2-dependent metabolic effects significantly associate with changes in gut microbiota, correlation analyses were performed between the two outcomes. PERMANOVA was used to perform an overall association test based on the Bray-Curtis distance. For metabolic measures where multiple samples within the same mouse were obtained, within-mouse permutations were done. Next, correlation tests were done to identify microbial taxa associated with metabolic measures. To control for the potential confounding effects due to diet and E2 treatment, residuals were taken by fitting regression models (linear mixed effects model) to the microbial taxa abundance (square-root transformed) and metabolic measures adjusting for diet and E2 effects. Spearman correlation tests were then performed on the residuals. To reduce multiple testing burden, correlation analyses were focused on the taxa associated with E2 treatment. The associations with an FDR-adjusted *p* value (or q value) < 0.1 were considered as statistically significant.

5. Conclusions

The present study provides compelling evidence that estrogens profoundly impact energy and metabolic homeostasis in female mice. Consistent with previous studies, the key metabolic changes, including food intake, energy expenditure, and glucose turnover, were improved by E2 in females fed HFD. Moreover, the present findings reveal that gut microbiota and gut barrier integrity are additional targets of E2-mediated protection against diet-induced metabolic disorders. Furthermore, the role of gut microbiota in metabolic health is supported by the present findings of strong correlations of multiple microbial taxa with specific metabolic measures and physical activity. In future studies, it will be important to perform shotgun metagenomics for the functional study of the gut microbiome and explore the potential beneficial effects of *Akkermansia* and other microbes identified in this study and their causative links with metabolic protection in females provided by E2. In addition, identification and characterization of microbial metabolites that contribute to the beneficial effects of E2 on metabolism will provide important insights for targeting gut microbiota to improve women's metabolic health.

Supplementary Materials: The following are available online at https://www.mdpi.com/article/10.3390/metabo11080499/s1. Figure S1. E2 alters water intake in female mice during STND (A) and HFD (B). Mice were kept in metabolic cages from days 11–13 after ovariectomy and E2 implant and the average 24-h data were obtained from 72-h data and used for statistical analysis. * indicates differences between E2 and Veh mice (n = 6/group) ($p < 0.05$, t-test), Figure S2. Estradiol decreases the respiratory exchange ratio (RER) in female mice during the day. Respiratory exchange ratios of mice on STND (A) or HFD (B) were measured in metabolic cages for 72 h. The average 24-h data were obtained from 72-h data and used for statistical analysis. * indicates differences between E2 and Veh mice (n = 6/group) ($p < 0.05$, t-test), Figure S3: Estradiol or high fat diet did not alter levels of the plasma cytokines, IL-6 and TNF-α in female mice. 5 h-fasting blood samples were used to measure IL-6 (A) and TNF-α (B) during STND (on D8) and HFD (on D23); n = 6/group, Figure S4. Estradiol does not alter hepatic insulin sensitivity and lipid production in female mice on HFD. Mice E2 (n = 6) and Veh (n = 5) underwent hyperinsulinemic-euglycemic clamp on days 43–45, a week following jugular vein surgery. Blood glucose (A) Whole-body glycolysis (B) Hepatic glucose production (C) Hepatic insulin action (D) Liver triglycerides (E), Figure S5. Estradiol does not alter zona occludens (ZO-1) immunoreactivity in the colonic epithelium in female mice fed a HFD. Percent area (A) and mean intensity (B) in the three subdivisions of the colon (n = 6/group), Figure S6. HFD alters gut microbiota α-diversity in female mice. HFD lowers richness (A) and increases evenness (B), (n = 12/group). * indicates a difference between STND vs. HFD ($p < 0.05$; "lme" in regression).

Author Contributions: Each author has made substantial contributions to the work: Conceptualization; formal analysis; writing—original draft; supervision, review and editing: K.D.A., J.K.K. and M.J.T. Methodology, review and editing: K.D.A., S.S., H.L.N. and R.H.F. Formal Analysis: K.D.A., C.C.G., A.E.R.P., and J.C. Project administration and funding acquisition: K.D.A., M.E.G., J.K.K. and M.J.T. All authors have read and agreed to the published version of the manuscript.

Funding: This work was funded by NIH 5U24DK076169-13 Subaward # 30835-64 (K.D.A.), NIH 5U2C-DK093000 (J.K.K.), and NIH R01 DK61935 and Wellesley College Jenkins Distinguished Chair in Neuroscience Funds (M.J.T.). Part of this study was performed at the National Mouse Metabolic Phenotyping Center (MMPC) at UMass Medical School.

Institutional Review Board Statement: The study was conducted in accordance with National Institutes of Health Animal Care and Use Guidelines, and approved by the Institutional Animal Care and Use Committees of University of Massachusetts Medical School (PROTO202000104, 11/01/20) and Wellesley College (#2101, 02/05/21).

Informed Consent Statement: Not applicable.

Data Availability Statement: Datasets generated during and/or analyzed during the current study are not publicly available but can be made available upon reasonable request.

Conflicts of Interest: The authors declare no conflict of interest.

References

1. Haslam, D.W.; James, W.P.T. Obesity. *Lancet* **2005**, *366*, 1197–1209. [CrossRef]
2. Richardson, S.; Hirsch, J.S.; Narasimhan, M.; Crawford, J.M.; McGinn, T.; Davidson, K.W.; Barnaby, D.P.; Becker, L.B.; Chelico, J.D.; Cohen, S.L.; et al. Presenting Characteristics, Comorbidities, and Outcomes Among 5700 Patients Hospitalized With COVID-19 in the New York City Area. *JAMA* **2020**, *323*, 2052–2059. [CrossRef]
3. Fryar, C.D.; Carroll, M.D.; Ogden, C.L. Prevalence of Overweight, Obesity, and Extreme Obesity Among Adults Aged 20 and Over: United States, 1960–1962 through 2017–2018; NCHS Health E-Stats: 2020. Available online: https://www.cdc.gov/nchs/data/hestat/obesity-adult-17-18/obesity-adult.htm (accessed on 30 July 2021).
4. Miranda, P.J.; DeFronzo, R.A.; Califf, R.M.; Guyton, J.R. Metabolic syndrome: Definition, pathophysiology, and mechanisms. *Am. Heart J.* **2005**, *149*, 33–45. [CrossRef] [PubMed]
5. Steinbaum, S.R. The metabolic syndrome: An emerging health epidemic in women. *Prog. Cardiovasc. Dis.* **2004**, *46*, 321–336. [CrossRef]
6. Carr, M.C. The Emergence of the Metabolic Syndrome with Menopause. *J. Clin. Endocrinol. Metab.* **2003**, *88*, 2404–2411. [CrossRef]
7. Schmiegelow, M.D.; Hedlin, H.; Mackey, R.H.; Martin, L.W.; Vitolins, M.Z.; Stefanick, M.L.; Perez, M.V.; Allison, M.; Hlatky, M.A. Race and Ethnicity, Obesity, Metabolic Health, and Risk of Cardiovascular Disease in Postmenopausal Women. *J. Am. Heart Assoc.* **2015**, *4*, e001695. [CrossRef]
8. Gurney, E.P.; Nachtigall, M.J.; Nachtigall, L.E.; Naftolin, F. The Women's Health Initiative trial and related studies: 10 years later: A clinician's view. *J. Steroid Biochem. Mol. Biol.* **2014**, *142*, 4–11. [CrossRef]
9. Hedlin, H.; Weitlauf, J.; Crandall, C.J.; Nassir, R.; Cauley, J.A.; Garcia, L.; Brunner, R.; Robinson, J.; Stefanick, M.L.; Robbins, J. Development of a comprehensive health-risk prediction tool for postmenopausal women. *Menopause* **2019**, *26*, 1385–1394. [CrossRef]
10. Santoro, N.; Sutton-Tyrrell, K. The SWAN Song: Study of Women's Health Across the Nation's Recurring Themes. *Obstet. Gynecol. Clin. N. Am.* **2011**, *38*, 417–423. [CrossRef]
11. Bless, E.P.; Reddy, T.; Acharya, K.D.; Beltz, B.S.; Tetel, M.J. Oestradiol and Diet Modulate Energy Homeostasis and Hypothalamic Neurogenesis in the Adult Female Mouse. *J. Neuroendocr.* **2014**, *26*, 805–816. [CrossRef]
12. Mamounis, K.J.; Hernandez, M.R.; Margolies, N.; Yasrebi, A.; Roepke, T.A. Interaction of 17beta-estradiol and dietary fatty acids on energy and glucose homeostasis in female mice. *Nutr. Neurosci.* **2017**, *21*, 715–728. [CrossRef]
13. Bryzgalova, G.; Lundholm, L.; Portwood, N.; Gustafsson, J.; Khan, A.; Efendic, S.; Dahlman-Wright, K. Mechanisms of antidiabetogenic and body weight-lowering effects of estrogen in high-fat diet-fed mice. *Am. J. Physiol. Metab.* **2008**, *295*, E904–E912. [CrossRef]
14. Camporez, J.P.G.; Jornayvaz, F.; Lee, H.-Y.; Kanda, S.; Guigni, B.; Kahn, M.; Samuel, V.T.; Carvalho, C.; Petersen, K.F.; Jurczak, M.; et al. Cellular Mechanism by Which Estradiol Protects Female Ovariectomized Mice From High-Fat Diet-Induced Hepatic and Muscle Insulin Resistance. *Endocrinology* **2013**, *154*, 1021–1028. [CrossRef] [PubMed]
15. Brown, L.M.; Clegg, D.J. Central effects of estradiol in the regulation of food intake, body weight, and adiposity. *J. Steroid Biochem. Mol. Biol.* **2010**, *122*, 65–73. [CrossRef] [PubMed]
16. Geary, N.; Asarian, L.; Korach, K.S.; Pfaff, D.W.; Ogawa, S. Deficits in E2-dependent control of feeding, weight gain, and cholecystokinin satiation in ER-alpha null mice. *Endocrinology* **2001**, *142*, 4751–4757. [CrossRef]
17. Heine, P.A.; Taylor, J.A.; Iwamoto, G.A.; Lubahn, D.B.; Cooke, P.S. Increased adipose tissue in male and female estrogen receptor-alpha knockout mice. *Proc. Natl. Acad. Sci. USA* **2000**, *97*, 12729–12734. [CrossRef]
18. Van Sinderen, M.L.; Steinberg, G.R.; Jorgensen, S.B.; Honeyman, J.; Chow, J.D.; Herridge, K.A.; Winship, A.; Dimitriadis, E.; Jones, M.; Simpson, E.R.; et al. Effects of Estrogens on Adipokines and Glucose Homeostasis in Female Aromatase Knockout Mice. *PLoS ONE* **2015**, *10*, e0136143. [CrossRef] [PubMed]
19. Tetel, M.J.; De Vries, G.J.; Melcangi, R.C.; Panzica, G.; O'Mahony, S.M. Steroids, stress and the gut microbiome-brain axis. *J. Neuroendocr.* **2017**, *30*, e12548. [CrossRef]
20. Le Chatelier, E.; Nielsen, T.; Qin, J.; Prifti, E.; Hildebrand, F.; Falony, G.; Almeida, M.; Arumugam, M.; Batto, J.-M.; Kennedy, S.; et al. Richness of human gut microbiome correlates with metabolic markers. *Nat. Cell Biol.* **2013**, *500*, 541–546. [CrossRef]
21. Hildebrandt, M.A.; Hoffmann, C.; Sherrill–Mix, S.A.; Keilbaugh, S.A.; Hamady, M.; Chen, Y.-Y.; Knight, R.; Ahima, R.S.; Bushman, F.; Wu, G.D. High-Fat Diet Determines the Composition of the Murine Gut Microbiome Independently of Obesity. *Gastroenterology* **2009**, *137*, 1716–1724.e2. [CrossRef]
22. Vrieze, A.; Van Nood, E.; Holleman, F.; Salojärvi, J.; Kootte, R.S.; Bartelsman, J.F.; Dallinga-Thie, G.; Ackermans, M.T.; Serlie, M.J.; Oozeer, R.; et al. Transfer of Intestinal Microbiota From Lean Donors Increases Insulin Sensitivity in Individuals With Metabolic Syndrome. *Gastroenterology* **2012**, *143*, 913–916.e7. [CrossRef]
23. Wostmann, B.S.; Larkin, C.; Moriarty, A.; Bruckner-Kardoss, E. Dietary intake, energy metabolism, and excretory losses of adult male germfree Wistar rats. *Lab. Anim. Sci.* **1983**, *33*, 46–50.
24. Bäckhed, F.; Ding, H.; Wang, T.; Hooper, L.V.; Koh, G.Y.; Andras, N.; Semenkovich, C.F.; Gordon, J.I. The gut microbiota as an environmental factor that regulates fat storage. *Proc. Natl. Acad. Sci. USA* **2004**, *101*, 15718–15723. [CrossRef]
25. Jumpertz, R.; Le, D.S.; Turnbaugh, P.J.; Trinidad, C.; Bogardus, C.; Gordon, J.I.; Krakoff, J. Energy-balance studies reveal associations between gut microbes, caloric load, and nutrient absorption in humans. *Am. J. Clin. Nutr.* **2011**, *94*, 58–65. [CrossRef]

26. Rezzi, S.; Ramadan, Z.; Martin, F.-P.J.; Fay, L.B.; Van Bladeren, P.; Lindon, J.; Nicholson, J.K.; Kochhar, S. Human Metabolic Phenotypes Link Directly to Specific Dietary Preferences in Healthy Individuals. *J. Proteome Res.* **2007**, *6*, 4469–4477. [CrossRef]
27. Walker, A.; Ince, J.; Duncan, S.H.; Webster, L.M.; Holtrop, G.; Ze, X.; Brown, D.; Stares, M.D.; Scott, P.; Bergerat, A.; et al. Dominant and diet-responsive groups of bacteria within the human colonic microbiota. *ISME J.* **2010**, *5*, 220–230. [CrossRef]
28. Ley, R.; Peterson, D.A.; Gordon, J.I. Ecological and Evolutionary Forces Shaping Microbial Diversity in the Human Intestine. *Cell* **2006**, *124*, 837–848. [CrossRef]
29. Turnbaugh, P.; Hamady, M.; Yatsunenko, T.; Cantarel, B.L.; Duncan, A.; Ley, R.; Sogin, M.L.; Jones, W.J.; Roe, B.A.; Affourtit, J.P.; et al. A core gut microbiome in obese and lean twins. *Nat. Cell Biol.* **2008**, *457*, 480–484. [CrossRef]
30. Devkota, S.; Wang, Y.; Musch, M.W.; Leone, V.; Fehlner-Peach, H.; Nadimpalli, A.; Antonopoulos, D.A.; Jabri, B.; Chang, E.B. Dietary-fat-induced taurocholic acid promotes pathobiont expansion and colitis in Il10$^{-/-}$ mice. *Nat. Cell Biol.* **2012**, *487*, 104–108. [CrossRef]
31. Cani, P.D.; Delzenne, N.M.; Amar, J.; Burcelin, R. Role of gut microflora in the development of obesity and insulin resistance following high-fat diet feeding. *Pathol. Biol.* **2008**, *56*, 305–309. [CrossRef]
32. Kim, K.-A.; Gu, W.; Lee, I.-A.; Joh, E.-H.; Kim, D.-H. High Fat Diet-Induced Gut Microbiota Exacerbates Inflammation and Obesity in Mice via the TLR4 Signaling Pathway. *PLoS ONE* **2012**, *7*, e47713. [CrossRef]
33. Qin, Y.; Roberts, J.D.; Grimm, S.A.; Lih, F.; Deterding, L.J.; Li, R.; Chrysovergis, K.; Wade, P.A. An obesity-associated gut microbiome reprograms the intestinal epigenome and leads to altered colonic gene expression. *Genome Biol.* **2018**, *19*, 1–14. [CrossRef]
34. Karl, J.P.; Meydani, M.; Barnett, J.B.; Vanegas, S.M.; Goldin, B.; Kane, A.; Rasmussen, H.; Saltzman, E.; Vangay, P.; Knights, D.; et al. Substituting whole grains for refined grains in a 6-wk randomized trial favorably affects energy-balance metrics in healthy men and postmenopausal women. *Am. J. Clin. Nutr.* **2017**, *105*, 589–599. [CrossRef] [PubMed]
35. Goedert, J.J.; Jones, G.; Hua, X.; Xu, X.; Yu, G.; Flores, R.; Falk, R.T.; Gail, M.H.; Shi, J.; Ravel, J.; et al. Investigation of the Association Between the Fecal Microbiota and Breast Cancer in Postmenopausal Women: A Population-Based Case-Control Pilot Study. *J. Natl. Cancer Inst.* **2015**, *107*, djv147. [CrossRef] [PubMed]
36. Guadamuro, L.; Delgado, S.; Redruello, B.; Flórez, A.B.; Suárez, A.; Camblor, P.M.; Mayo, B. Equol status and changes in fecal microbiota in menopausal women receiving long-term treatment for menopause symptoms with a soy-isoflavone concentrate. *Front. Microbiol.* **2015**, *6*, 777. [CrossRef]
37. Clavel, T.; Fallani, M.; Lepage, P.; Levenez, F.; Mathey, J.; Rochet, V.; Sérézat, M.; Sutren, M.; Henderson, G.; Bennetau-Pelissero, C.; et al. Isoflavones and Functional Foods Alter the Dominant Intestinal Microbiota in Postmenopausal Women. *J. Nutr.* **2005**, *135*, 2786–2792. [CrossRef]
38. Moreno-Indias, I.; Sánchez-Alcoholado, L.; Sánchez-Garrido, M.; Núñez, G.M.M.; Pérez-Jiménez, F.; Tena-Sempere, M.; Tinahones, F.J.; Queipo-Ortuño, M.I. Neonatal Androgen Exposure Causes Persistent Gut Microbiota Dysbiosis Related to Metabolic Disease in Adult Female Rats. *Endocrinology* **2016**, *157*, 4888–4898. [CrossRef]
39. Cox-York, K.A.; Sheflin, A.M.; Foster, M.T.; Gentile, C.L.; Kahl, A.; Koch, L.G.; Britton, S.L.; Weir, T.L. Ovariectomy results in differential shifts in gut microbiota in low versus high aerobic capacity rats. *Physiol. Rep.* **2015**, *3*, e12488. [CrossRef]
40. Choi, S.; Hwang, Y.-J.; Shin, M.-J.; Yi, H. Difference in the Gut Microbiome between Ovariectomy-Induced Obesity and Diet-Induced Obesity. *J. Microbiol. Biotechnol.* **2017**, *27*, 2228–2236. [CrossRef]
41. Kaliannan, K.; Robertson, R.C.; Murphy, K.; Stanton, C.; Kang, C.; Wang, B.; Hao, L.; Bhan, A.K.; Kang, J.X. Estrogen-mediated gut microbiome alterations influence sexual dimorphism in metabolic syndrome in mice. *Microbiome* **2018**, *6*, 1–22. [CrossRef]
42. Acharya, K.D.; Gao, X.; Bless, E.P.; Chen, J.; Tetel, M.J. Estradiol and high fat diet associate with changes in gut microbiota in female ob/ob mice. *Sci. Rep.* **2019**, *9*, 1–13. [CrossRef]
43. Saladin, R.; De Vos, P.; Guerre-Millot, M.; Leturque, A.; Girard, J.; Staels, B.; Auwerx, J. Transient increase in obese gene expression after food intake or insulin administration. *Nature* **1995**, *377*, 527–528. [CrossRef]
44. Yasrebi, A.; Rivera, J.A.; Krumm, E.A.; Yang, J.A.; Roepke, T.A. Activation of estrogen response element–independent er α signaling protects female mice from diet-induced obesity. *Endocrinology* **2017**, *158*, 319–334. [CrossRef]
45. Soto, M.; Herzog, C.; Pacheco, J.A.; Fujisaka, S.; Bullock, K.; Clish, C.; Kahn, C.R. Gut microbiota modulate neurobehavior through changes in brain insulin sensitivity and metabolism. *Mol. Psychiatry* **2018**, *23*, 2287–2301. [CrossRef]
46. Carmody, R.N.; Gerber, G.K.; Luevano, J.M.; Gatti, D.M.; Somes, L.; Svenson, K.L.; Turnbaugh, P.J. Diet Dominates Host Genotype in Shaping the Murine Gut Microbiota. *Cell Host Microbe* **2015**, *17*, 72–84. [CrossRef]
47. Daniel, H.; Gholami, A.M.; Berry, D.; Desmarchelier, C.; Hahne, H.; Loh, G.; Mondot, S.; Lepage, P.; Rothballer, M.; Walker, A.; et al. High-fat diet alters gut microbiota physiology in mice. *ISME J.* **2013**, *8*, 295–308. [CrossRef]
48. Turnbaugh, P.; Ridaura, V.K.; Faith, J.J.; Rey, F.E.; Knight, R.; Gordon, J.I. The Effect of Diet on the Human Gut Microbiome: A Metagenomic Analysis in Humanized Gnotobiotic Mice. *Sci. Transl. Med.* **2009**, *1*, 6ra14. [CrossRef]
49. Sau, L.; Olmstead, C.M.; Cui, L.J.; Chen, A.; Shah, R.S.; Kelley, S.T.; Thackray, V.G. Alterations in Gut Microbiota Do Not Play a Causal Role in Diet-independent Weight Gain Caused by Ovariectomy. *J. Endocr. Soc.* **2020**, *5*, bvaa173. [CrossRef]
50. Jensen, E.V.; Suzuki, T.; Kawashima, T.; Stumpf, W.E.; Jungblut, P.W.; DeSombre, E.R. A two-step mechanism for the interaction of estradiol with rat uterus. *Proc. Natl. Acad. Sci. USA* **1968**, *59*, 632–638. [CrossRef]
51. Kuiper, G.G.; Enmark, E.; Pelto-Huikko, M.; Nilsson, S.; Gustafsson, J.A. Cloning of a novel receptor expressed in rat prostate and ovary. *Proc. Natl. Acad. Sci. USA* **1996**, *93*, 5925–5930. [CrossRef]

52. Tetel, M.J.; Pfaff, D.W. Contributions of estrogen receptor-α and estrogen receptor-β to the regulation of behavior. *Biochim. Biophys. Acta (BBA) Bioenerg.* **2010**, *1800*, 1084–1089. [CrossRef]
53. Wada-Hiraike, O.; Imamov, O.; Hiraike, H.; Hultenby, K.; Schwend, T.; Omoto, Y.; Warner, M.; Gustafsson, J.-A. Role of estrogen receptor beta in colonic epithelium. *Proc. Natl. Acad. Sci. USA* **2006**, *103*, 2959–2964. [CrossRef]
54. Gulhane, M.; Murray, L.; Lourie, R.; Tong, H.; Sheng, Y.H.; Wang, R.; Kang, A.; Schreiber, V.; Wong, K.Y.; Magor, G.; et al. High Fat Diets Induce Colonic Epithelial Cell Stress and Inflammation that is Reversed by IL-22. *Sci. Rep.* **2016**, *6*, 28990. [CrossRef]
55. Régnier, M.; Van Hul, M.; Knauf, C.; Cani, P.D. Gut microbiome, endocrine control of gut barrier function and metabolic diseases. *J. Endocrinol.* **2021**, *248*, R67–R82. [CrossRef]
56. Turnbaugh, P.; Ley, R.; Mahowald, M.A.; Magrini, V.; Mardis, E.R.; Gordon, J.I. An obesity-associated gut microbiome with increased capacity for energy harvest. *Nat. Cell Biol.* **2006**, *444*, 1027–1031. [CrossRef]
57. Cani, P.; Bibiloni, R.; Knauf, C.; Waget, A.; Neyrinck, A.; Delzenne, N.; Burcelin, R. Changes in Gut Microbiota Control Metabolic Endotoxemia-Induced Inflammation in High-Fat Diet-Induced Obesity and Diabetes in Mice. *Diabetes* **2008**, *57*, 1470–1481. [CrossRef]
58. Hassan, A.; Mancano, G.; Kashofer, K.; Liebisch, G.; Farzi, A.; Zenz, G.; Claus, S.P.; Holzer, P. Anhedonia induced by high-fat diet in mice depends on gut microbiota and leptin. *Nutr. Neurosci.* **2020**, 1–14. [CrossRef]
59. Santos-Marcos, J.A.; Barroso, A.; Rangel-Zuñiga, O.A.; Perdices-Lopez, C.; Haro, C.; Sanchez-Garrido, M.A.; Molina-Abril, H.; Ohlsson, C.; Perez-Martinez, P.; Poutanen, M.; et al. Interplay between gonadal hormones and postnatal overfeeding in defining sex-dependent differences in gut microbiota architecture. *Aging* **2020**, *12*, 19979–20000. [CrossRef]
60. Torres, P.J.; Ho, B.S.; Arroyo, P.; Sau, L.; Chen, A.; Kelley, S.T.; Thackray, V.G. Exposure to a Healthy Gut Microbiome Protects Against Reproductive and Metabolic Dysregulation in a PCOS Mouse Model. *Endocrinology* **2019**, *160*, 1193–1204. [CrossRef]
61. Rizk, M.G.; Thackray, V.G. Intersection of Polycystic Ovary Syndrome and the Gut Microbiome. *J. Endocr. Soc.* **2020**, *5*, bvaa177. [CrossRef]
62. Han, Q.; Wang, J.; Li, W.; Chen, Z.-J.; Du, Y. Androgen-induced gut dysbiosis disrupts glucolipid metabolism and endocrinal functions in polycystic ovary syndrome. *Microbiome* **2021**, *9*, 1–16. [CrossRef] [PubMed]
63. Derrien, M.; Belzer, C.; de Vos, W.M. Akkermansia muciniphila and its role in regulating host functions. *Microb. Pathog.* **2017**, *106*, 171–181. [CrossRef]
64. Van Passel, M.W.J.; Kant, R.; Zoetendal, E.G.; Plugge, C.M.; Derrien, M.; Malfatti, S.A.; Chain, P.; Woyke, T.; Palva, A.; De Vos, W.M.; et al. The Genome of Akkermansia muciniphila, a Dedicated Intestinal Mucin Degrader, and Its Use in Exploring Intestinal Metagenomes. *PLoS ONE* **2011**, *6*, e16876. [CrossRef]
65. De Vos, W.M. Microbe Profile: Akkermansia muciniphila: A conserved intestinal symbiont that acts as the gatekeeper of our mucosa. *Microbiology* **2017**, *163*, 646–648. [CrossRef]
66. Dao, M.C.; Everard, A.; Aron-Wisnewsky, J.; Sokolovska, N.; Prifti, E.; Verger, E.O.; Kayser, B.; Levenez, F.; Chilloux, J.; Hoyles, L.; et al. Akkermansia muciniphila and improved metabolic health during a dietary intervention in obesity: Relationship with gut microbiome richness and ecology. *Gut* **2015**, *65*, 426–436. [CrossRef]
67. Yassour, M.; Lim, M.Y.; Yun, H.S.; Tickle, T.L.; Sung, J.; Song, Y.-M.; Lee, K.; Franzosa, E.A.; Morgan, X.C.; Gevers, D.; et al. Sub-clinical detection of gut microbial biomarkers of obesity and type 2 diabetes. *Genome Med.* **2016**, *8*, 1–14. [CrossRef]
68. Derrien, M.; Vaughan, E.E.; Plugge, C.M.; De Vos, W.M. Akkermansia muciniphila gen. nov., sp. nov., a human intestinal mucin-degrading bacterium. *Int. J. Syst. Evol. Microbiol.* **2004**, *54*, 1469–1476. [CrossRef]
69. Santacruz, A.; Collado, M.C.; García-Valdés, L.; Segura, M.T.; Martín-Lagos, J.A.; Anjos, T.; Martí-Romero, M.; Lopez, R.M.; Florido, J.; Campoy, C.; et al. Gut microbiota composition is associated with body weight, weight gain and biochemical parameters in pregnant women. *Br. J. Nutr.* **2010**, *104*, 83–92. [CrossRef]
70. Brahe, L.K.; Le Chatelier, E.; Prifti, E.; Pons, N.; Kennedy, S.B.; Hansen, T.; Pedersen, O.; Astrup, A.; Ehrlich, S.; Larsen, L.H. Specific gut microbiota features and metabolic markers in postmenopausal women with obesity. *Nutr. Diabetes* **2015**, *5*, e159. [CrossRef]
71. Depommier, C.; Everard, A.; Druart, C.; Plovier, H.; Van Hul, M.; Vieira-Silva, S.; Falony, G.; Raes, J.; Maiter, D.; Delzenne, N.; et al. Supplementation with Akkermansia muciniphila in overweight and obese human volunteers: A proof-of-concept exploratory study. *Nat. Med.* **2019**, *25*, 1096–1103. [CrossRef]
72. Everard, A.; Belzer, C.; Geurts, L.; Ouwerkerk, J.P.; Druart, C.; Bindels, L.B.; Guiot, Y.; Derrien, M.; Muccioli, G.; Delzenne, N.; et al. Cross-talk between Akkermansia muciniphila and intestinal epithelium controls diet-induced obesity. *Proc. Natl. Acad. Sci. USA* **2013**, *110*, 9066–9071. [CrossRef] [PubMed]
73. Zhao, S.; Liu, W.; Wang, J.; Shi, J.; Sun, Y.; Wang, W.; Ning, G.; Liu, R.-X.; Hong, J. Akkermansia muciniphila improves metabolic profiles by reducing inflammation in chow diet-fed mice. *J. Mol. Endocrinol.* **2017**, *58*, 1–14. [CrossRef]
74. Li, J.; Lin, S.; Vanhoutte, P.M.; Woo, C.W.H.; Xu, A. Akkermansia Muciniphila Protects Against Atherosclerosis by Preventing Metabolic Endotoxemia-Induced Inflammation in Apoe$^{-/-}$ Mice. *Circ.* **2016**, *133*, 2434–2446. [CrossRef]
75. Shin, N.-R.; Lee, J.-C.; Lee, H.-Y.; Kim, M.-S.; Whon, T.W.; Lee, M.-S.; Bae, J.-W. An increase in theAkkermansiaspp. population induced by metformin treatment improves glucose homeostasis in diet-induced obese mice. *Gut* **2013**, *63*, 727–735. [CrossRef]
76. Crost, E.H.; Tailford, L.E.; Le Gall, G.; Fons, M.; Henrissat, B.; Juge, N. Utilisation of Mucin Glycans by the Human Gut Symbiont Ruminococcus gnavus Is Strain-Dependent. *PLoS ONE* **2013**, *8*, e76341. [CrossRef]

77. Javurek, A.B.; Spollen, W.G.; Johnson, S.A.; Bivens, N.J.; Bromert, K.H.; Givan, S.; Rosenfeld, C.S. Effects of exposure to bisphenol A and ethinyl estradiol on the gut microbiota of parents and their offspring in a rodent model. *Gut Microbes* **2016**, *7*, 471–485. [CrossRef]
78. Neves, A.R.; Pool, W.A.; Kok, J.; Kuipers, O.P.; Santos, H. Overview on sugar metabolism and its control inLactococcus lactis—The input from in vivo NMR. *FEMS Microbiol. Rev.* **2005**, *29*, 531–554. [CrossRef]
79. Kwa, M.; Plottel, C.S.; Blaser, M.J.; Adams, S. The Intestinal Microbiome and Estrogen Receptor–Positive Female Breast Cancer. *J. Natl. Cancer Inst.* **2016**, *108*, djw029. [CrossRef]
80. Yano, J.M.; Yu, K.; Donaldson, G.P.; Shastri, G.G.; Ann, P.; Ma, L.; Nagler, C.R.; Ismagilov, R.F.; Mazmanian, S.K.; Hsiao, E.Y. Indigenous Bacteria from the Gut Microbiota Regulate Host Serotonin Biosynthesis. *Cell* **2015**, *161*, 264–276. [CrossRef]
81. Das, N.K.; Schwartz, A.J.; Barthel, G.; Inohara, N.; Liu, Q.; Sankar, A.; Hill, D.R.; Ma, X.; Lamberg, O.; Schnizlein, M.K.; et al. Microbial Metabolite Signaling Is Required for Systemic Iron Homeostasis. *Cell Metab.* **2020**, *31*, 115–130.e6. [CrossRef]
82. Massey, W.; Brown, J.M. The Gut Microbial Endocrine Organ in Type 2 Diabetes. *Endocrinology* **2020**, *162*, bqaa235. [CrossRef]
83. Zhu, L.; Brown, W.C.; Cai, Q.; Krust, A.; Chambon, P.; McGuinness, O.P.; Stafford, J.M. Estrogen Treatment After Ovariectomy Protects Against Fatty Liver and May Improve Pathway-Selective Insulin Resistance. *Diabetes* **2012**, *62*, 424–434. [CrossRef] [PubMed]
84. Pettersson, U.S.; Waldén, T.B.; Carlsson, P.-O.; Jansson, L.; Phillipson, M. Female Mice are Protected against High-Fat Diet Induced Metabolic Syndrome and Increase the Regulatory T Cell Population in Adipose Tissue. *PLoS ONE* **2012**, *7*, e46057. [CrossRef] [PubMed]
85. Santollo, J.; Edwards, A.A.; Howell, J.A.; Myers, K.E. Bidirectional effects of estradiol on the control of water intake in female rats. *Horm. Behav.* **2021**, *133*, 104996. [CrossRef] [PubMed]
86. Allard, C.; Morford, J.J.; Xu, B.; Salwen, B.; Xu, W.; Desmoulins, L.; Zsombok, A.; Kim, J.K.; Levin, E.R.; Mauvais-Jarvis, F. Loss of Nuclear and Membrane Estrogen Receptor-α Differentially Impairs Insulin Secretion and Action in Male and Female Mice. *Diabetes* **2018**, *68*, 490–501. [CrossRef]
87. Elmquist, J.K.; Bjorbaek, C.; Ahima, R.S.; Flier, J.S.; Saper, C.B. Distributions of leptin receptor mRNA isoforms in the rat brain. *J. Comp. Neurol.* **1998**, *395*, 535–547. [CrossRef]
88. Pelleymounter, M.; Cullen, M.; Baker, M.; Hecht, R.; Winters, D.; Boone, T.; Collins, F. Effects of the obese gene product on body weight regulation in ob/ob mice. *Science* **1995**, *269*, 540–543. [CrossRef]
89. González, M.; Del Mar Bibiloni, M.; Pons, A.; Llompart, I.; Tur, J.A. Inflammatory markers and metabolic syndrome among adolescents. *Eur. J. Clin. Nutr.* **2012**, *66*, 1141–1145. [CrossRef]
90. Bastard, J.-P.; Jardel, C.; Bruckert, E.; Blondy, P.; Capeau, J.; Laville, M.; Vidal, H.; Hainque, B. Elevated Levels of Interleukin 6 Are Reduced in Serum and Subcutaneous Adipose Tissue of Obese Women after Weight Loss. *J. Clin. Endocrinol. Metab.* **2000**, *85*, 3338–3342. [CrossRef]
91. Qi, Y.; Nie, Z.; Lee, Y.-S.; Singhal, N.; Scherer, P.E.; Lazar, M.A.; Ahima, R.S. Loss of Resistin Improves Glucose Homeostasis in Leptin Deficiency. *Diabetes* **2006**, *55*, 3083–3090. [CrossRef]
92. Banerjee, R.R.; Rangwala, S.M.; Shapiro, J.S.; Rich, A.S.; Rhoades, B.; Qi, Y.; Wang, J.; Rajala, M.W.; Pocai, A.; Scherer, P.E.; et al. Regulation of Fasted Blood Glucose by Resistin. *Science* **2004**, *303*, 1195–1198. [CrossRef] [PubMed]
93. Kim, J.H.; Lee, E.; Friedline, R.H.; Suk, S.; Jung, D.Y.; Dagdeviren, S.; Hu, X.; Inashima, K.; Noh, H.L.; Kwon, J.Y.; et al. Endoplasmic reticulum chaperone GRP78 regulates macrophage function and insulin resistance in diet-induced obesity. *FASEB J.* **2018**, *32*, 2292–2304. [CrossRef]
94. Xu, J.; Lloyd, D.; Hale, C.; Stanislaus, S.; Chen, M.; Sivits, G.; Vonderfecht, S.; Hecht, R.; Li, Y.-S.; Lindberg, R.A.; et al. Fibroblast Growth Factor 21 Reverses Hepatic Steatosis, Increases Energy Expenditure, and Improves Insulin Sensitivity in Diet-Induced Obese Mice. *Diabetes* **2008**, *58*, 250–259. [CrossRef] [PubMed]
95. Kim, J.K. Hyperinsulinemic–Euglycemic Clamp to Assess Insulin Sensitivity In Vivo. *Methods Mol. Biol.* **2009**, *560*, 221–238. [CrossRef]
96. Dagdeviren, S.; Jung, D.Y.; Friedline, R.H.; Noh, H.L.; Kim, J.H.; Patel, P.R.; Tsitsilianos, N.; Inashima, K.; Tran, D.A.; Hu, X.; et al. IL-10 prevents aging-associated inflammation and insulin resistance in skeletal muscle. *FASEB J.* **2016**, *31*, 701–710. [CrossRef]
97. RRID:AB_2756387. Available online: https://antibodyregistry.org/search.php?q=AB_2756387 (accessed on 30 July 2021).
98. Dagdeviren, S.; Jung, D.Y.; Lee, E.; Friedline, R.H.; Noh, H.L.; Kim, J.H.; Patel, P.R.; Tsitsilianos, N.; Tsitsilianos, A.V.; Tran, D.A.; et al. Altered Interleukin-10 Signaling in Skeletal Muscle Regulates Obesity-Mediated Inflammation and Insulin Resistance. *Mol. Cell. Biol.* **2016**, *36*, 2956–2966. [CrossRef]
99. Caporaso, J.G.; Lauber, C.L.; Walters, W.A.; Berg-Lyons, D.; Lozupone, C.A.; Turnbaugh, P.; Fierer, N.; Knight, R. Global patterns of 16S rRNA diversity at a depth of millions of sequences per sample. *Proc. Natl. Acad. Sci. USA* **2010**, *108*, 4516–4522. [CrossRef]
100. Chen, X.; Johnson, S.; Jeraldo, P.; Wang, J.; Chia, N.; Kocher, J.-P.A.; Chen, J. Hybrid-denovo: A de novo OTU-picking pipeline integrating single-end and paired-end 16S sequence tags. *GigaScience* **2017**, *7*, 1–7. [CrossRef] [PubMed]
101. Cole, J.R.; Wang, Q.; Fish, J.A.; Chai, B.; McGarrell, D.M.; Sun, Y.; Brown, C.T.; Porras-Alfaro, A.; Kuske, C.R.; Tiedje, J.M. Ribosomal Database Project: Data and tools for high throughput rRNA analysis. *Nucleic Acids Res.* **2013**, *42*, D633–D642. [CrossRef]
102. RRID:SCR_002830. Available online: https://scicrunch.org/resolver/RRID:SCR_002830 (accessed on 30 July 2021).

103. Price, M.N.; Dehal, P.S.; Arkin, A.P. FastTree: Computing Large Minimum Evolution Trees with Profiles instead of a Distance Matrix. *Mol. Biol. Evol.* **2009**, *26*, 1641–1650. [CrossRef]
104. Bialkowska, A.B.; Ghaleb, A.M.; Nandan, M.O.; Yang, V.W. Improved Swiss-rolling Technique for Intestinal Tissue Preparation for Immunohistochemical and Immunofluorescent Analyses. *J. Vis. Exp.* **2016**, e54161. [CrossRef]
105. Gumber, S.; Nusrat, A.; Villinger, F. Immunohistological characterization of intercellular junction proteins in rhesus macaque intestine. *Exp. Toxicol. Pathol.* **2014**, *66*, 437–444. [CrossRef]
106. RRID:AB_2744671. Available online: https://antibodyregistry.org/search.php?q=AB_2744671 (accessed on 30 July 2021).
107. RRID:AB_2890613. Available online: https://antibodyregistry.org/search.php?q=AB_2890613 (accessed on 30 July 2021).
108. Mohankumar, K.; Namachivayam, K.; Song, T.; Cha, B.J.; Slate, A.; Hendrickson, J.E.; Pan, H.; Wickline, S.A.; Oh, J.-Y.; Patel, R.P.; et al. A murine neonatal model of necrotizing enterocolitis caused by anemia and red blood cell transfusions. *Nat. Commun.* **2019**, *10*, 1–17. [CrossRef]
109. RRID:AB_2536183. Available online: https://antibodyregistry.org/search.php?q=AB_2536183 (accessed on 30 July 2021).
110. RRID:AB_2534102. Available online: https://antibodyregistry.org/search?q=AB_2534102 (accessed on 30 July 2021).
111. RRID:SCR_016555. Available online: https://scicrunch.org/resolver/RRID:SCR_016555 (accessed on 30 July 2021).
112. RRID:SCR_003070. Available online: https://scicrunch.org/resolver/SCR_003070 (accessed on 30 July 2021).
113. RRID:SCR_015655. Available online: https://scicrunch.org/resolver/SCR_015655 (accessed on 30 July 2021).
114. RRID:SCR_019096. Available online: https://scicrunch.org/resolver/RRID:SCR_019096 (accessed on 30 July 2021).
115. Weiss, S.; Xu, Z.Z.; Peddada, S.; Amir, A.; Bittinger, K.; Gonzalez, A.; Lozupone, C.; Zaneveld, J.R.; Vázquez-Baeza, Y.; Birmingham, A.; et al. Normalization and microbial differential abundance strategies depend upon data characteristics. *Microbiome* **2017**, *5*, 1–18. [CrossRef] [PubMed]
116. Hu, Y.-J.; Satten, G.A. Testing hypotheses about the microbiome using the linear decomposition model (LDM). *Bioinformatics* **2020**, *36*, 4106–4115. [CrossRef] [PubMed]
117. Walther-António, M.R.S.; Chen, J.; Multinu, F.; Hokenstad, A.; Distad, T.J.; Cheek, E.H.; Keeney, G.L.; Creedon, D.J.; Nelson, H.; Mariani, A.; et al. Potential contribution of the uterine microbiome in the development of endometrial cancer. *Genome Med.* **2016**, *8*, 1–15. [CrossRef] [PubMed]
118. Chen, L.; Reeve, J.; Zhang, L.; Huang, S.; Wang, X.; Chen, J. GMPR: A robust normalization method for zero-inflated count data with application to microbiome sequencing data. *PeerJ* **2018**, *6*, e4600. [CrossRef] [PubMed]

 metabolites

MDPI

Brief Report

Evidence That Agouti-Related Peptide May Directly Regulate Kisspeptin Neurons in Male Sheep

Christina M. Merkley, Sydney L. Shuping, Jeffrey R. Sommer and Casey C Nestor *

Department of Animal Science, North Carolina State University, Raleigh, NC 27695, USA; cmmerkle@ncsu.edu (C.M.M.); slshupin@ncsu.edu (S.L.S.); jrsommer@ncsu.edu (J.R.S.)
* Correspondence: ccnestor@ncsu.edu

Abstract: Agouti-related peptide (AgRP) neurons, which relay information from peripheral metabolic signals, may constitute a key central regulator of reproduction. Given that AgRP inhibits luteinizing hormone (LH) secretion and that nutritional suppression of LH elicits an increase in AgRP while suppressing kisspeptin expression in the arcuate nucleus (ARC) of the hypothalamus, we sought to examine the degree to which AgRP could directly regulate ARC kisspeptin neurons. Hypothalamic tissue was collected from four castrated male sheep (10 months of age) and processed for the detection of protein (AgRP input to kisspeptin neurons) using immunohistochemistry and mRNA for melanocortin 3 and 4 receptors (MC3R; MC4R) in kisspeptin neurons using RNAscope. Immunohistochemical analysis revealed that the majority of ARC kisspeptin neurons are contacted by presumptive AgRP terminals. RNAscope analysis revealed that nearly two thirds of the ARC kisspeptin neurons express mRNA for MC3R, while a small percentage (<10%) colocalize MC4R. Taken together, this data provides neuroanatomical evidence for a direct link between orexigenic AgRP neurons and reproductively critical kisspeptin neurons in the sheep, and builds upon our current understanding of the central link between energy balance and reproduction.

Keywords: kisspeptin; AgRP; sheep; reproduction; LH

Citation: Merkley, C.M.; Shuping, S.L.; Sommer, J.R.; Nestor, C.C Evidence That Agouti-Related Peptide May Directly Regulate Kisspeptin Neurons in Male Sheep. *Metabolites* **2021**, *11*, 138. https://doi.org/10.3390/metabo11030138

Academic Editor: Giancarlo Panzica

Received: 28 January 2021
Accepted: 23 February 2021
Published: 26 February 2021

1. Introduction

Proper nutritional balance is paramount to the capacity for reproduction in mammals. Favorable metabolic conditions allow for pubertal development and fertility, while an unfavorable metabolic state such as undernutrition, can suppress reproductive function. This, at least in part, occurs via central integration of metabolic signals given that negative energy balance inhibits gonadotropin-releasing hormone (GnRH), and subsequently luteinizing hormone (LH), release [1–3]. Although GnRH neurons serve as the final common conduit from the central nervous system controlling reproduction, they appear to lack the appropriate receptors to directly respond to a metabolic hormone such as leptin [4–6]. Thus, changes in peripheral metabolic signals reflective of undernutrition (i.e., lower circulating leptin concentrations) must recruit inhibitory afferents and/or block stimulatory inputs to GnRH neurons in order to ultimately suppress GnRH/LH release during a negative energy state. One such stimulatory afferent is the neuropeptide kisspeptin, which has perikarya located primarily in two areas of the ventral forebrain, the anterior ventral periventricular area (AVPV; rodents)/preoptic area (POA; non-rodents) and the arcuate nucleus (ARC) of the hypothalamus [7–9]. Indeed, the stimulatory action of kisspeptin on LH secretion has been confirmed in several species [10], and the vast majority of GnRH neurons colocalize Kiss1R [11–14], supporting the idea of a direct stimulatory action of kisspeptin on GnRH neurons. Arcuate kisspeptin neurons in particular are identified as playing a dominant role in pulsatile GnRH/LH release [15,16], and anatomical evidence has shown that up to 60% of GnRH cell bodies receive kisspeptin input arising from the ARC kisspeptin population [17]. Indeed, food restriction has been shown to reduce expression of ARC kisspeptin [18–22]. Furthermore, there is evidence to show that kisspeptin neurons express leptin receptors [23–25], but others have failed to demonstrate activation of leptin signaling in these neurons in vivo [26]. Thus, central regulation of GnRH/LH secretion during undernutrition likely incorporates other afferent inputs to ARC kisspeptin neurons.

Known for an orexigenic role in energy homeostasis, neurons that express agouti-related peptide (AgRP) may also play an important role to relay metabolic signals to reproductive axis. Located in the ARC, AgRP neurons express leptin receptors [27], and delayed puberty onset in leptin receptor-deficient mice can be rescued with select expression of leptin receptor within AgRP neurons [28]. With data demonstrating that central administration of AgRP inhibits LH secretion [29] and that undernutrition increases AgRP expression [30–32], AgRP signaling may represent a means whereby undernutrition inhibits GnRH/LH secretion. Interestingly, nearly all ARC AgRP neurons co-express neuropeptide Y (NPY) [32], and reciprocal connections between NPY neurons and kisspeptin cells have been reported [24]. However, given that other populations of NPY perikarya exist outside of the ARC [33], the degree to which AgRP neurons innervate ARC kisspeptin neurons is unclear, and remains to be examined. Furthermore, although classically known as the endogenous antagonist to α-melanocyte stimulating hormone (αMSH), a neuropeptide produced by anorexigenic pro-opiomelanocortin (POMC) neurons, AgRP alone has been shown to inhibit cells that express melanocortin 3 receptors (MC3R) and melanocortin 4 receptors (MC4R) via $G_{i/o}$ signaling [34,35]. Work in mice has revealed very few ARC kisspeptin neurons express MC4R [25], but our recent work in sheep [36] led us to investigate the potential for AgRP regulation of ARC kisspeptin neurons in the ovine model. In this study, we examine the degree to which ARC kisspeptin perikarya are innervated by presumptive AgRP terminals using dual-label immunofluorescence, and utilizing a relatively new fluorescent in situ hybridization technique, RNAscope, we characterize the degree to which arcuate kisspeptin neurons colocalize MC3R and MC4R.

2. Results

2.1. AgRP Inputs to Kisspeptin Cells

In ARC tissue from 10 month old castrated male sheep (wethers), AgRP-immunoreactive (ir) terminals were observed in frequent apposition to kisspeptin cell bodies and fibers (Figure 1). A total of 171 kisspeptin cells were analyzed for AgRP inputs in four animals, and the majority (72.31 ± 7.6%) of kisspeptin neurons showed at least one close contact by AgRP-ir terminals. A subset of kisspeptin neurons that showed at least one AgRP-ir input was examined further, and the total number of putative close contacts onto the cell body or proximal dendrite were analyzed from z-stack images (1.0 μm optical sections) taken throughout the extent of the cell. The results indicated that each kisspeptin perikaryon received multiple AgRP-ir close contacts (4.92 ± 1.0 contacts/cell). As reported previously in sheep [37], no colocalization of kisspeptin and AgRP in cell bodies or fibers was observed. There were, however, some instances of kisspeptin-ir fibers in apposition to AgRP perikarya (data not quantified).

2.2. Colocalization of Kisspeptin, MC3R, and MC4R

The extent of colocalization of kisspeptin and the melanocortin receptors, MC3R and MC4R, was characterized in the ARC of wethers, and various instances of single, dual and triple mRNA-expressing cells were observed (Figure 2A). Of the MC3R mRNA-expressing cells (436 ± 30.4 examined per wether) in the middle ARC nearly two thirds (63.79 ± 11.7%; Figure 2B) colocalized mRNA for kisspeptin, while 14.42 ± 1.1% (Figure 2B) colocalized mRNA for MC4R. Of the total kisspeptin mRNA-expressing cells examined (431.75 ± 36.0 per wether), nearly two thirds co-expressed mRNA for MC3R (62.33 ± 6.3%), while only 6.33 ± 1.2% (Figure 2B) colocalize MC4R. Moreover, of the total MC4R mRNA-expressing cells identified (101.25 ± 5.3 per wether), 26.05 ± 2.1% contained kisspeptin (Figure 2B), while 62.59 ± 7.4% (Figure 2B) expressed mRNA for MC3R. Finally, we examined the degree to cells co-expressed in all three transcripts to find that 15.85 ± 2.4% of MC3R/MC4R cells express mRNA for kisspeptin, 60.53 ± 6.1% of Kiss1/MC4R cells express mRNA for MC3R, and 3.64 ± 0.4% of Kiss/MC3R cells express mRNA for MC4R (Figure 2B).

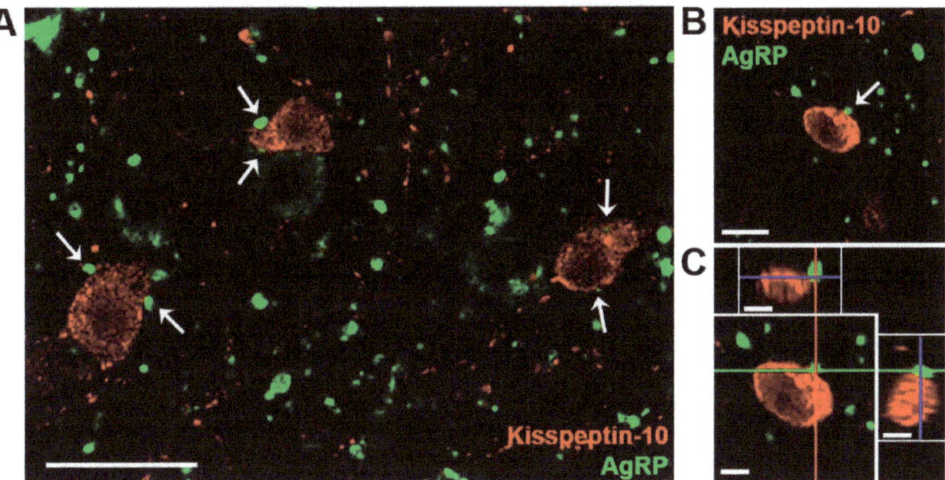

Figure 1. Confocal images (1.0 μm optical section; 40 × magnification) showing dual-label immunofluorescence for AgRP (green) and kisspeptin (red) in the arcuate nucleus of a castrated male sheep (wether). White arrows (**A**,**B**) indicate examples of AgRP-immunoreactive (ir) terminals in apposition to arcuate kisspeptin neurons, with AgRP-ir cell bodies and fibers seen in the vicinity. Orthogonal views (**C**) confirm close contact of an AgRP-labeled bouton to a kisspeptin cell body. Red, green, and blue lines in (**C**) indicate X, Y, and Z planes, respectively. Scale bars = 25 μm (**A**), 10 μm (**B**,**C**).

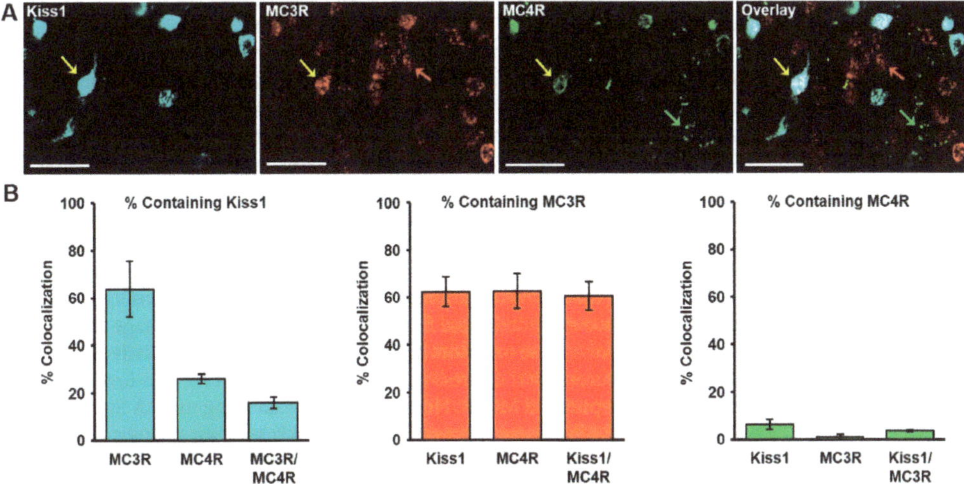

Figure 2. (**A**) Confocal image (1.0 μm optical section; 20 × magnification) of kisspeptin, MC3R, and MC4R mRNA-expressing cells in the arcuate nucleus (ARC) of a castrated male sheep. Yellow arrows indicate a Kiss1 cell expressing both MC3R and MC4R. Red and green arrows show MC3R and MC4R-expressing cells, respectively. Scale bar, 50 μm. (**B**) Mean (± SEM) percentage of kisspeptin (left), MC3R (middle), and MC4R (right) cells containing Kiss1, MC3R, and/or MC4R in the middle ARC.

PBS; Vector) for 1 hr each with four washes (5 min each) in PBS between incubations. Next, sections were incubated in biotinylated tyramide (BT; dilution 1:250 in PBS containing 3% H_2O_2 per mL; Perkin Elmer LAS, Inc., Boston, MA, USA) for 10 min. To visualize AgRP, sections were then incubated in Alexa 488- Streptavidin (dilution 1:100 in PBS; Invitrogen, Carlsbad, CA, USA, cat# S-32354) for 30 min. Following this step, sections were incubated in primary antibody rabbit anti-kisspeptin (1:10,000; Gift from I. Franceschini, cat# 566) diluted in PBS containing 0.4% Triton X-100 and 4% NGS for 17 h. On day four, sections were incubated in biotinylated goat anti-rabbit IgG Alexa 555 (1:100; Invitrogen, cat# A-21428) diluted in PBS, and then washed four times (5 min each) in PBS. Finally, sections were mounted onto Superfrost Plus microscope slides (Fisher Scientific, Waltham, MA, USA), coverslipped using ProLong Gold Antifade Mountant (Fisher Scientific), and stored at 4°C until imaging.

4.4. RNAscope In Situ Hybridization for Kisspeptin and Melanocortin Receptors (MC3R, MC4R)

To examine the extent of melanocortin receptor colocalization within ARC kisspeptin neurons of wethers, multi-plex RNAscope was performed for kisspeptin, MC3R, and MC4R. As described above, three sections in the middle ARC were selected, with the right hemisection of the brain used for RNAscope. In situ hybridization was performed based on instructions from Advanced Cell Diagnostics and technical recommendations with minor modifications using the RNAscope Multiplex Fluorescent Reagent Kit v2 (Advanced Cell Diagnostics, Newark, CA, USA; cat# 323100). All incubations between 40 and 60°C were conducted using an ACD HybEZ II Hybridization System with an EZ-Batch Slide System (Advanced Cell Diagnostics; cat# 321710). On day one, hemisections were washed overnight in 0.1 M PBS at 4 °C on a rocking shaker to remove excess cryoprotectant. On day two, sections were submerged in chilled 4% PFA (1 hr at 4 °C), and rinsed four times in 0.1 M PBS (5 min/rinse), followed by an incubation in Hydrogen Peroxide solution (10 min at RT; Advanced Cell Diagnostics, cat# 322335). Hemisections were then incubated with RNAscope Target Retrieval Solution (98 °C for 10 min; Advanced Cell Diagnostics, cat# 322001) and rinsed four times in 0.1M PBS (5min/rinse). Next, sections were mounted onto Superfrost Plus microscope slides (Fisher Scientific), a hydrophobic barrier was created around the tissue using an ImmEdge Pen (Advanced Cell Diagnostics; cat# 310018), and slides were stored overnight at 4 °C. On day three, slides were incubated in increasing concentrations of ethanol (50, 70, 100, and 100%; 5 min each) and allowed to air dry. Sections were then treated with RNAscope® Protease III (30 min at 40 °C; Advanced Cell Diagnostics, cat# 322337), and subsequently incubated with RNAscope target (kisspeptin, Oa-KISS1-C3, cat# 497471-C3; MC3R, Oa-MC3R-C2, cat# 537911-C2; MC4R, Oa-MC4R-C1, cat# 537921-C1) and control probes (positive controls, Oa-UBC-C3, cat#516181-C3; Oa-PPIB-C2, cat# 457031-C2, and Oa-POLR2A, cat# 516171; negative control, 3-plex Negative Control Probe, cat# 320871) for 2 h at 40 °C. Next, slides were washed twice with 1X Wash Buffer (Advanced Cell Diagnostics, cat# 310091; 2 min/rinse at RT) followed by sequential tissue application of 50 µl of the following, each for 30 min at 40 °C with 2 min washes using 1X Wash Buffer between applications: RNAscope Multiplex FL v2 Amp 1 (Advanced Cell Diagnostics, cat# 323101), RNAscope Multiplex FL v2 Amp 2 (Advanced Cell Diagnostics, cat# 323102), and RNAscope Multiplex FL v2 Amp 3 (Advanced Cell Diagnostics, cat# 323103). Following final incubation with Amp 3, slides were rinsed with 1X Wash Buffer twice (2 min/rinse at RT) followed by application of RNAscope Multiplex FL v2 HRP C1 (15 min at 40 °C; Advanced Cell Diagnostics, cat#323104). Slides were then washed with 1X Wash Buffer twice (2 min/rinse at RT), and incubated with 150 µL per slide of Opal 520 (Fisher Scientific; cat#NC1601877) diluted in RNAscope TSA buffer (Advanced Cell Diagnostics, cat# 322809) at a final concentration of 1:1500 for 30 min at 40 °C. Following a rinse with 1X Wash Buffer twice (2 min/rinse at RT), 50 µL of RNAscope® Multiplex FL v2 HRP Blocker (Advanced Cell Diagnostics, cat# 323107) was applied to tissue (15 min at 40 °C). Slides were then rinsed with 1X Wash Buffer twice (2 min/rinse at RT) followed by application of RNAscope Multiplex FL v2 HRP-C2 (15 min at 40 °C; Advanced Cell Diagnostics, cat# 323105). Next, following two rinses with 1X Wash Buffer (2 min/rinse at RT), slides were incubated with 150 μL per slide of Opal 570 (Fisher Scientific; cat# NC601878) diluted in RNAscope TSA buffer (Advanced Cell Diagnostics, cat# 322809) at a final concentration of 1:1500 for 30 min at 40 °C. Subsequently, after two rinses in 1X Wash Buffer (2 min/rinse at RT), 50 μL of RNAscope® Multiplex FL v2 HRP Blocker (Advanced Cell Diagnostics, cat# 323107) was applied to tissue for 15 min at 40 °C, followed by rinsing with 1X Wash Buffer twice (2 min/rinse at RT). Next, RNAscope Multiplex FL v2 HRP-C3 (15 min at 40 °C; Advanced Cell Diagnostics, cat# 323106) was applied to tissue. Slides were then rinsed with 1X Wash Buffer twice (2 min/rinse at RT), and incubated with 150 μL per slide of Opal 690 (Fisher Scientific; cat# NC1605064) in RNAscope TSA buffer (Advanced Cell Diagnostics, cat# 322809) at a final concentration of 1:1500 for 30 min at 40 °C. Next, slides were rinsed with 1X Wash Buffer twice (2 min/rinse at RT), followed by a final application of 50 μL of RNAscope® Multiplex FL v2 HRP Blocker (Advanced Cell Diagnostics, cat# 323107) for 15 min at 40 °C, and two rinses with 1X Wash Buffer (2 min/rinse at RT). Finally, slides were incubated with 50 μL DAPI, coverslipped with ProLong Gold Antifade Mountant (Fisher Scientific, cat# P36930), and stored at 4 °C until image acquisition.

4.5. *Confocal Analyses*

4.5.1. Immunofluorescence

Confocal imaging of immunostained hemisections was conducted using an LSM710 laser-scanning confocal microscope (Zeiss, Thornwood, NY, USA), with a Plan Achromat 40x/1.1 objective. For each section (three sections per wether), z-stacks (1.0-μm optical sections) were captured in the ARC using consistent acquisition settings used for all images. Confocal images were imported into Zen 3.0SR software (Black edition; Carl Zeiss Microscopy GmBH, Jena, Germany), where AgRP close contacts onto kisspeptin neurons were analyzed by a single observer. Based on the approximate size of each kisspeptin neuron, eight to eleven images (1.0- μm optical section) were analyzed per neuron through the z-plane. In each animal, 40–50 kisspeptin cells in which complete cell bodies (with visible nuclei) were imaged in the z-stack were selected for analysis. A close contact was defined as an immunolabeled bouton in direct apposition (no intervening pixels) to a kisspeptin cell body or proximal dendrite [17]. First, the percentage of kisspeptin neurons receiving one or more AgRP-positive close contact was calculated. Next, a subset of kisspeptin neurons (10 cells per animal) that showed at least one putative AgRP close contact was randomly selected in each wether to quantify all putative AgRP-positive inputs onto the cell body or proximal dendrite. Markers placed on putative contacts during analysis ensured that appositions flanking optical sections were not counted twice. In addition, putative terminals were viewed in orthogonal planes (X, Y, and Z), and only those contacting the kisspeptin neuron in all dimensions were quantified.

4.5.2. RNAscope

Hemisections processed for detection of mRNA for kisspeptin, MC3R, and MC4R were analyzed using an LSM 880 laser scanning confocal microscope (Zeiss). For each section (three sections per wether), z-stacks (1.0-μm optical sections) were captured in the ARC using a Plan Achromat 20x/0.8 objective, with consistent acquisition settings used for all images. Confocal images were imported into Zen 3.0 software (Carl Zeiss Microscopy), where the total number of cells expressing kisspeptin, MC3R, or MC4R mRNA were identified by a single observer. Markers placed on cells ensured that the same cell was not counted twice, and cells in which complete cell bodies were visible were included in the analysis, with DAPI used to visualize nuclear area. Images containing marked cells were imported in to FIJI/ImageJ software, where the numbers of cells were quantified. The degree of double or triple labeling was calculated as a percentage of the total number of kisspeptin cells expressing MC3R and/or MC4R, the total number of MC3R cells expressing kisspeptin and/or MC4R mRNA, and the total number of MC4R-expressing cells containing kisspeptin and/or MC3R mRNA.

Author Contributions: Project design, C.M.M., C.C.N.; project management, J.R.S., C.C.N.; collection of brain tissue, J.R.S., C.C.N.; implementation of IHC, C.M.M., C.C.N.; implementation of RNAscope, S.L.S.; data analysis and writing of this manuscript, C.M.M., S.L.S., C.C.N. All authors have read and agreed to the published version of the manuscript.

Funding: This work was supported by USDA National Institute of Food and Agriculture Hatch Project 1012905 and by National Institute of Food and Agriculture Grant no. 2019-67016-29408.

Institutional Review Board Statement: The study was approved by the North Carolina State University Animal Care and Use Committee (#17-020-B) and followed the National Institutes of Health guidelines for use of animals in research.

Informed Consent Statement: Not applicable.

Data Availability Statement: The data presented in this study are available on request from the corresponding author. The data are not public due to small samples size.

Acknowledgments: We thank Tabatha Wilson (North Carolina State University Metabolic Education Unit) for care of the animals, and Isabelle Franceschini for the kind gift of kisspeptin antisera. In addition, we thank Maggie Cummings, KaLynn Harlow, and Allison Renwick for their assistance with conducting this research (animal care and tissue collection). We also acknowledge the use of the Cellular and Molecular Imaging Facility (CMIF) at North Carolina State University, which is supported by the State of North Carolina and the National Science Foundation.

Conflicts of Interest: The authors declare that there is no conflict of interest that could be perceived as prejudicing the impartiality of the research reported.

References

1. Ebling, F.; Wood, R.; Karsch, F.; Vannerson, L.; Suttie, J.; Bucholtz, D.; Schall, R.; Foster, D. Metabolic interfaces between growth and reproduction. Iii. Central mechanisms controlling pulsatile luteinizing hormone secretion in the nutritionally growth-limited female lamb. *Endocrinology* **1990**, *126*, 2719–2727. [CrossRef]
2. Prasad, B.; Conover, C.; Sarkar, D.; Rabii, J.; Advis, J. Feed restriction in prepubertal lambs: Effect on puberty onset and on in vivo release of luteinizing-hormone-releasing hormonE.; neuropeptide y and beta-endorphin from the posterior-lateral median eminence. *Neuroendocrinology* **1993**, *57*, 1171–1181. [CrossRef]
3. I'anson, H.; Manning, J.; Herbos, C.; Pelt, J.; Friedman, C.; Wood, R.; Bucholtz, D.; Foster, D. Central inhibition of gonadotropin-releasing hormone secretion in the growth-restricted hypogonadotropic female sheep. *Endocrinology* **2000**, *141*, 520–527. [CrossRef]
4. Finn, P.; Cunningham, M.; Pau, K.; Spies, H.; Clifton, D.; Steiner, R. The stimulatory effect of leptin on the neuroendocrine reproductive axis of the monkey. *Endocrinology* **1998**, *139*, 4652–4662. [CrossRef]
5. Hakansson, M.; Brown, H.; Ghilardi, N.; Skoda, R.; Meister, B. Leptin receptor immunoreactivity in chemically defined target neurons of the hypothalamus. *J. Neurosci.* **1998**, *18*, 559–572. [CrossRef]
6. Quennell, J.; Mulligan, A.; Tups, A.; Liu, X.; Phipps, S.; Kemp, C.; Herbison, A.; Grattan, D.; Anderson, G. Leptin indirectly regulates gonadotropin-releasing hormone neuronal function. *Endocrinology* **2009**, *150*, 2805–2812. [CrossRef] [PubMed]
7. Gottsch, M.; Cunningham, M.; Smith, J.; Popa, S.; Acohido, B.; Crowley, W.; Seminara, S.; Clifton, D.; Steiner, R. A role for kisspeptins in the regulation of gonadotropin secretion in the mouse. *Endocrinology* **2004**, *145*, 4073–4077. [CrossRef] [PubMed]
8. Franceschini, I.; Lomet, D.; Cateau, M.; Delsol, G.; Tillet, Y.; Caraty, A. Kisspeptin immunoreactive cells of the ovine preoptic area and arcuate nucleus co-express estrogen receptor alpha. *Neurosci. Lett.* **2006**, *401*, 225–230. [CrossRef] [PubMed]
9. Lehman, M.; Merkley, C.; Coolen, L.; Goodman, R. Anatomy of the kisspeptin neural network in mammals. *Brain Res.* **2010**, *1364*, 90–102. [CrossRef] [PubMed]
10. Abbara, A.; Ratnasabapathy, R.; Jayasena, C.; Dhillo, W. The effects of kisspeptin on gonadotropin release in non-human mammals. *Adv. Exp. Med. Biol.* **2013**, *784*, 63–87. [PubMed]
11. Irwig, M.; Fraley, G.; Smith, J.; Acohido, B.; Popa, S.; Cunningham, M.; Gottsch, M.; Clifton, D.; Steiner, R. Kisspeptin activation of gonadotropin releasing hormone neurons and regulation of kiss-1 mrna in the male raT. *Neuroendocrinology* **2004**, *80*, 264–272. [CrossRef]
12. Herbison, A.; De Tassigny, X.; Doran, J.; Colledge, W. Distribution and postnatal development of gpr54 gene expression in mouse brain and gonadotropin-releasing hormone neurons. *Endocrinology* **2010**, *151*, 312–321. [CrossRef] [PubMed]
13. Smith, J.; Li, Q.; Yap, K.; Shahab, M.; Roseweir, A.; Millar, R.; Clarke, I. Kisspeptin is essential for the full preovulatory lh surge and stimulates gnrh release from the isolated ovine median eminence. *Endocrinology* **2011**, *152*, 1001–1012. [CrossRef] [PubMed]
14. Bosch, M.; Tonsfeldt, K.; Ronnekleiv, O. Mrna expression of ion channels in gnrh neurons: Subtype-specific regulation by 17beta-estradiol. *Mol. Cell Endocrinol.* **2013**, *367*, 85–97. [CrossRef] [PubMed]
15. Nestor, C.; Bedenbaugh, M.; Hileman, S.; Coolen, L.; Lehman, M.; Goodman, R. Regulation gnrh pulsatility in ewes. *Reproduction* **2018**, *156*, R83–R99. [CrossRef]

16. Lehman, M.; Coolen, L.; Goodman, R. Importance of neuroanatomical data from domestic animals to the development and testing of the kndy hypothesis for gnrh pulse generation. *Domest. Anim. Endocrinol.* **2020**. [CrossRef] [PubMed]
17. Merkley, C.; Coolen, L.; Goodman, R.; Lehman, M. Evidence for changes in numbers of synaptic inputs onto kndy and gnrh neurones during the preovulatory lh surge in the ewe. *J. Neuroendocrinol.* **2015**, *27*, 624–635. [CrossRef] [PubMed]
18. Castellano, J.; Navarro, V.; Fernandez-Fernandez, R.; Nogueiras, R.; Tovar, S.; Roa, J.; Vazquez, M.; Vigo, E.; Casanueva, F.; Aguilar, E.; et al. Changes in hypothalamic kiss-1 system and restoration of pubertal activation of the reproductive axis by kisspeptin in undernutrition. *Endocrinology* **2005**, *146*, 3917–3925. [CrossRef]
19. True, C.; Kirigiti, M.; Kievit, P.; Grove, K.; Smith, M. Leptin is not the critical signal for kisspeptin or luteinising hormone restoration during exit from negative energy balance. *J. Neuroendocrinol.* **2011**, *23*, 1099–1112. [CrossRef]
20. Navarro, V.; Ruiz-Pino, F.; Sanchez-Garrido, M.; Garcia-Galiano, D.; Hobbs, S.; Manfredi-Lozano, M.; Leon, S.; Sangiao-Alvarellos, S.; Castellano, J.; Clifton, D.; et al. Role of neurokinin b in the control of female puberty and its modulation by metabolic status. *J. Neurosci.* **2012**, *32*, 2388–2397. [CrossRef]
21. Polkowska, J.; Cieslak, M.; Wankowska, M.; Wojcik-Gladysz, A. The effect of short fasting on the hypothalamic neuronal system of kisspeptin in peripubertal female lambs. *Anim. Reprod. Sci.* **2015**, *159*, 184–190. [CrossRef] [PubMed]
22. Merkley, C.; Renwick, A.; Shuping, S.; Harlow, K.; Sommer, J.; Nestor, C. Undernutrition reduces kisspeptin and neurokinin b expression in castrated male sheep. *Reprod. Fertil.* **2020**, *1*, 21–33. [CrossRef]
23. Smith, J.; Acohido, B.; Clifton, D.; Steiner, R. Kiss-1 neurones are direct targets for leptin in the ob/ob mouse. *J. Neuroendocr.* **2006**, *18*, 298–303. [CrossRef]
24. Backholer, K.; Smith, J.; Rao, A.; Pereira, A.; Iqbal, J.; Ogawa, S.; Li, Q.; Clarke, I. Kisspeptin cells in the ewe brain respond to leptin and communicate with neuropeptide y and proopiomelanocortin cells. *Endocrinology* **2010**, *151*, 2233–2243. [CrossRef] [PubMed]
25. Cravo, R.; Margatho, L.; Osborne-Lawrence, S.; Donato, J.; Atkin, S.; Bookout, A.; Rovinsky, S.; Frazao, R.; Lee, E.; Gautron, L.; et al. Characterization of kiss1 neurons using transgenic mouse models. *Neuroscience* **2011**, *173*, 37–56. [CrossRef] [PubMed]
26. Louis, G.; Greenwald-Yarnell, M.; Phillips, R.; Coolen, L.; Lehman, M.; Myers, M. Molecular mapping of the neural pathways linking leptin to the neuroendocrine reproductive axis. *Endocrinology* **2011**, *152*, 2302–2310. [CrossRef]
27. Donato, J.; Cravo, R.; Frazao, R.; Gautron, L.; Scott, M.; Lachey, J.; Castro, I.; Margatho, L.; Lee, S.; Lee, C.; et al. Leptin's effect on puberty in mice is relayed by the ventral premammillary nucleus and does not require signaling in kiss1 neurons. *J. Clin. Invest.* **2011**, *121*, 355–368. [CrossRef] [PubMed]
28. Egan, O.; Inglis, M.; Anderson, G. Leptin signaling in agrp neurons modulates puberty onset and adult fertility in mice. *J. Neurosci.* **2017**, *37*, 3875–3886. [CrossRef]
29. Vulliemoz, N.; Xiao, E.; Xia-Zhang, L.; Wardlaw, S.; Ferin, M. Central infusion of agouti-related peptide suppresses pulsatile luteinizing hormone release in the ovariectomized rhesus monkey. *Endocrinology* **2005**, *146*, 784–789. [CrossRef] [PubMed]
30. Adam, C.; Archer, Z.; Findlay, P.; Thomas, L.; Marie, M. Hypothalamic gene expression in sheep for cocaine- and amphetamine-regulated transcript, pro-opiomelanocortin, neuropeptide y, agouti-related peptide and leptin receptor and responses to negative energy balance. *Neuroendocrinology* **2002**, *75*, 250–256. [CrossRef] [PubMed]
31. Wagner, C.; Mcmahon, C.; Marks, D.; Daniel, J.; Steele, B.; Sartin, J. A role for agouti-related protein in appetite regulation in a species with continuous nutrient delivery. *Neuroendocrinology* **2004**, *80*, 210–218. [CrossRef]
32. Hahn, T.; Breininger, J.; Baskin, D.; Schwartz, M. Coexpression of agrp and npy in fasting-activated hypothalamic neurons. *Nat. Neurosci.* **1998**, *1*, 271–272. [CrossRef]
33. Chaillou, E.; Baumont, R.; Chilliard, Y.; Tillet, T. Several subpopulations of neuropeptide y-containing neurons exist in the infundibular nucleus of sheep: An immunohistochemical study of animals on different diets. *J. Comp. Neurol.* **2002**, *444*, 129–143. [CrossRef]
34. Büch, T.; Heling, D.; Damm, E.; Gudermann, T.; Breit, A. Pertussis toxin-sensitive signaling of melanocortin-4 receptors in hypothalamic gt1–7 cells defines agouti-related protein as a biased agonisT. *J. Biol. Chem.* **2009**, *284*, 26411–26420. [CrossRef] [PubMed]
35. Yang, Z.; Tao, Y. Biased signaling initiated by agouti-related peptide through human melanocortin-3 and-4 receptors. *Biochim. Biophys. Acta* **2016**, *1862*, 1485–1494. [CrossRef] [PubMed]
36. Merkley, C.; Shuping, S.; Nestor, C. Neuronal networks that regulate gonadotropin-releasing hormone/luteinizing hormone secretion during undernutrition: Evidence from sheep. *Domest. Anim. Endocrinol.* **2020**. [CrossRef]
37. Goodman, R.; Lehman, M.; Smith, J.; Coolen, L.; De Oliveira, C.; Jafarzadehshirazi, M.; Pereira, A.; Iqbal, J.; Caraty, A.; Ciofi, P.; et al. Kisspeptin neurons in the arcuate nucleus of the ewe express both dynorphin a and neurokinin b. *Endocrinology* **2007**, *148*, 5752–5760. [CrossRef] [PubMed]
38. Broberger, C.; Johansen, J.; Johnasson, C.; Schalling, M.; Hokfelt, T. The neuropeptide y/agouti gene-related protein (agrp) brain circuitry in normal, anorectiC.; and monosodium glutamate-treated mice. *Proc. Natl. Acad. Sci. USA* **1998**, *95*, 15043–15048. [CrossRef]
39. Sheppard, K.; Padmanabhan, V.; Coolen, L.; Lehman, M. Prenatal programming by testosterone of hypothalamic metabolic control neurones in the ewe. *J. Neuroendocrinol.* **2011**, *23*, 401–411. [CrossRef]
40. Padilla, S.; Qiu, J.; Nestor, C.; Zhang, C.; Smith, A.; Whiddon, B.; Ronnekleiv, O.; Kelly, M.; Palmiter, R. Agrp to kiss1 neuron signaling links nutritional state and fertility. *Proc. Natl. Acad. Sci. USA* **2017**, *114*, 2413–2418. [CrossRef]

41. Mercer, R.; Chee, M.; Colmers, W. The role of npy in hypothalamic mediated food intake. *Front. Neuroendocrinol.* **2011**, *32*, 398–415. [CrossRef]
42. Anderson, E.; Cakir, I.; Carrington, S.; Cone, R.; Ghamari-Langroudi, M.; Gillyard, T.; Gimenez, L.; Litt, M. 60 years of pomc: Regulation of feeding and energy homeostasis by alpha-msh. *J. Mol. Endocrinol.* **2016**, *56*, t157–t174. [CrossRef]
43. Stanley, S.; Small, C.; Kim, M.; Heath, M.; Seal, L.; Russell, S.; Ghatei, M.; Bloom, S. Agouti related peptide (agrp) stimulates the hypothalamo pituitary gonadal axis in vivo and in vitro in male rats. *Endocrinology* **1999**, *140*, 5459–5462. [CrossRef]
44. Israel, D.; Sheffer-Babila, S.; De Luca, C.; Jo, Y.; Liu, S.; Xia, G.; Spergel, D.; Dun, S.; Dun, N.; Chua, S.J. Effects of leptin and melanocortin signaling interactions on pubertal development and reproduction. *Endocrinology* **2012**, *153*, 2408–2419. [CrossRef] [PubMed]
45. Roa, J.; Herbison, A. Direct regulation of gnrh neuron excitability by arcuate nucleus pomc and npy neuron neuropeptides in female mice. *Endocrinology* **2012**, *153*, 5587–5599. [CrossRef]
46. Yang, Y.-K.; Thompson, D.A.; Dickinson, C.J.; Wilken, J.; Barsh, G.S.; Kent, S.B.; Gantz, I. Characterization of agouti-related protein binding to melanocortin receptors. *Mol. Endocrinol.* **1999**, *13*, 148–155. [CrossRef] [PubMed]
47. Ollmann, M.M.; Wilson, B.D.; Yang, Y.-K.; Kerns, J.A.; Chen, Y.; Gantz, I.; Barsh, G.S. Antagonism of Central Melanocortin Receptors in Vitro and in Vivo by Agouti-Related Protein. *Sci.* **1997**, *278*, 135–138. [CrossRef]
48. Backholer, K.; Smith, J.; Clarke, I. Melanocortins may stimulate reproduction by activating orexin neurons in the dorsomedial hypothalamus and kisspeptin neurons in the preoptic area of the ewe. *Endocrinology* **2009**, *150*, 5488–5497. [CrossRef]
49. Manfredi-Lozano, M.; Roa, J.; Ruiz-Pino, F.; Piet, R.; Garcia-Galiano, D.; Pineda, R.; Zamora, A.; Leon, S.; Sanchez-Garrido, M.; Romero-Ruiz, A.; et al. Defining a novel leptin-melanocortin-kisspeptin pathway involved in the metabolic control of puberty. *Mol. Metab.* **2016**, *5*, 844–857. [CrossRef] [PubMed]
50. Sarvestani, F.; Tamadon, A.; Hematzadeh, A.; Jahanara, M.; Shirazi, M.; Moghadam, A.; Niazi, A.; Moghiminasr, R. Expression of melanocortin-4 receptor and agouti-related peptide mrnas in arcuate nucleus during long term malnutrition of female ovariectomized rats. *Iran. J. Basic Med. Sci.* **2015**, *18*, 104–107.
51. Mercer, A.; Stuart, R.; Attard, C.; Otero-Corchon, V.; Nillni, E.; Low, M. Temporal changes in nutritional state affect hypothalamic pomc peptide levels independently of leptin in adult male mice. *Am. J. Physiol. Endocrinol. Metab.* **2014**, *306*, e904–e915. [CrossRef] [PubMed]
52. Foradori, C.; Amstalden, M.; Goodman, R.; Lehman, M. Colocalisation of dynorphin a and neurokinin b immunoreactivity in the arcuate nucleus and median eminence of the sheep. *J. Neuroendocrinol.* **2006**, *18*, 534–541. [CrossRef] [PubMed]
53. Merkley, C.; Porter, K.; Coolen, L.; Hileman, S.; Billings, H.; Drews, S.; Goodman, R.; Lehman, M. Kndy (kisspeptin/neurokinin b/dynorphin) neurons are activated during both pulsatile and surge secretion of lh in the ewe. *Endocrinology* **2012**, *153*, 5406–5414. [CrossRef] [PubMed]

Article

Hypothalamic Expression of Neuropeptide Y (NPY) and Pro-OpioMelanoCortin (POMC) in Adult Male Mice Is Affected by Chronic Exposure to Endocrine Disruptors

Marilena Marraudino [1,2,†], Elisabetta Bo [1,2,†], Elisabetta Carlini [1,2], Alice Farinetti [1,2], Giovanna Ponti [2], Isabella Zanella [3], Diego Di Lorenzo [4], Gian Carlo Panzica [1,2] and Stefano Gotti [1,2,*]

1 Laboratory of Neuroendocrinology, Department of Neuroscience, University of Torino, 10126 Torino, Italy; marilena.marraudino@unito.it (M.M.); bo.elisabetta@gmail.com (E.B.); carlinielisabetta@hotmail.it (E.C.); alice.farinetti@unito.it (A.F.); giancarlo.panzica@unito.it (G.C.P.)
2 Neuroscience Institute Cavalieri-Ottolenghi (NICO), 10043 Orbassano, Italy; gponti2@gmail.com
3 Department of Molecular and Translational Medicine, University of Brescia, 25121 Brescia, Italy; isabella.zanella@unibs.it
4 Clinical Chemistry Laboratory, Diagnostic Department, ASST Spedali Civili di Brescia, 25123 Brescia, Italy; diego.dilorenzo@asst-spedalicivili.it
* Correspondence: stefano.gotti@unito.it; Tel.: +39-011-6706610; Fax: +39-011-2367054
† Equally contributed and should be considered as joint first authors.

Citation: Marraudino, M.; Bo, E.; Carlini, E.; Farinetti, A.; Ponti, G.; Zanella, I.; Di Lorenzo, D.; Panzica, G.C.; Gotti, S. Hypothalamic Expression of Neuropeptide Y (NPY) and Pro-OpioMelanoCortin (POMC) in Adult Male Mice Is Affected by Chronic Exposure to Endocrine Disruptors. *Metabolites* **2021**, *11*, 368. https://doi.org/10.3390/metabo11060368

Academic Editor: Amedeo Lonardo

Received: 24 March 2021
Accepted: 6 June 2021
Published: 9 June 2021

Publisher's Note: MDPI stays neutral with regard to jurisdictional claims in published maps and institutional affiliations.

Abstract: In the arcuate nucleus, neuropeptide Y (NPY) neurons, increase food intake and decrease energy expenditure, and control the activity of pro-opiomelanocortin (POMC) neurons, that decrease food intake and increase energy expenditure. Both systems project to other hypothalamic nuclei such as the paraventricular and dorsomedial hypothalamic nuclei. Endocrine disrupting chemicals (EDCs) are environmental contaminants that alter the endocrine system causing adverse health effects in an intact organism or its progeny. We investigated the effects of long-term exposure to some EDCs on the hypothalamic NPY and POMC systems of adult male mice that had been previously demonstrated to be a target of some of these EDCs after short-term exposure. Animals were chronically fed for four months with a phytoestrogen-free diet containing two different concentrations of bisphenol A, diethylstilbestrol, tributyltin, or E_2. At the end, brains were processed for NPY and POMC immunohistochemistry and quantitatively analyzed. In the arcuate and dorsomedial nuclei, both NPY and POMC immunoreactivity showed a statistically significant decrease. In the paraventricular nucleus, only the NPY system was affected, while the POMC system was not affected. Finally, in the VMH the NPY system was affected whereas no POMC immunoreactive material was observed. These results indicate that adult exposure to different EDCs may alter the hypothalamic circuits that control food intake and energy metabolism.

Keywords: endocrine disrupting chemicals; bisphenol A; diethylstilbestrol; tributyltin; neuropeptide Y; pro-opiomelanocortin

1. Introduction

Two neurochemically distinct sets of hypothalamic neurons controlling food intake are located in the arcuate nucleus (ARC). One group expresses neuropeptide Y (NPY) and agouti-related protein (AgRP). The NPY release by these neurons results in increased food intake and decreased energy expenditure. The other group expresses cocaine- and amphetamine-regulated transcript (CART) and pro-opiomelanocortin POMC, which is processed to melanocortin peptides, such as α-melanocyte-stimulating hormone (α-MSH). The activation of these neurons decreases food intake and increases energy expenditure [1] with an opposite effect of the NPY/AgRP system. Interactions between these two populations allow the NPY neurons to control the activity of the POMC cells. NPY/AgRP and POMC/CART neuronal projections reach hypothalamic nuclei such as the paraventricular nucleus (PVN), dorsomedial hypothalamic nucleus (DMH), and perifornical area [2]. These secondary centers process information regarding energy homeostasis.

Many factors can influence the activity of this system (for example the secretion of leptin by adipocytes), but estrogenic signaling may intersect at several levels with the hypothalamic circuits controlling food intake [3]. In fact, estradiol is involved in the regulation of metabolism through the modulation of food intake, body weight, glucose/insulin balance, body fat distribution, lipogenesis, lipolysis, and energy consumption [4]. The estradiol regulates neuroendocrine circuits controlling the metabolism [5] by acting on the POMC neurons through the estrogen receptor α (ERα) and on the NPY cells through an estrogen-activated membrane receptor, Gq-mER [6]. Indeed, estradiol has an inhibitor function on food intake, repressing the synthesis of NPY and AgRP [7]. Moreover, it seems that the leptin (secreted by adipocytes in proportion to fat mass and the activator of anorexigenic signals) has a common pathway with estradiol to regulate energy metabolism, namely the STAT3 pathway in POMC neurons [7]. Peripherally E_2 increases both leptin mRNA expression in 3T3 adipocytes and leptin secretion in omental adipose tissue [8]. Alternatively, lack of E_2 after ovariectomy may affect body weight regulation at a central level and mice deficient in ERα show a marked increase of adipose tissue [9]. There is also some evidence that ovariectomy increases hypothalamic NPY expression and decreases CRH immunoreactivity, promoting hyperphagia [10]. Moreover, E_2 deficiency causes central leptin insensitivity [9].

Endocrine-disrupting chemicals (EDCs) are industrial pollutants or natural molecules, which can be found as contaminants in the environment. They can interact with natural hormones by mimicking, antagonizing, or altering their actions [11] and may interfere with several brain circuits [12]. Recent evidence from many laboratories has shown that a variety of environmental EDCs (now called metabolic disrupting chemicals, MDCs) can influence adipogenesis and obesity and these effects may be partly mediated by sex steroid dysregulation due to the exposure to these substances and by alterations of nervous circuits involved in the control of food intake and energy metabolism [13,14].

In the present study, we analyzed three widely diffused MDCs—bisphenol A (BPA), diethylstilbestrol (DES), and tributyltin (TBT).

The BPA, one of the most diffused chemicals in the world, is a xenoestrogen present in a very large number of products and may affect multiple endocrine pathways, due to its ability to bind classical estrogen receptors (particularly ER-α) and non-classical ones (membrane receptors) [15], as well as the G-protein-coupled receptor 30 (GPR30) [16]. BPA can also act through non-genomic pathways [17] and bind to a variety of other hormone receptors (e.g., androgen receptor, thyroid hormone receptor, glucocorticoid receptor, and PPARγ) [18]. In vitro experiments have demonstrated that BPA may dysregulate NPY, AgRP, and POMC expression in hypothalamic immortalized cell lines [19–21].

The DES is a powerful nonsteroidal synthetic estrogen (pharmaceutical) used until the early 70s to prevent miscarriage in pregnant women. Later this compound was recognized as a cause of reproductive cancers, genital malformations, and infertility in sons or daughters that had been exposed to this drug in utero [22], but it is still in use for veterinary purposes in some countries and is bioaccumulated in the environment [23]. DES exerts an agonistic effect against ER-α and an antagonistic effect against estrogen-related receptor-γ (ERR-γ) [24]. In ovariectomized female rats exposed to an isoflavone-rich diet, DES had no effect on hypothalamic NPY mRNA and increased POMC mRNA [25].

TBT belongs to the EDC family of organotins, it has been employed primarily as an antifouling agent in paint for boats. Other uses are as a fungicide on food crops, and an antifungal agent in wood treatments and industrial and textile water systems [26]. Due to its use in paint for boats, TBT has exerted toxicological effects on marine organisms. For example, TBT can induce masculinization in fish species [27]. Humans are exposed to TBT largely through contaminated dietary sources (seafood and shellfish [28]). In mammals TBT can increase body weight [29], alter hypothalamic NPY and POMC systems in short-term (4 weeks) exposed adult mice [30,31], and may also alter behavior—exposure to a low dose of TBT induced lower activity, high level of anxiety, and fear in mice [32]. TBT binds with high affinity to steroid receptors; in particular, it binds androgen receptor [33] and interferes with the expression of brain aromatase and estrogen receptors [34]. TBT can act as an agonist of retinoid X receptor (RXR) and peroxisome proliferator-activated receptor-γ (PPARγ) [35]. This inappropriate receptor activation could lead to disruption of the normal developmental and homeostatic controls over adipogenesis and energy balance, especially under the influence of the typical high-fat Western diet [36]. In addition, changes in the microbiome are associated with TBT exposure [37].

As previously reported, studies on the action of EDC on hypothalamic neurons related to eating behavior and energy control used a variety of experimental conditions (exposure to isoflavones, in vitro experiments, and short-term exposure). For this reason, in the present study, we exposed, for a longer time period (4 months), adult male mice to phytoestrogen-free food containing different putative MDCs to understand if the central neuroendocrine, orexinergic, and anorexinergic circuits are differentially affected by these compounds. Due to the alleged xenoestrogenic activity of some of them we also included, as a positive control, a group of animals treated with E_2.

2. Results

2.1. Body Weight

At the end of the experiment the animals were weighted. Data collected showed a global effect of treatment on the body weight of exposed animals ($p < 0.05$, $F_{(8)} = 2.185$). In particular, the post-hoc analysis with Fisher's LSD test showed a reduction in body weight for mice treated with the higher dose of DES ($p < 0.05$) and for those treated with both doses of E_2 ($p < 0.05$). No statistically significant effects were observed in the other groups (see Table 1).

Table 1. Summary of statistical analysis of body weight data. The values (in grams) are indicated as mean ± standard error of the mean (SEM). Bold numbers and asterisks indicate significant differences among the differently treated groups: * $p < 0.05$, different from control ($p < 0.05$, Fisher's test).

Groups	Body Weight (g) Mean +/− SEM	*p* Value
CRL	31.2 ± 2.92	
TBT 0.5	31 ± 0.89	0.912
TBT 500	31.2 ± 0.80	1.000
DES 0.05	29.4 ± 1.21	0.323
DES 50	26.6 ± 0.93	0.015 *
BPA 5	28.6 ± 0.75	0.156
BPA 500	29.6 ± 0.93	0.379
E2 5	26.75 ± 0.48	0.025 *
E2 50	27.17 ± 0.65	0.025 *

2.2. Immunohistochemistry

2.2.1. NPY System

A preliminary qualitative analysis showed a distribution similar to those already reported in previous contributions [30,38–41]. In particular, we did not observe positive cell bodies (confirming previous reports that NPY cell bodies in ARC are visible only after colchicine treatment [42]), whereas a large number of positive fibers was observed along the entire hypothalamus. These were particularly dense within the PVN (Figure 1) and the ARC (Figure 2) nuclei, but they were also abundant within the suprachiasmatic, supraoptic, and DMH (Figure 2) nuclei. Other regions displayed less dense innervations, as for example, the VMH (Figure 2). In the experimental groups, we observed a qualitative decrease of the NPY immunoreactivity (ir) in all the considered nuclei for all the different treatments.

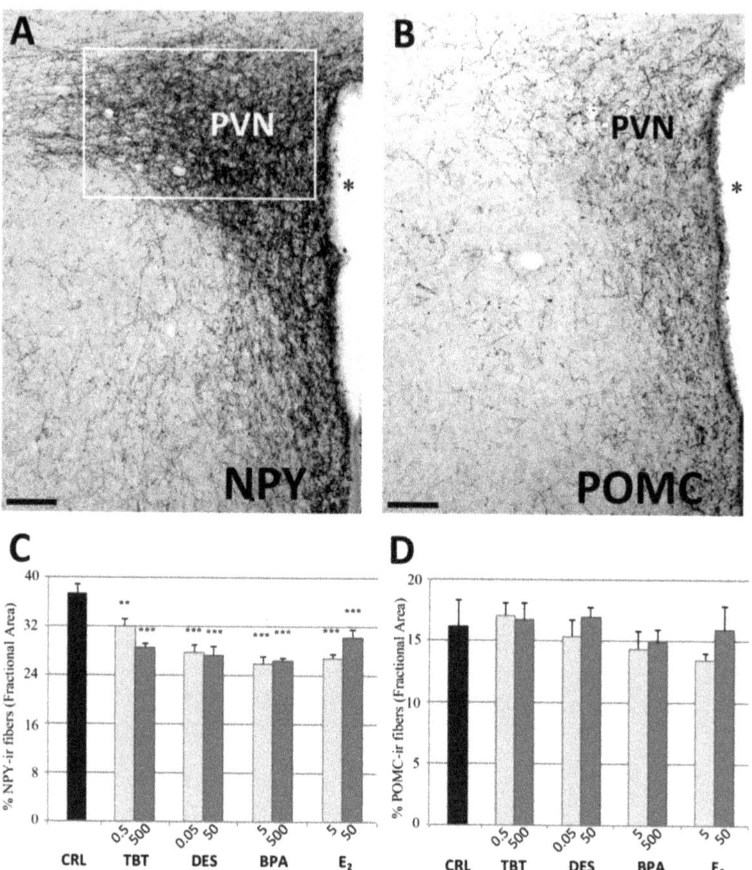

Figure 1. NPY and POMC immunohistochemistry in the PVN. Microphotographs and histograms illustrating the immunohistochemical immunoreactivity for NPY and POMC in the paraventricular nucleus (PVN). (**A**) Low magnification of a control mouse (CRL) illustrating the NPY immunoreactivity in PVN. The white box represent the ROI selected for the quantitative analysis. (**B**) Low magnification of a control mouse (CRL) illustrating the POMC immunoreactivity in PVN. Scale bar = 100 μm. * = Third ventricle. (**C**,**D**) Histograms illustrating the quantitative analysis of the fractional area covered by NPY (C) and POMC (D) immunoreactivity in the PVN. Bars represent the mean and the standard error of the mean (SEM). Asterisks indicate significant differences (Fisher's test) of the experimental groups in comparison to controls (CRL): ** $p < 0.01$, *** $p < 0.001$.

This qualitative impression was confirmed by the statistical analysis. For all nuclei we found a statistically significant effect of treatment (PVN: $p < 0.001$, $F_{(8)} = 10.672$; ARC: $p < 0.01$, $F_{(8)} = 3.566$; DMH: $p < 0.01$, $F_{(8)} = 3.767$; VMH: $p < 0.001$, $F_{(8)} = 5.780$).

The post-hoc analysis with Fisher's LSD test showed a significant decrease of NPYir in all nuclei and for almost all the treatments. In PVN, all groups showed a significantly lower NPYir than controls ($p < 0.01$, Figure 1). In ARC, we did not observe statistically significant differences for the lowest dose of TBT and the highest dose of BPA, while all the other treatments induced a significant decrease of NPY expression ($p < 0.05$, Figure 2). In DMH we observed a significant reduction of NPYir in all treated groups ($p < 0.05$; Figure 2), except for the lowest dose of DES. Finally, in VMH we observed a strong reduction of NPYir due to the treatments ($p < 0.01$) except for the highest dose of TBT (for details see Table S2, Supplementary Materials).

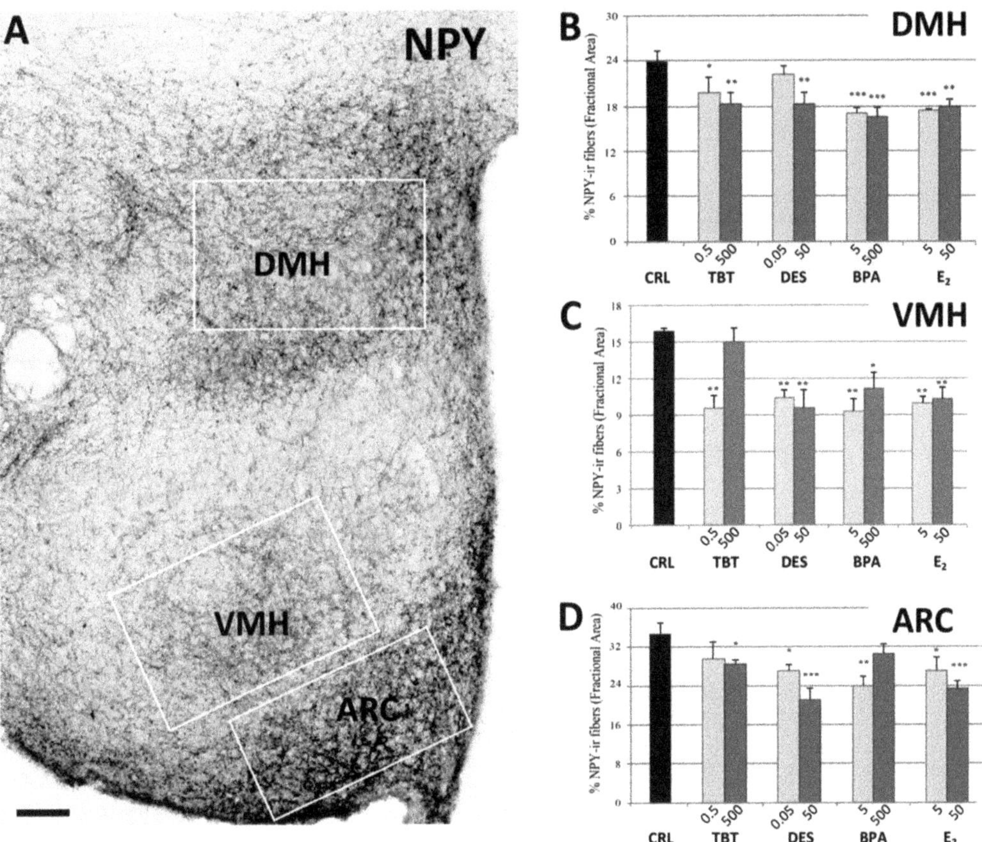

Figure 2. NPY immunohistochemistry. Microphotograph and histograms illustrating the immunohistochemical immunoreactivity for NPY in the dorsomedial (DMH), ventromedial (VMH), and arcuate (ARC) nuclei. (**A**) Low magnification of the hypothalamic region of a control mouse (CRL) illustrating the NPY immunoreactivity in DMH, VMH, and ARC nuclei. The white boxes represent the ROI selected for each nucleus in the quantitative analysis. Scale bar = 100 μm. (**B–D**) Histograms illustrating the quantitative analysis of the fractional area covered by NPY immunoreactivity in the DMH (B), VMH (C), and ARC (D) nuclei in the different experimental groups. Bars represent the mean and the standard error of the mean (SEM). Asterisks indicate significant differences (Fisher's test) of the experimental groups in comparison to controls (CRL): * $p < 0.05$, ** $p < 0.01$, *** $p < 0.001$.

2.2.2. POMC System

The distribution of POMCir in control mice was in agreement with the few previous studies that described this system in rats [43–45] and mice [31,46]. Contrary to NPY, hypothalamic POMC cell bodies are clearly visible, and they were fully included within the rostrocaudal extent of the ARC (Figure 3) and periarcuate area, which also showed a local dense innervation of ir fibers. Two major targets of this system are the PVN and the DMH. In the PVN, POMCir fibers outlined the entire nucleus, starting from its rostral portion up to the more caudal levels. The distribution of these fibers was not homogeneous, in particular they were denser in the medial PVN (corresponding to the parvocellular regions of this nucleus) compared to the lateral PVN (corresponding to the magnocellular region) (Figure 1). The DMH nucleus (Figure 3) showed a denser innervation in the caudal part of the nucleus compared with the rostral part. Other hypothalamic nuclei, such as the VMH, did not show a significant number of positive fibers.

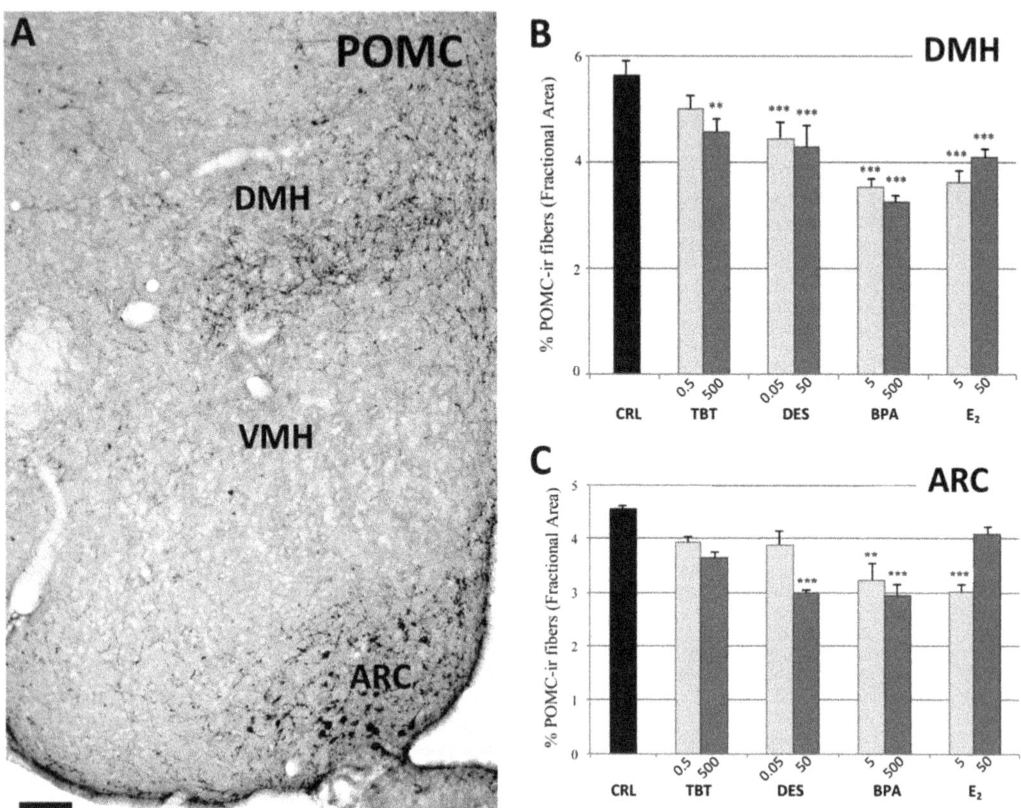

Figure 3. POMC immunohistochemistry. Microphotograph and histograms illustrating the immunohistochemical immunoreactivity for POMC in the dorsomedial (DMH), and arcuate (ARC) nuclei. (**A**) Low magnification of the hypothalamic region of a control mouse (CRL) illustrating the POMC immunoreactivity in DMH, and ARC nuclei. Due to the extreme paucity of immunoreactive structures, it was not possible to measure POMC immunoreactivity in the VMH. Scale bar = 100 μm. (**B**,**C**) Histograms illustrating the quantitative analysis of the fractional area covered by POMC immunoreactivity in the DMH (B), and ARC (C) nuclei in the different experimental groups. Bars represent the mean and the standard error of the mean (SEM). Asterisks indicate significant differences (Fisher's test) of the experimental groups in comparison to controls (CRL): ** $p < 0.01$, *** $p < 0.001$.

In the PVN we did not observe variations due to treatment. In fact, the statistical analysis showed no effect of treatment (F =1.097, $p = 0.396$, Figure 1). On the contrary, data collected in the ARC showed a decrease of the POMCir (including positive cell bodies and fibers), following the different treatments ($F_{(8)} = 8.289$, $p < 0.001$). The Fisher LSD test showed a statistically significant decrease in the groups treated with the highest dose of DES ($p < 0.001$), the lowest of E_2 ($p < 0.001$) and in both groups treated with BPA ($p < 0.001$, Figure 3).

The quantitative analysis also showed a decrease of the POMCir in the DMH following different treatments ($F_{(8)} = 19.563$, $p < 0.001$). The Fisher LSD test showed a decrease of

POMC ir in all groups compared to controls, except for the lowest dose of TBT (Figure 3; for details see Table S2, Supplementary Materials).

3. Discussion

The control of energy metabolism and food intake is in part dependent on central neuroendocrine circuits that have been detailed in the introduction. Among the various systems, the NPY and the POMC systems (both located in the hypothalamic arcuate nucleus and sending their axons to other hypothalamic nuclei) exert orexigenic (NPY) and anorexigenic (POMC) effects. Several studies (recently reviewed by [14]) demonstrated that these neural circuits are altered when the animals are exposed to some environmental compounds that are now classified as metabolism-disrupting chemicals (MDCs) [13,47].

In the present study, we showed that some of the putative MDCs, when chronically administered through a phytoestrogen-free diet (reported in the literature as inducing body weight gain [48]), affected the expression of both NPY and POMC in the hypothalamic circuits of adult male mice. For comparison, we included two additional groups, one without any treatment (control group) and the second one exposed to E_2 (added to the diet), which has a well-known anti-adipogenic effect [49,50].

As expected, in the present experiment, both doses of E_2 induced a significant reduction of the body weight in comparison to the control group. On the contrary, male mice fed with the same diet but with different concentrations of three different EDCs, except the group treated with the highest dose of DES, did not show any significant reduction of the body weight. These results suggest that BPA, DES, and TBT are not able, in adult male mice, to counteract the consequence of an exposure to a phytoestrogen-free diet on the body weight, whereas E_2 is able to do this. Therefore, whereas E_2 has an anti-obesogenic effect, the EDCs considered in this study do not show this property. It is possible that the lack of effect on body weight is due to the fact that the reduction of the activity of the orexinergic circuits originated by the reduction of NPY is compensated by the reduction of the activity of the anorexinergic circuits caused by the reduction of the expression of POMC.

Our data show that the NPY expression in male mice hypothalamic nuclei involved in food intake regulation is reduced by E_2 as well as by all tested EDCs at almost all doses. Therefore DES, BPA, and TBT have the same effect of E_2 on the NPY system. In particular, DES and BPA have a well-known strong xenoestrogenic activity because they specifically bind to ERs [51]. On the contrary, TBT does not bind ERs, but it also has xenoandrogenic or antiandrogenic activity [52]. The reduction of NPY expression in the hypothalamus after TBT treatment confirms our previous results [30] and may be due to the activation of other pathways, not directly regulated by E_2.

The effects of treatments on the POMC system of male mice are less homogeneous. In fact, we observed significant effects on ARC and DMH, while in the PVN we have not detected significant effects. DES, BPA, and also E_2 significantly decreased the POMC expression in ARC, while TBT showed no significant effect. It is important to note that 30% of POMC cells in ARC colocalize with ERα while they do not express ERβ [53], thus suggesting a possible direct role of ERs in regulating part of this system that represents, consequently, a putative target for xenoestrogens, like BPA and DES. The lack of TBT effect is also in line with our recent results that showed no effects of TBT on the POMC system in adult male mice [31].

The POMC neurons of the ARC send axons to two main targets, the DMH and the PVN. All treatments (including TBT at the highest dose) induced a significant decrease in the immunoreactivity in the DMH, whereas no effect was detected in the PVN, even when the quantitative analysis was performed on the different parts of the PVN, according to the method detailed in [54] (results summarized in Figure S3 of Supplementary Material). It is still possible that the paucity of POMC fibers in the PVN (compared to the NPY ones) has prevented the detection of small differences in the present experimental material.

In a limited number of experimental groups, the tested EDCs showed a significant effect in reducing immunoreactivity at the low dose and not at the high dose, for example see the effect of TBT on VMH NPY immunoreactivity, the effect of BPA on ARC NPY immunoreactivity, or the effect of E_2 on ARC POMC immunoreactivity. These results confirm the nonmonotonic dose response described in many experimental situations for several EDCs [55]. The differences of the results of EDC treatments on NPY and POMC immunoreactivity with those obtained with E_2, are probably due to the activation of pathways not directly or indirectly regulated by E_2. For example, it has been found that intracerebroventricular injections of oxytocin (OT) in adult fasted male rats decreases food intake [56]. Moreover, a retrograde tracer study revealed OT projection from PVN and SON to ARC, demonstrating that oxytocinergic signaling may regulate feeding [57]. OT cells, expressing ER-β, of the PVN [58], are a possible target for xenoestrogen that binds ER-β, like the phytoestrogen genistein [59]. This suggests that some EDCs may alter POMC expression via the OT system. However, the physiological significance of the OT neuronal projections from PVN and SON to ARC POMC neurons, still remains unclear, and further studies are required to clarify it.

One of the most important regulators of the NPY [60] and POMC [61] systems is represented by the cannabinoid receptor CB1. Some EDCs may modulate the expression of this receptor: prolonged exposure to DES produced a reduction in the mRNA for CB1 receptor in the rat pituitary [62], while BPA caused a downregulation of CB1 receptor in the mice hypothalamus [63]. No data are yet available for an action of TBT on the expression of CB1 receptor. Therefore, it is possible that present results on the alterations of NPY and POMC systems are partly due to an effect of the EDCs on the expression of CB1 receptor and a consequent functional alteration of these two systems. Future work should clarify this aspect.

The levels of circulating glucose are also important in controlling the NPY and POMC circuits, through glucose sensitive neurons located in the VMH and LH (for a recent review see [64]). All the three EDCs analyzed in this study disrupt glucose homeostasis by acting on pancreatic islets [37,65,66]. Even if in the present study we have not detected glucose blood levels, it is therefore possible that part of the dysregulation of the NPY and POMC systems is due to alterations of glucose homeostasis.

Another crucial point is that we do not know, at the moment, if we are observing an activational or an organizational effect of these EDCs. In the first case we may expect that the differences in the expression of immunoreactivity are due to an increase or a decrease in the production of neuropeptides in stable circuits (see the effects of BPA on NPY mRNA in neuronal cell cultures [20]). In the second case the hypothesis is that the exposure to the EDCs may induce permanent (or long-term) changes in the observed circuits. In fact, it has been demonstrated that gonadal hormones produced during puberty are inducing neurogenesis in some hypothalamic [67] or extrahypothalamic [68] nuclei and that this process is necessary to stabilize the sexual differences evidenced in these nuclei. A recent review [69] analyzed the available data for the development of hypothalamic circuits that control food intake and energy balance. In summary, in these circuits neurogenesis is only present during the prenatal period [70], but the full maturation of the connections ARC–PVN is reached during the postnatal days 28–35 [71]. However, more recent studies demonstrated that adult neurogenesis of NPY and POMC neurons in mice ARC is stimulated by changes among high fat–low fat diets [72]. Being our animal was three weeks old, it is therefore possible that exposure to EDCs had altered the connection of ARC towards VMH, DMH, and PVN, or even determined a change in the number of NPY and POMC neurons (BPA may induce apoptosis in hippocampal cells [73]). According to this hypothesis the observed changes in the immunoreactivity could be linked to an alteration (plasticity) of fibers' system reaching these nuclei. At the moment it is impossible to know if NPY and POMC circuits, after such a long exposure to EDCs, when provided with EDCs-free food, may recover to a status comparable to the non-treated animals (this is compatible with an activational effect). Future studies should elucidate this point, in particular not only if there is a recovery, but also how long it will take to recover.

In conclusion, these data, together with those already present in the literature, suggest that EDCs may alter energy metabolism not only at the level of peripheral tissues [13], but also in neuroendocrine circuits involved in the control of food intake, in particular, the NPY and POMC systems. The control of physiological processes by these systems is highly complex, making the understanding of neuroendocrine disruption a particular challenge.

4. Materials and Methods

4.1. Animals and Treatment

The procedures involving animals and their care were performed in Brescia according to the Union Council Directive of 22 September 2010 (2010/63/UE). The study was approved by the Ethical Committee of Animal Experimentation of the Hospital and the Italian Minister of Health (407/2018-PR). All care was taken to use the minimum number of animals.

C57BJ/6 male mice (Harlan, Udine) were housed in same-sex groups of 4 per cage on a 12:12-h light/dark cycle; animal rooms were maintained at a temperature of 23 °C. Estrogen-free diet was purchased from Dottori Piccioni S.r.L. Via Guglielmo Marconi, 29/31 Gessate (MI, Italy) (https://totofood.it/, assessed on 7 June 2021). The diet was prepared in pellets (the composition is reported in Table S1, Supplementary Materials).

The treatment started when mice were three weeks old and lasted for four months. Animals were divided randomly in nine experimental groups: control mice were fed with the base diet (estrogen-free diet) while experimental groups were fed with the base diet enriched with two different concentrations of E_2, BPA, DES, or TBT (according to previous studies [74]). All the chemicals were obtained from Sigma-Aldrich, Milano, Italy, dissolved in DMSO and further diluted before their addition to the diet, for homogeneous preparations. These are the doses used: E_2 (stock solution 97%, cat. number E8515; 5 or 50 μg/kg diet); BPA (stock solution 99%, cat. number 239658; 5 or 500 μg/kg diet); DES (stock solution 99%, cat. number D-4628; 0.05 or 50 μg/kg diet); and TBT (stock solution 96%, cat. number T50202; 0.5 or 500 μg/kg diet).

The normal food consumption in adult mice corresponds to 15g/100g body weight/day [75]; since mice used in this experiment had a mean body weight of 30g, it was considered an approximate consumption of 4.5 g food/day was appropriate. Accordingly, in this case mice were exposed daily to approximately 0.15–1.5 μg/g body weight of E_2, 0.15–15 μg/g body weight of BPA, 0.0015–1.5 μg/g body weight of DES, and 0.015–15 μg/g body weight of TBT.

Body weights were recorded at the end of the experiment, before sacrifice (see Table 1).

Food consumption was monitored every two days as the difference between the weight of the pellets supplied and that consumed. Spilled food, if any, was collected in apposite trays underneath the food containers, measured, and taken into account.

4.2. Tissue Sampling and Histological Examination

Four months after the beginning of treatment adult mice were deeply anesthetized with an intraperitoneal injection of a mixture of ketamine (100 mg/kg of body weight, Ketavet, Gelling, Italy) and xylazine (10 mg/kg of body weight, Rompun, Bayer, Germany) solution, monitored until the pedal reflex was abolished and killed by cervical dislocation. Animals were decapitated, brains were quickly dissected and placed into acrolein (5% in 0.01 M saline phosphate buffer, PBS) for 150 min at room temperature. Brains were rinsed several times in PBS, placed overnight in a 30% sucrose solution in PBS at 4 °C, frozen in liquid isopentane at −40 °C and stored in a deep freezer at −80 °C until sectioning.

Brains (N = 4 for each group) were serially cut in the coronal plane with a cryostat (Leica CM 1900) at 25 μm of thickness. Sections were collected in four series for free-floating procedure in multiwell dishes, filled with a cryoprotectant solution [76] and stored at −20 °C until used for immunohistochemistry. One series of sections was stained for NPY immunohistochemistry and another for POMC immunohistochemistry. Brain sections were always stained in groups containing each treatment, so that between-assay variance could not cause systematic group differences.

After overnight washing in PBS, sections were incubated in 0.01% sodium borohydride for 20 min to remove the acrolein and rinsed in PBS several times. Then, sections were exposed to Triton X-100 (0.2% in PBS) for 30 min and treated for blocking endogenous peroxidase activity with PBS solution containing methanol/hydrogen peroxide for 20 min. Sections were afterwards incubated with normal goat serum (Vector Laboratories, Burlingame, CA, USA) for 30 min. One series was incubated overnight at 4 °C with the rabbit polyclonal antibody against synthetic porcine NPY (gift by Professor Vaudry, France) diluted 1:5000 in 0.2% PBS-Triton X-100, pH 7.3–7.4 and another with the rabbit polyclonal antibody against POMC (Phoenix Pharmaceuticals, Inc.,Burlingame, CA USA) [31,77,78] diluted 1:5000 in 0.2% PBS-Triton X-100 and 1% of BSA, pH 7.3–7.4. The next day, sections were incubated for 60 min in biotinylated goat anti-rabbit IgG (Vector Laboratories, Burlingame, CA, USA) 1:200. The antigen–antibody reaction was revealed by 60 min incubation with the biotin–avidin system (Vectastain ABC Kit Elite, Vector Laboratories, Burlingame, CA, USA). The peroxidase activity was visualized with a solution containing 0.400 mg/mL of 3,3′-diamino-benzidine (DAB, Sigma–Aldrich, Milano, Italy) and 0.004% hydrogen peroxide in 0.05 M Tris–HCl buffer, pH 7.6. Sections were mounted on chromallum-coated slides, air-dried, cleared in xylene, and cover slipped with Entellan (Merck, Milano, Italy).

The production and characterization of NPY polyclonal antibody has been previously reported [79,80] and it has been employed to detect the NPY system in a wide range of species [40].

The POMC antibody from Phoenix Pharmaceuticals recognizes a sequence corresponding to N terminal amino acids 27–52 of Pig POMC precursor and has often been used in mouse and rat studies [31,42,81].

We performed the following additional controls in our material: (a) the primary antibody was omitted or replaced with an equivalent concentration of normal serum (negative controls) and (b) the secondary antibody was omitted. In these conditions, cells and fibers were completely unstained.

4.3. Quantitative Analysis

All sections were acquired with a NIKON Digital Sight DS-Fi1 video camera connected to a NIKON Eclipse 80i microscope (Nikon Italia S.p.S., Firenze, Italy). The staining density of NPY- and POMC-immunoreactive (ir)-containing structures was measured in selected nuclei with the freeware ImageJ (version 1.49b, Wayne Rasband, NIH, Bethesda, MD, USA) by calculating in binary transformations of the images (threshold function) the fractional area (percentages of pixels) covered by immunoreactive structures in predetermined fields (area of interest, ROI). Due to differences in the immunostaining, according to our previous reports [50,63], the range of the threshold was individually adjusted for each section.

For quantification of NPY and POMC systems we selected four hypothalamic nuclei involved in controlling food intake—ARC, DMH, PVN, and ventromedial hypothalamic nucleus (VMH). For each nucleus, we measured the density of immunoreactive structures on three consecutive sections identified by the Mouse Brain Atlas (ARC, VMH, DMH: bregma −1.46mm, −1.58mm, −1.70mm; PVN: bregma −0.70mm, −0.82mm, −0.94mm [82,83].

The ROI selected for each nucleus was a box of fixed size and shape, selected to cover immunoreactive material only within the boundaries of each nucleus (about 140,000 μm^2 for VMH and DMH, 110,000 μm^2 for ARC, and 200,000 μm^2 for PVN). Due to the extreme paucity of immunoreactive structures, it was not possible to measure POMC-immunoreactivity in the VMH.

4.4. Statistical Analysis

Collected data were analyzed with the program SPSS 24.0 (SPSS Inc., Chicago, IL, USA); the p values and the significance threshold were set at $p \leq 0.05$. Data collected for the body weight were analyzed by one-way ANOVA followed by post-hoc analysis with a Fisher LSD test. Data collected for the immunohistochemistry were analyzed by repeated-measure one-way ANOVA. When the analysis did not show significant differences between different levels of the same nucleus, we calculated a mean value for each nucleus that was used to assess variations due to the treatment. When statistically significant, the ANOVA analysis was followed by a Fisher LSD test.

Supplementary Materials: The following are available online at https://www.mdpi.com/article/10.3390/metabo11060368/s1, Figure S1: NPY immunoreactivity. Immunohistochemical comparison of NPY immunoreactivity among control animals (CRL) and the different treated groups (in all case it was shown to be the lowest dose used) in the dorsomedial (DMH), ventromedial (VMH), arcuate (ARC), and paraventricular (PVN) nuclei. Estradiol, E2; tributyltin, TBT; diethylstilbestrol, DES; bisphenol A, BPA. Scale bar = 100 µm, Figure S2: POMC immunoreactivity. Immunohistochemical comparison of POMC immunoreactivity among control animals (CRL) and the different treated groups (in all case it was shown to be the lowest dose used) in the dorsomedial (DMH), arcuate (ARC), and paraventricular (PVN) nuclei. Estradiol, E2; tributyltin, TBT; diethylstilbestrol, DES; bisphenol A, BPA. Scale bar = 100 µm, Figure S3: Regional analysis of POMC immunoreactivity in the PVN. To further confirm the absence of variations in the POMC expression within the PVN, we measured the immunoreactivity, according to our previous studies [54], by dividing the PVN into four quadrants: dorsomedial (DM), dorsolateral (DL), ventromedial (VM), and ventrolateral (VL). The results of this analysis reported no significant differences for all the analyzed subregions and are summarized in the histograms (B—E). Scale bar = 100 µm, Table S1. Composition of the soy-free diet (SFSD), Table S2. Summary of quantitative analysis of the fractional area. Fractional area data in the different nuclei and in the different groups analyzed in this study. The values reported are the mean and standard error of the mean (SEM). Bold numbers and asterisks indicate significant differences (Fisher's test) among the differently treated groups: * $p < 0.05$, ** $p < 0.01$, *** $p < 0.001$ different from control.

Author Contributions: M.M. and E.B. performed experiments, analyzed data, and wrote the paper. E.C., A.F., G.P. (Giovanna Ponti), and I.Z. performed the experiments. D.D.L., G.C.P. (GianCarlo Panzica), and S.G. designed experiments, and wrote and supervised the paper. D.D.L. and I.Z. hosted, maintained, and treated the mice until sacrifice. All authors have read and agreed to the published version of the manuscript.

Funding: This work was supported by European Union Grants QLK4-CT-2002-02221 (EDERA) and LSHB-CT-2006-037168 (EXERA), Ministero dell'Istruzione, dell'Università e della Ricerca–MIUR project Dipartimenti di Eccellenza 2018–2022 to Department of Neuroscience Rita Levi Montalcini; University of Torino, Ricerca locale to GP, GCP, SG, and Cavalieri-Ottolenghi Foundation, Orbassano, Italy.

Institutional Review Board Statement: The study was performed in Brescia according to the Union Council Directive of 22 September 2010 (2010/63/UE); The study was approved by the Ethical Committee of Animal Experimentation of the Hospital and the Italian Minister of Health (407/2018-PR, 02/06/2018).

Informed Consent Statement: Not applicable.

Data Availability Statement: All the data are available from the authors upon reasonable request. The data presented in this study are available in this supplementary.

Acknowledgments: MM was a fellow of G.C. Bergui (2020) and is now a fellow of the Fondazione Umberto Veronesi (2021).

Conflicts of Interest: The authors declare no conflict of interest.

References

1. Broberger, C. Brain regulation of food intake and appetite: Molecules and networks. *J. Intern. Med.* **2005**, *258*, 301–327. [CrossRef]
2. Kalra, S.P.; Dube, M.G.; Pu, S.; Xu, B.; Horvath, T.L.; Kalra, P.S. Interacting appetite-regulating pathways in the hypothalamic regulation of body weight. *Endocr. Rev.* **1999**, *20*, 68–100. [CrossRef]
3. Grun, F.; Blumberg, B. Endocrine disrupters as obesogens. *Mol. Cell Endocrinol.* **2009**, *304*, 19–29. [CrossRef]
4. Clegg, D.J. Minireview: The year in review of estrogen regulation of metabolism. *Mol. Endocrinol.* **2012**, *26*, 1957–1960. [CrossRef]

5. Stincic, T.L.; Ronnekleiv, O.K.; Kelly, M.J. Diverse actions of estradiol on anorexigenic and orexigenic hypothalamic arcuate neurons. *Horm. Behav.* **2018**, *104*, 146–155. [CrossRef] [PubMed]
6. Roepke, T.A.; Bosch, M.A.; Rick, E.A.; Lee, B.; Wagner, E.J.; Seidlova-Wuttke, D.; Wuttke, W.; Scanlan, T.S.; Ronnekleiv, O.K.; Kelly, M.J. Contribution of a membrane estrogen receptor to the estrogenic regulation of body temperature and energy homeostasis. *Endocrinology* **2010**, *151*, 4926–4937. [CrossRef] [PubMed]
7. Gao, Q.; Horvath, T.L. Cross-talk between estrogen and leptin signaling in the hypothalamus. *Am. J. Physiol. Endocrinol. Metab.* **2008**, *294*, E817–E826. [CrossRef]
8. Casabiell, X.; Pineiro, V.; Peino, R.; Lage, M.; Camina, J.; Gallego, R.; Vallejo, L.G.; Dieguez, C.; Casanueva, F.F. Gender differences in both spontaneous and stimulated leptin secretion by human omental adipose tissue in vitro: Dexamethasone and estradiol stimulate leptin release in women, but not in men. *J. Clin. Endocrinol. Metab.* **1998**, *83*, 2149–2155. [CrossRef]
9. Ainslie, D.A.; Morris, M.J.; Wittert, G.; Turnbull, H.; Proietto, J.; Thorburn, A.W. Estrogen deficiency causes central leptin insensitivity and increased hypothalamic neuropeptide Y. *Int. J. Obes.* **2001**, *25*, 1680–1688. [CrossRef]
10. Shimizu, H.; Ohtani, K.; Kato, Y.; Tanaka, Y.; Mori, M. Withdrawal of estrogen increases hypothalamic neuropeptide Y (NPY) mRNA expression in ovariectomized obese rat. *Neurosci. Lett.* **1996**, *204*, 81–84. [CrossRef]
11. La Merrill, M.A.; Vandenberg, L.N.; Smith, M.T.; Goodson, W.; Browne, P.; Patisaul, H.B.; Guyton, K.Z.; Kortenkamp, A.; Cogliano, V.J.; Woodruff, T.J.; et al. Consensus on the key characteristics of endocrine-disrupting chemicals as a basis for hazard identification. *Nat. Rev. Endocrinol.* **2020**, *16*, 45–57. [CrossRef]
12. Panzica, G.C.; Bo, E.; Martini, M.A.; Miceli, D.; Mura, E.; Viglietti-Panzica, C.; Gotti, S. Neuropeptides and enzymes are targets for the action of endocrine disrupting chemicals in the vertebrate brain. *J. Toxicol. Environ. Health Part B* **2011**, *14*, 449–472. [CrossRef] [PubMed]
13. Heindel, J.J.; Blumberg, B.; Cave, M.; Machtinger, R.; Mantovani, A.; Mendez, M.A.; Nadal, A.; Palanza, P.; Panzica, G.; Sargis, R.; et al. Metabolism disrupting chemicals and metabolic disorders. *Reprod. Toxicol.* **2017**, *68*, 3–33. [CrossRef] [PubMed]
14. Marraudino, M.; Bonaldo, B.; Farinetti, A.; Panzica, G.; Ponti, G.; Gotti, S. Metabolism Disrupting Chemicals and Alteration of Neuroendocrine Circuits Controlling Food Intake and Energy Metabolism. *Front. Endocrinol.* **2018**, *9*, 766. [CrossRef]
15. Alonso-Magdalena, P.; Laribi, O.; Ropero, A.B.; Fuentes, E.; Ripoll, C.; Soria, B.; Nadal, A. Low doses of bisphenol A and diethylstilbestrol impair Ca2+ signals in pancreatic alpha-cells through a nonclassical membrane estrogen receptor within intact islets of Langerhans. *Environ. Health Perspect.* **2005**, *113*, 969–977. [CrossRef]
16. Thomas, P.; Dong, J. Binding and activation of the seven-transmembrane estrogen receptor GPR30 by environmental estrogens: A potential novel mechanism of endocrine disruption. *J. Steroid Biochem. Mol. Biol.* **2006**, *102*, 175–179. [CrossRef]
17. Ropero, A.B.; Alonso-Magdalena, P.; Ripoll, C.; Fuentes, E.; Nadal, A. Rapid endocrine disruption: Environmental estrogen actions triggered outside the nucleus. *J. Steroid Biochem. Mol. Biol.* **2006**, *102*, 163–169. [CrossRef]
18. MacKay, H.; Abizaid, A. A plurality of molecular targets: The receptor ecosystem for bisphenol-A (BPA). *Horm. Behav.* **2017**. [CrossRef] [PubMed]
19. Loganathan, N.; Salehi, A.; Chalmers, J.A.; Belsham, D.D. Bisphenol A Alters Bmal1, Per2, and Rev-Erba mRNA and Requires Bmal1 to Increase Neuropeptide Y Expression in Hypothalamic Neurons. *Endocrinology* **2019**, *160*, 181–192. [CrossRef]
20. Loganathan, N.; McIlwraith, E.K.; Belsham, D.D. BPA Differentially Regulates NPY Expression in Hypothalamic Neurons Through a Mechanism Involving Oxidative Stress. *Endocrinology* **2020**, *161*. [CrossRef] [PubMed]
21. Loganathan, N.; McIlwraith, E.K.; Belsham, D.D. Bisphenol A induces Agrp gene expression in hypothalamic neurons through a mechanism involving ATF3. *Neuroendocrinology* **2020**. [CrossRef] [PubMed]
22. Goldberg, J.M.; Falcone, T. Effect of diethylstilbestrol on reproductive function. *Fertil. Steril.* **1999**, *72*, 1–7. [CrossRef]
23. Tapiero, H.; Nguyen Ba, G.; Tew, K.D. Estrogens and environmental estrogens. *Biomed. Pharmacother.* **2002**, *56*, 36–44. [CrossRef]
24. Nam, K.; Marshall, P.; Wolf, R.M.; Cornell, W. Simulation of the different biological activities of diethylstilbestrol (DES) on estrogen receptor alpha and estrogen-related receptor gamma. *Biopolymers* **2003**, *68*, 130–138. [CrossRef] [PubMed]
25. Zhang, Y.B.; Zhang, Y.; Li, L.N.; Zhao, X.Y.; Na, X.L. Soy isoflavone and its effect to regulate hypothalamus and peripheral orexigenic gene expression in ovariectomized rats fed on a high-fat diet. *Biomed. Environ. Sci.* **2010**, *23*, 68–75. [CrossRef]
26. Zuo, Z.; Chen, S.; Wu, T.; Zhang, J.; Su, Y.; Chen, Y.; Wang, C. Tributyltin causes obesity and hepatic steatosis in male mice. *Environ. Toxicol.* **2011**, *26*, 79–85. [CrossRef]
27. McAllister, B.G.; Kime, D.E. Early life exposure to environmental levels of the aromatase inhibitor tributyltin causes masculinisation and irreversible sperm damage in zebrafish (Danio rerio). *Aquat. Toxicol.* **2003**, *65*, 309–316. [CrossRef]
28. Golub, M.; Doherty, J. Triphenyltin as a potential human endocrine disruptor. *J. Toxicol. Environ. Health B Crit. Rev.* **2004**, *7*, 281–295. [CrossRef]
29. Si, J.; Wu, X.; Wan, C.; Zeng, T.; Zhang, M.; Xie, K.; Li, J. Peripubertal exposure to low doses of tributyltin chloride affects the homeostasis of serum T, E2, LH, and body weight of male mice. *Environ. Toxicol.* **2011**, *26*, 307–314. [CrossRef]
30. Bo, E.; Farinetti, A.; Marraudino, M.; Sterchele, D.; Eva, C.; Gotti, S.; Panzica, G. Adult exposure to tributyltin affects hypothalamic neuropeptide Y, Y1 receptor distribution, and circulating leptin in mice. *Andrology* **2016**, *4*, 723–734. [CrossRef] [PubMed]
31. Farinetti, A.; Marraudino, M.; Ponti, G.; Panzica, G.; Gotti, S. Chronic treatment with tributyltin induces sexually dimorphic alterations in the hypothalamic POMC system of adult mice. *Cell Tissue Res.* **2018**, *374*, 587–594. [CrossRef]

32. Ohtaki, K.; Aihara, M.; Takahashi, H.; Fujita, H.; Takahashi, K.; Funabashi, T.; Hirasawa, T.; Ikezawa, Z. Effects of tributyltin on the emotional behavior of C57BL/6 mice and the development of atopic dermatitis-like lesions in DS-Nh mice. *J. Dermatol. Sci.* **2007**, *47*, 209–216. [CrossRef]
33. Yamabe, Y.; Hoshino, A.; Imura, N.; Suzuki, T.; Himeno, S. Enhancement of androgen-dependent transcription and cell proliferation by tributyltin and triphenyltin in human prostate cancer cells. *Toxicol. Appl. Pharmacol.* **2000**, *169*, 177–184. [CrossRef]
34. Zhang, J.; Zuo, Z.; Zhu, W.; Sun, P.; Wang, C. Sex-different effects of tributyltin on brain aromatase, estrogen receptor and retinoid X receptor gene expression in rockfish (Sebastiscus marmoratus). *Mar. Environ. Res.* **2013**, *90*, 113–118. [CrossRef] [PubMed]
35. Nakanishi, T. Endocrine disruption induced by organotin compounds; organotins function as a powerful agonist for nuclear receptors rather than an aromatase inhibitor. *J. Toxicol. Sci.* **2008**, *33*, 269–276. [CrossRef]
36. Grun, F.; Blumberg, B. Environmental obesogens: Organotins and endocrine disruption via nuclear receptor signaling. *Endocrinology* **2006**, *147*, S50–S55. [CrossRef]
37. Zhan, J.; Ma, X.; Liu, D.; Liang, Y.; Li, P.; Cui, J.; Zhou, Z.; Wang, P. Gut microbiome alterations induced by tributyltin exposure are associated with increased body weight, impaired glucose and insulin homeostasis and endocrine disruption in mice. *Environ. Pollut.* **2020**, *266*, 115276. [CrossRef] [PubMed]
38. Oberto, A.; Mele, P.; Zammaretti, F.; Panzica, G.C.; Eva, C. Evidence of Altered Neuropeptide Y Content and Neuropeptide Y1 Receptor Gene Expression in the Hypothalamus of Pregnant Transgenic Mice. *Endocrinology* **2003**, *144*, 4826–4830. [CrossRef]
39. Chronwall, B.M. Anatomical distribution of NPY and NPY messenger RNA in rat brain. In *Neuropeptide Y*; Mutt, V., Ed.; Raven Press: New York, NY, USA, 1989; pp. 51–59.
40. Danger, J.M.; Tonon, M.C.; Basille, C.; Jenks, B.G.; Saint Pierre, S.; Martel, J.C.; Fasolo, A.; Quirion, R.; Pelletier, G.; Vaudry, H. Neuropeptide Y: Localization in the central nervous system and neuroendocrine functions. *Fundam. Clin. Pharmacol.* **1990**, *4*, 307–340. [CrossRef]
41. Morris, B.J. Neuronal localisation of neuropeptide Y gene expression in rat brain. *J. Comp. Neurol.* **1989**, *290*, 358–368. [CrossRef] [PubMed]
42. Gumbs, M.C.R.; Vuuregge, A.H.; Eggels, L.; Unmehopa, U.A.; Lamuadni, K.; Mul, J.D.; la Fleur, S.E. Afferent neuropeptide Y projections to the ventral tegmental area in normal-weight male Wistar rats. *J. Comp. Neurol.* **2019**, *527*, 2659–2674. [CrossRef] [PubMed]
43. Watson, S.J.; Richard, C.W., 3rd; Barchas, J.D. Adrenocorticotropin in rat brain: Immunocytochemical localization in cells and axons. *Science* **1978**, *200*, 1180–1182. [CrossRef]
44. Jacobowitz, D.M.; O'Donohue, T.L. alpha-Melanocyte stimulating hormone: Immunohistochemical identification and mapping in neurons of rat brain. *Proc. Natl. Acad. Sci. USA* **1978**, *75*, 6300–6304. [CrossRef]
45. Bagnol, D.; Lu, X.Y.; Kaelin, C.B.; Day, H.E.; Ollmann, M.; Gantz, I.; Akil, H.; Barsh, G.S.; Watson, S.J. Anatomy of an endogenous antagonist: Relationship between Agouti-related protein and proopiomelanocortin in brain. *J. Neurosci.* **1999**, *19*, RC26. [CrossRef]
46. King, C.M.; Hentges, S.T. Relative number and distribution of murine hypothalamic proopiomelanocortin neurons innervating distinct target sites. *PLoS ONE* **2011**, *6*, e25864. [CrossRef]
47. Heindel, J.J.; Vom Saal, F.S.; Blumberg, B.; Bovolin, P.; Calamandrei, G.; Ceresini, G.; Cohn, B.A.; Fabbri, E.; Gioiosa, L.; Kassotis, C.; et al. Parma consensus statement on metabolic disruptors. *Environ. Health Glob. Access Sci. Source* **2015**, *14*, 54. [CrossRef] [PubMed]
48. Cederroth, C.; Vinciguerra, M.; Kühne, F.; Madani, R.; Doerge, D.; Visser, T.; Foti, M.; Rohner-Jeanrenaud, F.; Vassalli, J.-D.; Nef, S. A phytoestrogen-rich diet increases energy expenditure and decreases adiposity in mice. *Environ. Health Perspect.* **2007**, *115*, 1467–1473. [CrossRef] [PubMed]
49. Murata, Y.; Robertson, K.M.; Jones, M.E.; Simpson, E.R. Effect of estrogen deficiency in the male: The ArKO mouse model. *Mol. Cell Endocrinol.* **2002**, *193*, 7–12. [CrossRef]
50. Heine, P.A.; Taylor, J.A.; Iwamoto, G.A.; Lubahn, D.B.; Cooke, P.S. Increased adipose tissue in male and female estrogen receptor-alpha knockout mice. *Proc. Natl. Acad. Sci. USA* **2000**, *97*, 12729–12734. [CrossRef]
51. Welshons, W.V.; Nagel, S.C.; vom Saal, F.S. Large effects from small exposures. III. Endocrine mechanisms mediating effects of bisphenol A at levels of human exposure. *Endocrinology* **2006**, *147*, S56–S69. [CrossRef]
52. Allera, A.; Lo, S.; King, I.; Steglich, F.; Klingmuller, D. Impact of androgenic/antiandrogenic compounds (AAC) on human sex steroid metabolizing key enzymes. *Toxicology* **2004**, *205*, 75–85. [CrossRef]
53. De Souza, F.S.; Nasif, S.; Lopez-Leal, R.; Levi, D.H.; Low, M.J.; Rubinsten, M. The estrogen receptor alpha colocalizes with proopiomelanocortin in hypothalamic neurons and binds to a conserved motif present in the neuron-specific enhancer nPE2. *Eur. J. Pharmacol.* **2011**, *660*, 181–187. [CrossRef] [PubMed]
54. Marraudino, M.; Miceli, D.; Farinetti, A.; Ponti, G.; Panzica, G.; Gotti, S. Kisspeptin innervation of the hypothalamic paraventricular nucleus: Sexual dimorphism and effect of estrous cycle in female mice. *J. Anat.* **2017**, *230*, 775–786. [CrossRef] [PubMed]
55. Vandenberg, L.N.; Colborn, T.; Hayes, T.B.; Heindel, J.J.; Jacobs, D.R., Jr.; Lee, D.H.; Shioda, T.; Soto, A.M.; vom Saal, F.S.; Welshons, W.V.; et al. Hormones and endocrine-disrupting chemicals: Low-dose effects and nonmonotonic dose responses. *Endocr. Rev.* **2012**, *33*, 378–455. [CrossRef] [PubMed]
56. Arletti, R.; Benelli, A.; Bertolini, A. Oxytocin inhibits food and fluid intake in rats. *Physiol. Behav.* **1990**, *48*, 825–830. [CrossRef]

57. Maejima, Y.; Sakuma, K.; Santoso, P.; Gantulga, D.; Katsurada, K.; Ueta, Y.; Hiraoka, Y.; Nishimori, K.; Tanaka, S.; Shimomura, K.; et al. Oxytocinergic circuit from paraventricular and supraoptic nuclei to arcuate POMC neurons in hypothalamus. *FEBS Lett.* **2014**, *588*, 4404–4412. [CrossRef]
58. Hrabovszky, E.; Kallo, I.; Steinhauser, A.; Merchenthaler, I.; Coen, C.W.; Petersen, S.L.; Liposits, Z. Estrogen receptor-beta in oxytocin and vasopressin neurons of the rat and human hypothalamus: Immunocytochemical and in situ hybridization studies. *J. Comp. Neurol.* **2004**, *473*, 315–333. [CrossRef]
59. Forsling, M.L.; Kallo, I.; Hartley, D.E.; Heinze, L.; Ladek, R.; Coen, C.W.; File, S.E. Oestrogen receptor-beta and neurohypophysial hormones: Functional interaction and neuroanatomical localisation. *Pharmacol. Biochem. Behav.* **2003**, *76*, 535–542. [CrossRef]
60. Gamber, K.M.; Macarthur, H.; Westfall, T.C. Cannabinoids augment the release of neuropeptide Y in the rat hypothalamus. *Neuropharmacology* **2005**, *49*, 646–652. [CrossRef]
61. Koch, M.; Varela, L.; Kim, J.G.; Kim, J.D.; Hernandez-Nuno, F.; Simonds, S.E.; Castorena, C.M.; Vianna, C.R.; Elmquist, J.K.; Morozov, Y.M.; et al. Hypothalamic POMC neurons promote cannabinoid-induced feeding. *Nature* **2015**, *519*, 45–50. [CrossRef] [PubMed]
62. Gonzalez, S.; Mauriello-Romanazzi, G.; Berrendero, F.; Ramos, J.A.; Franzoni, M.F.; Fernandez-Ruiz, J. Decreased cannabinoid CB1 receptor mRNA levels and immunoreactivity in pituitary hyperplasia induced by prolonged exposure to estrogens. *Pituitary* **2000**, *3*, 221–226. [CrossRef]
63. Suglia, A.; Chianese, R.; Migliaccio, M.; Ambrosino, C.; Fasano, S.; Pierantoni, R.; Cobellis, G.; Chioccarelli, T. Bisphenol A induces hypothalamic down-regulation of the the cannabinoid receptor 1 and anorexigenic effects in male mice. *Pharmacol. Res.* **2016**, *113*, 376–383. [CrossRef]
64. Garcia, S.M.; Hirschberg, P.R.; Sarkar, P.; Siegel, D.M.; Teegala, S.B.; Vail, G.M.; Routh, V.H. Insulin actions on hypothalamic glucose-sensing neurones. *J. Neuroendocrinol.* **2021**, *33*, e12937. [CrossRef]
65. Farrugia, F.; Aquilina, A.; Vassallo, J.; Pace, N.P. Bisphenol A and Type 2 Diabetes Mellitus: A Review of Epidemiologic, Functional, and Early Life Factors. *Int. J. Environ. Res. Public. Health* **2021**, *18*, 716. [CrossRef] [PubMed]
66. Hao, C.J.; Cheng, X.J.; Xia, H.F.; Ma, X. The endocrine disruptor diethylstilbestrol induces adipocyte differentiation and promotes obesity in mice. *Toxicol. Appl. Pharmacol.* **2012**, *263*, 102–110. [CrossRef] [PubMed]
67. Ahmed, E.I.; Zehr, J.L.; Schulz, K.M.; Lorenz, B.H.; DonCarlos, L.L.; Sisk, C.L. Pubertal hormones modulate the addition of new cells to sexually dimorphic brain regions. *Nat. Neurosci.* **2008**, *11*, 995–997. [CrossRef] [PubMed]
68. Morishita, M.; Maejima, S.; Tsukahara, S. Gonadal Hormone-Dependent Sexual Differentiation of a Female-Biased Sexually Dimorphic Cell Group in the Principal Nucleus of the Bed Nucleus of the Stria Terminalis in Mice. *Endocrinology* **2017**, *158*, 3512–3525. [CrossRef] [PubMed]
69. Bouret, S.G. Development of Hypothalamic Circuits That Control Food Intake and Energy Balance. In *Appetite and Food Intake: Central Control*; Harris, R.B.S., Ed.; CRC Press: Boca Raton, FL, USA, 2017; pp. 135–154.
70. Ishii, Y.; Bouret, S.G. Embryonic birthdate of hypothalamic leptin-activated neurons in mice. *Endocrinology* **2012**, *153*, 3657–3667. [CrossRef] [PubMed]
71. Melnick, I.; Pronchuk, N.; Cowley, M.A.; Grove, K.L.; Colmers, W.F. Developmental switch in neuropeptide Y and melanocortin effects in the paraventricular nucleus of the hypothalamus. *Neuron* **2007**, *56*, 1103–1115. [CrossRef]
72. Safahani, M.; Aligholi, H.; Noorbakhsh, F.; Djalali, M.; Pishva, H.; Modarres Mousavi, S.M.; Alizadeh, L.; Gorji, A.; Koohdani, F. Switching from high-fat diet to foods containing resveratrol as a calorie restriction mimetic changes the architecture of arcuate nucleus to produce more newborn anorexigenic neurons. *Eur. J. Nutr.* **2019**, *58*, 1687–1701. [CrossRef]
73. Yirun, A.; Ozkemahli, G.; Balci, A.; Erkekoglu, P.; Zeybek, N.D.; Yersal, N.; Kocer-Gumusel, B. Neuroendocrine disruption by bisphenol A and/or di(2-ethylhexyl) phthalate after prenatal, early postnatal and lactational exposure. *Environ. Sci. Pollut. Res. Int.* **2021**. [CrossRef] [PubMed]
74. Penza, M.; Jeremic, M.; Marrazzo, E.; Maggi, A.; Ciana, P.; Rando, G.; Grigolato, P.G.; Di Lorenzo, D. The environmental chemical tributyltin chloride (TBT) shows both estrogenic and adipogenic activities in mice which might depend on the exposure dose. *Toxicol. Appl. Pharmacol.* **2011**, *255*, 65–75. [CrossRef]
75. Harkness, J.E.; Wagner, J.E. *The Biology and Medicine of Rabbits and Rodents*, 3rd ed.; Lea and Febiger: Philadelphia, PA, USA, 1989.
76. Watson, R.E.; Wiegand, S.J.; Clough, R.W.; Hoffman, G.E. Use of cryoprotectant to maintain long-term peptide immunoreactivity and tissue morphology. *Peptides* **1986**, *7*, 155–159. [CrossRef]
77. Gouaze, A.; Brenachot, X.; Rigault, C.; Krezymon, A.; Rauch, C.; Nedelec, E.; Lemoine, A.; Gascuel, J.; Bauer, S.; Penicaud, L.; et al. Cerebral cell renewal in adult mice controls the onset of obesity. *PLoS ONE* **2013**, *8*, e72029. [CrossRef] [PubMed]
78. Kabra, D.G.; Pfuhlmann, K.; Garcia-Caceres, C.; Schriever, S.C.; Casquero Garcia, V.; Kebede, A.F.; Fuente-Martin, E.; Trivedi, C.; Heppner, K.; Uhlenhaut, N.H.; et al. Hypothalamic leptin action is mediated by histone deacetylase 5. *Nat. Commun.* **2016**, *7*, 10782. [CrossRef]
79. Pelletier, G.; Desy, L.; Kerkerian, L.; Cote, J. Immunocytochemical localization of neuropeptide Y (NPY) in the human hypothalamus. *Cell Tissue Res.* **1984**, *238*, 203–205. [CrossRef]
80. Pelletier, G.; Guy, J.; Allen, Y.S.; Polak, J.M. Electron microscopic immunocytochemical localization of neuropeptide Y (NPY) in the rat brain. *Neuropeptides* **1984**, *4*, 319–324. [CrossRef]

81. Lemus, M.B.; Bayliss, J.A.; Lockie, S.H.; Santos, V.V.; Reichenbach, A.; Stark, R.; Andrews, Z.B. A stereological analysis of NPY, POMC, Orexin, GFAP astrocyte, and Iba1 microglia cell number and volume in diet-induced obese male mice. *Endocrinology* **2015**, *156*, 1701–1713. [CrossRef]
82. Marraudino, M.; Martini, M.; Trova, S.; Farinetti, A.; Ponti, G.; Gotti, S.; Panzica, G. Kisspeptin system in ovariectomized mice: Estradiol and progesterone regulation. *Brain Res.* **2018**, *1688*, 8–14. [CrossRef] [PubMed]
83. Paxinos, G.; Franklin, K.B.J. *The Mouse Brain in Stereotaxic Coordinates*, 2nd ed.; Academic Press: San Diego, CA, USA, 2001.

 *metabolites*

MDPI

Article

Genistein during Development Alters Differentially the Expression of POMC in Male and Female Rats

Jose Manuel Fernandez-Garcia [1,2], Beatriz Carrillo [1,2], Patricia Tezanos [3], Paloma Collado [1,2] and Helena Pinos [1,2,*]

1 Departamento de Psicobiología, Facultad Psicología, Universidad Nacional de Educación a Distancia (UNED), 28040 Madrid, Spain; jmferdande@psi.uned.es (J.M.F.-G.); bcarrillo@psi.uned.es (B.C.); pcollado@psi.uned.es (P.C.)
2 Instituto Mixto de Investigación Escuela Nacional de Sanidad-UNED (IMIENS), 28040 Madrid, Spain
3 Departamento de Neurociencia Traslacional, Instituto Cajal, Consejo Superior de Investigaciones Científicas (CSIC), 28002 Madrid, Spain; ptezanos@cajal.csic.es
* Correspondence: hpinos@psi.uned.es

Abstract: Phytoestrogens are considered beneficial for health, but some studies have shown that they may cause adverse effects. This study investigated the effects of genistein administration during the second week of life on energy metabolism and on the circuits regulating food intake. Two different genistein doses, 10 or 50 μg/g, were administered to male and female rats from postnatal day (P) 6 to P13. Physiological parameters, such as body weight and caloric intake, were then analyzed at P90. Moreover, proopiomelanocortin (POMC) expression in the arcuate nucleus (Arc) and orexin expression in the dorsomedial hypothalamus (DMH), perifornical area (PF) and lateral hypothalamus (LH) were studied. Our results showed a delay in the emergence of sex differences in the body weight in the groups with higher genistein doses. Furthermore, a significant decrease in the number of POMC-immunoreactive (POMC-ir) cells in the Arc in the two groups of females treated with genistein was observed. In contrast, no alteration in orexin expression was detected in any of the structures analyzed in either males or females. In conclusion, genistein can modulate estradiol's programming actions on the hypothalamic feeding circuits differentially in male and female rats during development.

Keywords: genistein; proopiomelanocortin; arcuate nucleus; sex differences; rats

Citation: Fernandez-Garcia, J.M.; Carrillo, B.; Tezanos, P.; Collado, P.; Pinos, H. Genistein during Development Alters Differentially the Expression of POMC in Male and Female Rats. *Metabolites* **2021**, *11*, 293. https://doi.org/10.3390/metabo11050293

Academic Editor: Amedeo Lonardo

Received: 28 March 2021
Accepted: 29 April 2021
Published: 2 May 2021

1. Introduction

Genistein is a phytoestrogen that belongs to the group of isoflavones. It is present in a wide variety of legumes, mainly soybeans and their derivatives, which makes it one of the most consumed phytoestrogens by humans [1]. Soy is a typical ingredient in the traditional Asian diet, and it is also a widely consumed food in Western countries, being one of the most common milk substitutes, mostly in children [2–4].

Phytoestrogens exert their actions principally through the estradiol receptor (ER) α and ERβ, with higher reported affinity to the latter [5–7], but also through the G protein-coupled estrogen receptor (GPER) [8,9]. Specifically, genistein has structural similarities with estradiol, which allows it to bind estrogen receptors, acting as a potential agonist or antagonist of the estrogens, depending on the estradiol levels or the tissue [10–14].

Phytoestrogens are considered endocrine disruptors (EDCs), and although most of the EDCs have been demonstrated to have harmful effects on the organism (e.g., pesticides, bisphenol A) [15], phytoestrogens, in general, seem to have beneficial effects on health. Some practical advantages include the prevention of cardiovascular diseases [16], decreased inflammatory response in microglia [17], prevention of denervation-induced muscle atrophy [18], prevention of different types of cancer [19,20] and improvement in menopausal symptoms [21,22]. However, not all results evidence beneficial actions on the organism due to reports of damaging effects arising from phytoestrogens' exposure. In addition, a proportion of clinical studies do not demonstrate a clear health improvement (for review, see [23–25]).

The basal sexual genetic differences have an impact on structural, metabolic and behavioral differences between male and female rats that are more evident in adulthood [26–28]. Some authors have recently shown that genistein produces some alterations in various neural systems, such as vasopressinergic or dopaminergic systems, mainly when administered during development, and that those effects are sexually dimorphic in males and females [29,30]. Moreover, previous results of our group have demonstrated that estradiol administered from postnatal day (P) 6 to P13 has a modulatory role in rats in the early stages of life due to under- or overnutrition [31–33], specifically in the programming of the body weight in males and the mRNA POMC hypothalamic levels in females [34]. Estradiol conveyed mainly through ERα [35–39] is involved in the regulation of energy metabolism inhibiting food intake [40]. Considering these results and the possible estrogenic/antiestrogenic effects of genistein, it seems reasonable to assume that exposure to genistein during the early postnatal period may produce some alteration to the development of energy metabolism and on hypothalamic circuits that regulate food intake. Numerous orexigenic and anorexigenic peptides are involved in the regulation of body weight and feeding. Among them are the orexin and POMC peptides. The former are synthesized and released from LH neurons and increase food intake in response to the release of neuropeptide Y (NPY) from Arc neurons [41,42]. The latter, anorexigenic in their activity, are expressed by Arc neurons that send satiety signals to the LH and paraventricular hypothalamic nuclei (PVH) [43].

Given the increase in soy consumption in the general population and in children in particular, it is necessary to determine the effects of its main component, genistein. In the present work, we analyzed genistein treatment's effect during the second week of life on physiological parameters such as body weight and food consumption and neurohormonal parameters, such as POMC and orexin hypothalamic expression, in male and female rats.

2. Results

2.1. Differences in Body Weight and Caloric Intake

In the evolution of body weight, a main effect of sex ($F1{,}50 = 253.559$; $p < 0.001$) was found. Neither the treatment ($F2{,}50 = 0.170$; $p = 0.844$) nor the interaction ($F2{,}50 = 0.520$; $p = 0.598$) showed a significant effect.

As shown in Figure 1A, sex differences in body weight appeared on P41 in control and G10 groups ($p < 0.05$ in all cases), with males heavier than females. A delay in the appearance of sex differences can be observed in G50 groups from P48 onwards ($p < 0.05$ in all cases). No significant differences were observed when males and females were analyzed separately.

Food intake was measured in grams. A main effect of sex ($F1{,}50 = 297.788$; $p < 0.001$) was found. Treatment ($F2{,}50 = 0.183$; $p = 0.833$) and an interaction between these two factors ($F2{,}50 = 0.696$; $p = 0.504$) were not significant. As can be seen in Figure 1B, sex differences appeared on P41 and continued until P83. In all groups, males ate more than the corresponding females.

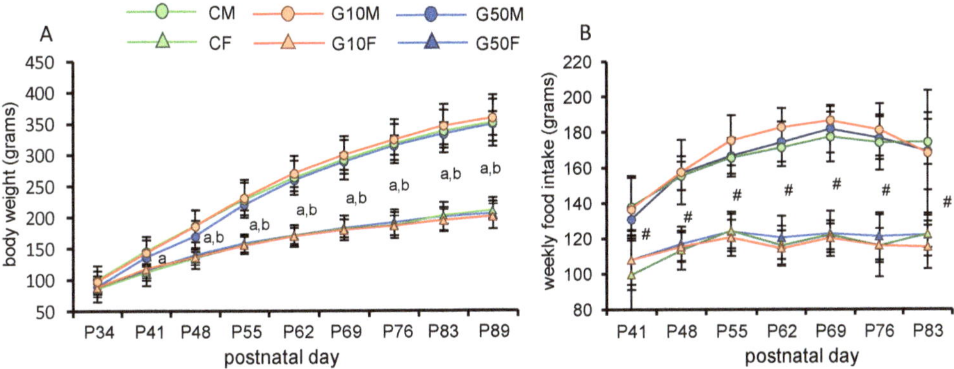

Figure 1. (**A**) Body weight evolution in all groups. (**B**) Weekly food intake in all groups (repeated measured ANOVA). Statistically significant differences ($p < 0.05$) are labelled as follows: a = sex differences in C and G10 groups; b = sex differences in G50 groups. # = sex differences in all groups studied. All values are expressed as means ± S.D. CM: control males: CF: control females; G10M: genistein treated males, dose 10 µg/g; G10F: genistein treated females, dose 10 µg/g; G50M: genistein treated males, dose 50 µg/g; G50F: genistein treated females, dose 50 µg/g.

2.2. Orexin-ir and POMC-ir Cell Analysis

No orexin-ir cells were detected in the ventromedial (VMH) or paraventricular hypothalamic nuclei (PVH). Orexin-ir cells were observed in the medial-ventral area of the lateral hypothalamus (LHmv). Likewise, a small population of orexin-ir cells was detected in the lateral edge of the dorsomedial hypothalamic nucleus (DMH) adjacent to the orexin-ir cells in the perifornical nucleus (PF). Together, these two areas were considered as the DMH-PF continuum for counting and analysis. In all nuclei studied, the cells that expressed orexin were easily detectable because the cell body was heavily labelled (Figure 2D,E).

No differences between the hemispheres were found in the PF-DMH continuum ($F1,54 = 0.926$; $p = 0.402$) or the LH ($F1,54 = 1.196$; $p = 0.310$). Moreover, no main effect of sex ($F1,30 = 0.598$; $p = 0.445$), treatment ($F2,30 = 1.614$; $p = 0.216$) or interaction between the factors ($F2,30 = 0.079$; $p = 0.924$) was found in this same continuum, and similar results were also observed in the LH with no main effect of sex ($F1,30 = 0.002$; $p = 0.964$), treatment ($F2,30 = 0.218$; $p = 0.805$) or interaction between the factors ($F2,30 = 0.234$; $p = 0.793$).

POMC-ir cells were easily distinguishable because the cell body was heavily labelled (Figure 2B,C). Cells expressing POMC were detected in the medial (ArcM), lateral (ArcL) and posteromedial (ArcPM) subdivisions of the arcuate nucleus (Arc) but not in the Arc-Dorsal subdivision.

No differences between the hemispheres were found in the anterior arcuate ($F1,56 = 0.001$; $p = 0.979$) or the posterior arcuate ($F1,56 = 0.021$; $p = 0.884$).

In the ArcM, a main effect of sex was found ($F1,36 = 5.347$; $p < 0.001$) but no effect of treatment ($F2,36 = 0.196$; $p = 0.823$) or interaction ($F2,36 = 1.051$; $p = 0.164$). Moreover, in the ArcL and the ArcPM, no effect of sex ($F1,36 = 0.001$; $p = 0.973$; $F1,36 = 0.795$; $p = 0.379$, respectively), treatment ($F2,36 = 1.514$; $p = 0.234$; $F2,36 = 0.724$; $p = 0.492$, respectively) or interaction ($F2,36 = 1.905$; $p = 0.164$; $F1,36 = 0.467$; $p = 0.631$, respectively) were found.

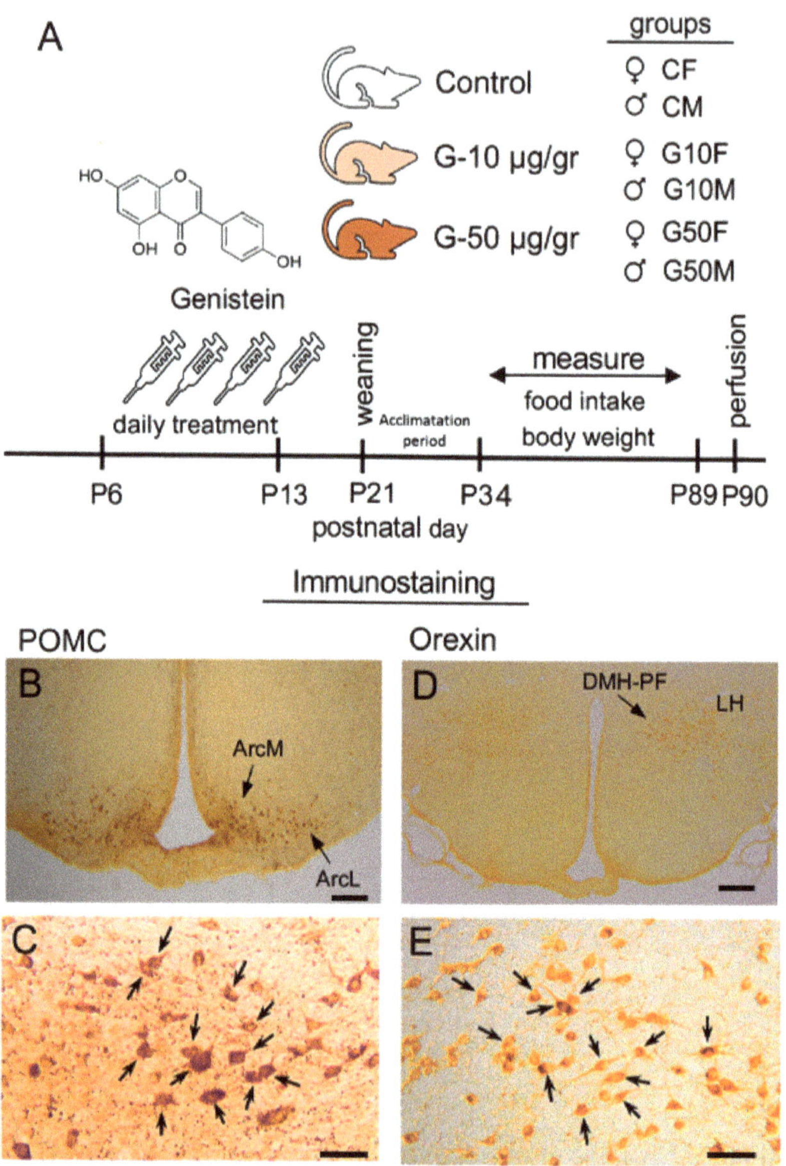

Figure 2. (**A**) Schematic representation of the procedure used from treatment to immunostaining. From P6 to P13, daily injections of synthetic genistein (10 or 50 µg/g) or vehicle according to experimental group were administered. From weaning on P21 to P34, an acclimatization period was implemented. Food intake and body weight were measured weekly from P34 until P89. Animals were sacrificed on P90. (**B**,**C**) Photomicrographs showing the distribution of immunostaining of POMC-ir positive cells in Arc nucleus. (**D**,**E**) Orexin-ir positive cells in the PF and LH. Arrows show orexin-ir and POMC-ir positive cells counted. ArcM: arcuate medial subdivision, ArcL: arcuate lateral subdivision; LH: lateral nucleus of the hypothalamus; DMH-PF: dorsomedial-perifornical nucleus B Bar = 200 µm; C Bar = 50 µm ; D Bar = 300 µm; E Bar = 75 µm [44].

Analysis of each sex separately showed a difference in the ArcM in females. The CF exhibited a greater number of POMC-ir cells than the females treated with low or high genistein doses ($p < 0.05$, in both cases). In contrast, males did not show significant differences among the three groups studied (Figure 3).

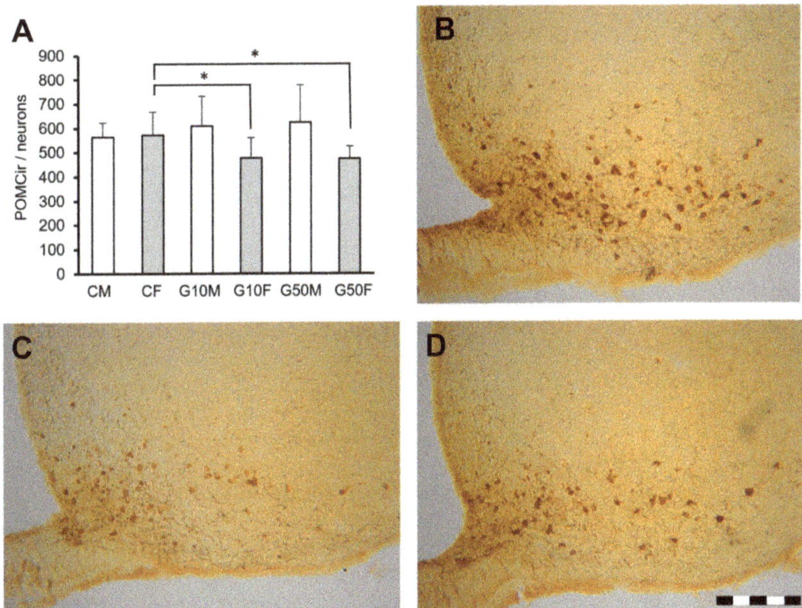

Figure 3. Graph (**A**) showing the number of POMC-ir neurons in ArcM in all experimental groups (two-way ANOVA) * indicates differences between groups ($p < 0.05$ in all cases). The error bars indicate standard error of the mean. CM: control males; CF: control females; G10M: genistein treated males, dose 10 µg/g; G10F: genistein treated females, dose 10 µg/g; G50M: genistein treated males, dose 50 µg/g; G50F: genistein treated females, dose 50 µg/g. $n = 6$ in all groups. All treatments, either injection of genistein or vehicle, were administered from P6 to P13. (**B–D**) photomicrographs showing the distribution of POMC-ir positive cells in Arc nucleus. **B** = control female, **C** = G10 female, **D** = G50 female. Bar = 200 µm [44].

3. Discussion

The present study results showed that the exposure to genistein in the early stages of development modifies hypothalamic POMC neurons' long-term expression in the arcuate nucleus of female but not male Wistar rats. Moreover, high doses of genistein produced a delay in the emergence of sex differences in body weight. In contrast, caloric intake or orexin expression were not altered in either sex.

Treatment with genistein from P6 to P13 did not affect the caloric intake because there were no differences between the groups or both sexes' developmental pattern. The differences appeared from P48 onwards in all groups. Concerning the body weight, we detected differences in the evolution of normal sexual dimorphism, an observation resulting from the one week delay in the emergence of sex differences in the genistein groups with high dose. Sex differences in control and G10 groups were observed from P41 onwards but at P48 in G50 groups. Similar results in the emergence of sex differences in control groups were reported in a previous study by our group [32].

Although the effects of genistein during development do not significantly alter physiological parameters such as caloric intake or body weight, a relevant effect has been detected in the brain. Specifically, the number of cells expressing POMC decreased in the medial subdivision of the arcuate nucleus of female rats when low or high doses of genistein were administered from P6 to P13. However, no effect of genistein was detected on POMC expression in this same nucleus in males.

The fact that genistein treatment altered the expression of POMC in the ArcM in female but not male rats is not surprising if previous results from our group are considered. In the same postnatal period, the administration of estradiol modulated the levels of hypothalamic POMC mRNA in females on a low-protein or high-fat diet, but no effect was detected in the male rats or in any other feeding-related peptide studied [31,32]. Furthermore, when the activity of ERα, ERβ and GPER receptors was blocked from P5 to P13, there was a decrease in hypothalamic POMC mRNA levels in female rats, but again, this effect was not detected in males, and no other alterations were shown in other peptides studied [34]. In line with these data, the results of the present study show first that genistein through an agonist or antagonist estrogenic activity alters the long-term expression of POMC during development in female but not male rats, and therefore this phytoestrogen could interfere with the programming activity of estradiol early in life. Secondly, in all cases, a misbalance in estrogenic activity affects POMC but no other peptides related to food intake in the hypothalamus [31,32] or specifically orexin in the DMH, PF or LH nuclei.

Gao et al. [45] demonstrated the existence of a direct relation between estradiol and POMC since this hormone can increase excitatory inputs on POMC neurons and POMC tone in the Arc in rats and mice [45]. This fact can explain the consistent response of POMC to the changes in the estradiol activity during development and how important the activity of this hormone is to the long-term expression of POMC in the hypothalamus. Our results show that the influence of genistein on estrogenic activity during development also results in an alteration of the melanocortin system in females in the long term.

At this point, it is important to note that the administration of estradiol in control animals in the same postnatal period did not alter hypothalamic POMC mRNA levels either in males or females in adulthood [32], suggesting a specific antagonist effect of genistein on the activity of estradiol in the programming of feeding circuits during development. On the one hand, it is worth bearing in mind that the effects of estradiol on food intake are mainly via the ERα [35–39] and that genistein has a higher affinity for ERβ, which could lead to differential effects of the phytoestrogen compared to estradiol. To our knowledge, no direct effect of genistein on POMC has been reported, but some results suggest a possible indirect effect of genistein on the downregulation of POMC expression in the Arc. It has been demonstrated that dietary soy produces an increase in hypothalamic neuropeptide Y (NPY) or agouti-related protein (AgRP) levels [46–48] and the inhibitory action of NPY/AgRP neurons on POMC neurons, possibly through GABA, has been soundly demonstrated [49–53]. Further research is therefore needed to determine the specific action of genistein during the programming period on ERs and whether genistein acts directly on POMC neurons or whether its effects on this system are due to an indirect action via NPY.

Numerous studies have reported that soy proteins may exert beneficial health effects by improving metabolic parameters and preventing obesity and diabetes, mainly in adult pre-and postmenopausal women [54–56]. However, many studies have shown that exposure to phytoestrogens early in development has adverse effects on reproductive function (see [57], for review) and alters various neurotransmitter systems [29,30,58]. These data, along with the results of the present work, demonstrate that exposure to genistein during the most sensitive stages of development alters neurotransmitter and neuropeptidergic systems involved in reproductive or feeding neurohormonal systems.

Very few studies have paid attention to the effects of phytoestrogens during development on the functions of the hypothalamic circuits regulating energy metabolism and/or feeding disorders. Data from other authors reported that phytoestrogens alter the reproductive system [57], and now our results reveal that phytoestrogens can also differentially modulate some actions of the estradiol during development in male and female rats. Taking into account that soy is the main substitute food for milk in children and that the lactation period is a sensitive period for the optimal development of brain circuits, further research will be needed to unravel the mechanisms through which genistein during development may alter the intake system in order to detect adverse effects on energy metabolism and feeding.

4. Materials and Methods

4.1. Animals

All experiments were designed according to the guidelines published in the "NIH Guide for the care and use of laboratory animals", the principles presented in the "Guidelines for the Use of Animals in Neuroscience Research" by the Society for Neuroscience, the European Union legislation (Council Directives 86/609/EEC and 2010/63/UE) and the Spanish Government Directive (R.D. 1201/2005). Experimental procedures were approved by our Institutional Bioethical Committee (UNED, Madrid). Special care was taken to minimize animal suffering and reduce the number of animals used to the minimum necessary. Wistar rats were reared under stable temperature, humidity and light conditions (22 ± 2 °C; 55 ± 10% humidity; 12 h light/12 h dark cycle, lights on from 08:00 to 20:00) with food and water ad libitum. For mating, a male was placed in a cage with two females for one week. Pregnant females were housed individually in plastic maternity cages with wood shavings as the nesting material. On postnatal day 1 (P1), pups born on the same day were weighed, sexed, and randomly distributed (five females and five males/dam). From P6 to P13, pups were treated with a daily s.c. injection of vehicle (corn oil), or synthetic genistein (Genistein Synthetic, ≥98%, Sigma-Aldrich St. Luois, MO, USA) in two doses: a low dose of genistein (10 µg/g) or a high dose of genistein 50 µg/g. The 10 µg dose was previously used by other authors who obtained differential effects in males and females on several physiological and brain parameters [59–61] This experimental design resulted in the following groups: control male (n = 10, CM), control female (n = 9, CF), G10 male (n = 10, G10M), G10 female (n = 9, G10F), G50 male (n = 10, G50M) and G50 female (n = 8, G50F). n = 6 and n = 7 in each group were used to study orexin and POMC expression, respectively. From weaning on P21 to P34, an acclimatization period was implemented. From P33 to P89, body weight and food intake were measured every 7 days, except for the week prior to perfusion when body weight was recorded at 6 days (Figure 2A).

4.2. Tissue Preparation

On P90, animals were deeply anaesthetized with an overdose of tribromoethanol in saline (1 mL/kg). Then, the animals were transcardially perfused with saline followed by 4% paraformaldehyde (PAF). The brains were removed, stored in a freshly prepared PAF solution for two hours at 4 °C and then washed several times in phosphate-buffered saline (PBS). Next, the brains were stored in a 30% sucrose solution in PBS at 4 °C until they were examined. The brains were then frozen on dry ice and serially sectioned along the coronal plane at a thickness of 40 µm. Serial sections were collected in four series, two of which were used in this study processed as free-floating sections for orexin and POMC immunostaining.

4.3. Orexin and POMC Immunostaining

The sections were incubated in PBS overnight. Endogenous peroxidase activity was blocked by incubation with H_2O_2 in 0.5% Triton X-100 in PBS for 30 min. After a brief wash in PBS, the sections were incubated in normal goat serum (diluted 1:5 in PBS; Vector, Burlingame, CA, USA) for 30 min at room temperature. Then, the sections were incubated for 48 h at 4 °C in a rabbit anti-orexin A primary antibody (Calbiochem, San Diego, CA, USA) or in a rabbit anti-POMC primary antibody (Phoenix Pharmaceuticals Inc., Burlingame, CA, USA); 1:2000 in both cases. After several brief washes in PBS, the sections were incubated with biotinylated anti-rabbit IgG serum (Vector, 1:200) for 90 min and then in avidin-peroxidase complex (Immunopure ABC Vector Burlingame, CA, USA) for 60 min at room temperature. Finally, the presence of peroxidase activity was visualized with a solution containing 0.02 g/mL diaminobenzidine (DAB; Aldrich, Madrid, Spain) and 0.025% hydrogen peroxidase in Tris–HCl, pH 7.6. The sections were mounted on gelatin-coated slides, dehydrated in ethanol, washed in xylene and coverslipped with DPX (Surgipath Europe Ltd., Peterborough, UK).

The number of orexin-ir cells in the dorsomedial hypothalamus (DMH), perifornical area (PF) and lateral hypothalamus (LH) and the number of POMC-ir cells in the subdivisions of the arcuate nucleus (the dorsal [ArcD], medial [ArcM], lateral [ArcL] and medial posterior [ArcPM]) [44] were estimated. Briefly, a microphotograph of each section was acquired using a scanner (Nikon Collscope Eclipse Net-VSL, Tokyo, Japan) with a monitor (Digital Sight DS-L1, Tokyo, Japan). The number of orexin-ir or POMC-ir cells in each section was estimated using ImageJ (ImageJ bundled with 64-bit Java 1.8.0; National Institutes of Health, Bethesda, MD, USA) following the Königsmark cell counting procedure [62]. The scattered orexin-ir cells on the lateral edge of the DMH were considered to be continuous with the PF nucleus. The orexin-ir or POMC-ir cells included within the boundaries of the different nuclei were counted.

4.4. Statistical Analysis

The evolution of body weight and caloric intake during the experimental procedure was analyzed using repeated-measures ANOVA with treatment as the within-subject factor and body weight and caloric intake as the between-subject factors. To determine the differences among the groups, one-way ANOVA was performed when appropriate. Post hoc comparisons were performed with Student–Newman–Keuls tests. The significance level was set at $p < 0.05$.

The number of orexin-ir and POMC-ir cells in both hemispheres was estimated. The data were subjected to one-way ANOVA with the hemisphere as a factor to determine the potential differences between the right and left hemispheres. Once the effect of the hemisphere was discarded, the mean value of the two hemispheres was used for statistical analysis performed by one-way ANOVA followed by Student–Newman–Keuls tests when appropriate, and the significance level was $p < 0.05$.

Author Contributions: Each author has made substantial contributions to the work: Conceptualization; formal analysis; writing—original draft; supervision, review and editing: B.C., P.C. and H.P. Methodology, review and editing: J.M.F.-G. and P.T. Project administration and funding acquisition: P.C. and H.P. All authors have read and agreed to the published version of the manuscript.

Funding: The present work was supported by grants PSI2017-86396-P and IMIENS (PC/HP).

Institutional Review Board Statement: All experiments were designed according to the guidelines published in the "NIH Guide for the care and use of laboratory animals", the principles presented in the "Guidelines for the Use of Animals in Neuroscience Research" by the Society for Neuroscience, the European Union legislation (Council Directives 86/609/EEC and 2010/63/UE) and the Spanish Government Directive (R.D. 1201/2005). Experimental procedures were approved by our Institutional Bioethical Committee (UNED, Madrid). Special care was taken to minimize animal suffering and to reduce the number of animals used to the minimum necessary.

Informed Consent Statement: Not applicable.

Data Availability Statement: The data presented in this study are available on request from the corresponding author.

Acknowledgments: We are grateful to L. Carrillo, A. Marcos and G. Moreno for their technical assistance. We are grateful to Stephen Morgan for his editorial help.

Conflicts of Interest: The authors declare no conflict of interest.

References

1. Cederroth, C.R.; Vinciguerra, M.; Kühne, F.; Madani, R.; Doerge, D.R.; Visser, T.J.; Foti, M.; Rohner-Jeanrenaud, F.; Vassalli, J.D.; Nef, S. A phytoestrogen-rich diet increases energy expenditure and decreases adiposity in mice. *Environ. Health Perspect.* **2007**, *115*, 1467–1473. [CrossRef] [PubMed]
2. Badger, T.M.; Ronis, M.J.J.; Hakkak, R.; Rowlands, J.C.; Korourian, S. The health consequences of early soy consumption. *J. Nutr.* **2002**, *132*. [CrossRef] [PubMed]
3. Barrett, J.R. The science of soy: What do we really know? *Environ. Health Perspect.* **2006**, *114*. [CrossRef] [PubMed]
4. Verduci, E.; Di Profio, E.; Cerrato, L.; Nuzzi, G.; Riva, L.; Vizzari, G.; D'Auria, E.; Giannì, M.L.; Zuccotti, G.; Peroni, D.G. Use of soy-based formulas and cow's milk allergy: Lights and shadows. *Front. Pediatr.* **2020**, *8*. [CrossRef]
5. Kuiper, G.G.J.M.; Lemmen, J.G.; Carlsson, B.; Corton, J.C.; Safe, S.H.; Van Der Saag, P.T.; Van Der Burg, B.; Gustafsson, J.Å. Interaction of estrogenic chemicals and phytoestrogens with estrogen receptor β. *Endocrinology* **1998**, *139*. [CrossRef]
6. Casanova, M. Developmental effects of dietary phytoestrogens in Sprague-Dawley rats and interactions of genistein and daidzein with rat estrogen receptors alpha and beta in vitro. *Toxicol. Sci.* **1999**, *51*. [CrossRef]
7. Patisaul, H.B.; Melby, M.; Whitten, P.L.; Young, L.J. Genistein affects ERβ- but not ERα-dependent gene expression in the hypothalamus. *Endocrinology* **2002**, *143*. [CrossRef]
8. Thomas, P.; Dong, J. Binding and activation of the seven-transmembrane estrogen receptor GPR30 by environmental estrogens: A potential novel mechanism of endocrine disruption. *J. Steroid Biochem. Mol. Biol.* **2006**, *102*. [CrossRef]
9. Du, Z.-R.; Feng, X.-Q.; Li, N.; Qu, J.-X.; Feng, L.; Chen, L.; Chen, W.-F. G protein-coupled estrogen receptor is involved in the anti-inflammatory effects of genistein in microglia. *Phytomedicine* **2018**, *43*. [CrossRef]
10. Dixon, R.; Ferreira, D. Genistein. Glycosides from stenochlaena palustris. *Phytochemistry* **2002**, *60*. [CrossRef]
11. Mueller, S.O. Overview of in vitro tools to assess the estrogenic and antiestrogenic activity of phytoestrogens. *J. Chromatogr. B* **2002**, *777*. [CrossRef]
12. Chen, A.-C.; Donovan, S.M. Genistein at a concentration present in soy infant formula inhibits Caco-2BBe cell proliferation by causing G2/M cell cycle arrest. *J. Nutr.* **2004**, *134*. [CrossRef] [PubMed]
13. Hwang, C.S.; Kwak, H.S.; Lim, H.J.; Lee, S.H.; Kang, Y.S.; Choe, T.B.; Hur, H.G.; Han, K.O. Isoflavone metabolites and their in vitro dual functions: They can act as an estrogenic agonist or antagonist depending on the estrogen concentration. *J. Steroid Biochem. Mol. Biol.* **2006**, *101*. [CrossRef]
14. Cederroth, C.R.; Nef, S. Soy, phytoestrogens and metabolism: A review. *Mol. Cell. Endocrinol.* **2009**, *304*. [CrossRef] [PubMed]
15. Gore, A.C.; Chappell, V.A.; Fenton, S.E.; Flaws, J.A.; Nadal, A.; Prins, G.S.; Toppari, J.; Zoeller, R.T. EDC-2: The endocrine society's second scientific statement on endocrine-disrupting chemicals. *Endocr. Rev.* **2015**, *36*. [CrossRef]
16. Fusi, F.; Trezza, A.; Tramaglino, M.; Sgaragli, G.; Saponara, S.; Spiga, O. The beneficial health effects of flavonoids on the cardiovascular system: Focus on K+ channels. *Pharm. Res.* **2020**, *152*. [CrossRef]
17. Cooke, P.S.; Selvaraj, V.; Yellayi, S. Genistein, estrogen receptors, and the acquired immune response. *J. Nutr.* **2006**, *136*. [CrossRef]
18. Aoyama, S.; Jia, H.; Nakazawa, K.; Yamamura, J.; Saito, K.; Kato, H. Dietary genistein prevents denervation-induced muscle atrophy in male rodents via effects on estrogen receptor-α. *J. Nutr.* **2016**, *146*. [CrossRef]
19. Carbonel, A.A.F.; Calió, M.L.; Santos, M.A.; Bertoncini, C.R.A.; da Silva Sasso, G.; Simões, R.S.; Simões, M.J.; Soares, J.M. Soybean isoflavones attenuate the expression of genes related to endometrial cancer risk. *Climacteric* **2015**, *18*. [CrossRef]
20. de Oliveira Andrade, F.; Liu, F.; Zhang, X.; Rosim, M.P.; Dani, C.; Cruz, I.; Wang, T.T.Y.; Helferich, W.; Li, R.W.; Hilakivi-Clarke, L. Genistein reduces the risk of local mammary cancer recurrence and ameliorates alterations in the gut microbiota in the offspring of obese dams. *Nutrients* **2021**, 13. [CrossRef]
21. Chen, L.-R.; Ko, N.-Y.; Chen, K.-H. Isoflavone supplements for menopausal women: A systematic review. *Nutrients* **2019**, 11. [CrossRef] [PubMed]
22. Thangavel, P.; Puga-Olguín, A.; Rodríguez-Landa, J.F.; Zepeda, R.C. Genistein as potential therapeutic candidate for menopausal symptoms and other related diseases. *Molecules* **2019**, *24*, 13892. [CrossRef] [PubMed]
23. Crain, D.A.; Janssen, S.J.; Edwards, T.M.; Heindel, J.; Ho, S.; Hunt, P.; Iguchi, T.; Juul, A.; McLachlan, J.A.; Schwartz, J.; et al. Female reproductive disorders: The roles of endocrine-disrupting compounds and developmental timing. *Fertil. Steril.* **2008**, *90*. [CrossRef]
24. Patisaul, H.B.; Jefferson, W. The pros and cons of phytoestrogens. *Front. Neuroendocr.* **2010**, *31*. [CrossRef] [PubMed]
25. Lephart, E.D.; Adlercreutz, H.; Lund, T.D. Dietary soy phytoestrogen effects on brain structure and aromatase in Long-Evans rats. *Neuroreport* **2001**, *12*. [CrossRef]
26. Yanai, J. Strain and sex differences in the rat brain. *Cells Tissues Organs* **1979**, *103*. [CrossRef]
27. Ngun, T.C.; Ghahramani, N.; Sánchez, F.J.; Bocklandt, S.; Vilain, E. The genetics of sex differences in brain and behavior. *Front. Neuroendocr.* **2011**, *32*. [CrossRef]
28. Ruoppolo, M.; Caterino, M.; Albano, L.; Pecce, R.; Di Girolamo, M.G.; Crisci, D.; Costanzo, M.; Milella, L.; Franconi, F.; Campesi, I. Targeted metabolomic profiling in rat tissues reveals sex differences. *Sci. Rep.* **2018**, *8*. [CrossRef]
29. Ponti, G.; Rodriguez-Gomez, A.; Farinetti, A.; Marraudino, M.; Filice, F.; Foglio, B.; Sciacca, G.; Panzica, G.C.; Gotti, S. Early postnatal genistein administration permanently affects nitrergic and vasopressinergic systems in a sex-specific way. *Neuroscience* **2017**, *346*. [CrossRef]

30. Ponti, G.; Farinetti, A.; Marraudino, M.; Panzica, G.; Gotti, S. Postnatal genistein administration selectively abolishes sexual dimorphism in specific hypothalamic dopaminergic system in mice. *Brain Res.* **2019**, *1724*. [CrossRef]
31. Carrillo, B.; Collado, P.; Díaz, F.; Chowen, J.A.; Pinos, H. Exposure to increased levels of estradiol during development can have long-term effects on the response to undernutrition in female rats. *Nutr. Neurosci.* **2016**, *19*. [CrossRef]
32. Carrillo, B.; Collado, P.; Díaz, F.; Chowen, J.A.; Pérez-Izquierdo, M.Á.; Pinos, H. Physiological and brain alterations produced by high-fat diet in male and female rats can be modulated by increased levels of estradiol during critical periods of development. *Nutr. Neurosci.* **2019**, *22*. [CrossRef]
33. Pinos, H.; Carrillo, B.; Díaz, F.; Chowen, J.A.; Collado, P. Differential vulnerability to adverse nutritional conditions in male and female rats: Modulatory role of estradiol during development. *Front. Neuroendocr.* **2018**, *48*, 13–22. [CrossRef]
34. Carrillo, B.; Collado, P.; Díaz, F.; Chowen, J.A.; Grassi, D.; Pinos, H. Blocking of estradiol receptors ERα, ERβ and GPER During development, differentially alters energy metabolism in male and female rats. *Neuroscience* **2020**, *426*. [CrossRef]
35. Roepke, T.A. Oestrogen modulates hypothalamic control of energy homeostasis through multiple mechanisms. *J. Neuroendocr.* **2009**, *21*. [CrossRef]
36. Meyer, M.R.; Clegg, D.J.; Prossnitz, E.R.; Barton, M. Obesity, insulin resistance and diabetes: Sex differences and role of oestrogen receptors. *Acta Physiol.* **2011**, *203*. [CrossRef] [PubMed]
37. Mauvais-Jarvis, F.; Clegg, D.J.; Hevener, A.L. The role of estrogens in control of energy balance and glucose homeostasis. *Endocr. Rev.* **2013**, *34*. [CrossRef] [PubMed]
38. Frank, A.; Brown, L.M.; Clegg, D.J. The role of hypothalamic estrogen receptors in metabolic regulation. *Front. Neuroendocr.* **2014**, *35*. [CrossRef] [PubMed]
39. Santollo, J.; Daniels, D. Multiple estrogen receptor subtypes influence ingestive behavior in female rodents. *Physiol. Behav.* **2015**, *152*. [CrossRef] [PubMed]
40. López, M.; Tena-Sempere, M. Estrogens and the control of energy homeostasis: A brain perspective. *Trends Endocrinol. Metab.* **2015**, *26*. [CrossRef] [PubMed]
41. Schwartz, M.W.; Woods, S.C.; Porte, D.; Seeley, R.J.; Baskin, D.G. Central nervous system control of food intake. *Nature* **2000**, *404*. [CrossRef]
42. Stuber, G.D.; Wise, R.A. Lateral hypothalamic circuits for feeding and reward. *Nat. Neurosci.* **2016**, *19*. [CrossRef] [PubMed]
43. Toda, C.; Santoro, A.; Kim, J.D.; Diano, S. POMC neurons: From birth to death. *Annu. Rev. Physiol.* **2017**, *79*. [CrossRef] [PubMed]
44. Paxinos, G.; Watson, C. *The Rat Brain in Stereotaxic Coordinates*, 4th ed.; Academic Press: San Diego, CA, USA, 2013.
45. Gao, Q.; Mezei, G.; Nie, Y.; Rao, Y.; Choi, C.S.; Bechmann, I.; Leranth, C.; Toran-Allerand, D.; Priest, C.A.; Roberts, J.L.; et al. Anorectic estrogen mimics leptin's effect on the rewiring of melanocortin cells and Stat3 signaling in obese animals. *Nat. Med.* **2007**, *13*. [CrossRef]
46. Lephart, E.D.; Porter, J.P.; Lund, T.D.; Bu, L.; Setchell, K.D.; Ramoz, G.; Crowley, W.R. Dietary isoflavones alter regulatory behaviors, metabolic hormones and neuroendocrine function in long-evans male rats. *Nutr. Metab. (Lond.)* **2004**, *1*. [CrossRef]
47. Andreoli, M.F.; Stoker, C.; Rossetti, M.F.; Alzamendi, A.; Castrogiovanni, D.; Luque, E.H.; Ramos, J.G. Withdrawal of dietary phytoestrogens in adult male rats affects hypothalamic regulation of food intake, induces obesity and alters glucose metabolism. *Mol. Cell. Endocrinol.* **2015**, *401*. [CrossRef]
48. Andreoli, M.F.; Stoker, C.; Lazzarino, G.P.; Canesini, G.; Luque, E.H.; Ramos, J.G. Dietary whey reduces energy intake and alters hypothalamic gene expression in obese phyto-oestrogen-deprived male rats. *Br. J. Nutr.* **2016**, *116*. [CrossRef]
49. Roseberry, A.G.; Liu, H.; Jackson, A.C.; Cai, X.; Friedman, J.M. Neuropeptide Y-mediated inhibition of proopiomelanocortin neurons in the arcuate nucleus shows enhanced desensitization in ob/ob mice. *Neuron* **2004**, *41*. [CrossRef]
50. Tong, Q.; Ye, C.-P.; Jones, J.E.; Elmquist, J.K.; Lowell, B.B. Synaptic release of GABA by AgRP neurons is required for normal regulation of energy balance. *Nat. Neurosci.* **2008**, *11*. [CrossRef] [PubMed]
51. Atasoy, D.; Betley, J.N.; Su, H.H.; Sternson, S.M. Deconstruction of a neural circuit for hunger. *Nature* **2012**, *488*. [CrossRef]
52. Mercer, A.J.; Hentges, S.T.; Meshul, C.K.; Low, M.J. Unraveling the central proopiomelanocortin neural circuits. *Front. Neurosci.* **2013**, *7*. [CrossRef] [PubMed]
53. Newton, A.J.; Hess, S.; Paeger, L.; Vogt, M.C.; Fleming Lascano, J.; Nillni, E.A.; Brüning, J.C.; Kloppenburg, P.; Xu, A.W. AgRP Innervation onto POMC neurons increases with age and is accelerated with chronic high-fat feeding in male mice. *Endocrinology* **2013**, *154*. [CrossRef] [PubMed]
54. Gardner, C.D.; Newell, K.A.; Cherin, R.; Haskell, W.L. The effect of soy protein with or without isoflavones relative to milk protein on plasma lipids in hypercholesterolemic postmenopausal women. *Am. J. Clin. Nutr.* **2001**, *73*. [CrossRef]
55. Jayagopal, V.; Albertazzi, P.; Kilpatrick, E.S.; Howarth, E.M.; Jennings, P.E.; Hepburn, D.A.; Atkin, S.L. Beneficial effects of soy phytoestrogen intake in postmenopausal women with type 2 diabetes. *Diabetes Care* **2002**, *25*. [CrossRef] [PubMed]
56. Greany, K.A.; Nettleton, J.A.; Wangen, K.E.; Thomas, W.; Kurzer, M.S. Probiotic consumption does not enhance the cholesterol-lowering effect of soy in postmenopausal women. *J. Nutr.* **2004**, *134*. [CrossRef]
57. Jefferson, W.N.; Patisaul, H.B.; Williams, C.J. Reproductive consequences of developmental phytoestrogen exposure. *Reproduction* **2012**, *143*. [CrossRef] [PubMed]
58. Losa, S.M.; Todd, K.L.; Sullivan, A.W.; Cao, J.; Mickens, J.A.; Patisaul, H.B. Neonatal exposure to genistein adversely impacts the ontogeny of hypothalamic kisspeptin signaling pathways and ovarian development in the peripubertal female rat. *Reprod. Toxicol.* **2011**, *31*. [CrossRef]

59. Bateman, H.L.; Patisaul, H.B. Disrupted female reproductive physiology following neonatal exposure to phytoestrogens or estrogen specific ligands is associated with decreased GnRH activation and kisspeptin fiber density in the hypothalamus. *Neurotoxicology* **2008**, *29*. [CrossRef]
60. Patisaul, H.B.; Todd, K.L.; Mickens, J.A.; Adewale, H.B. Impact of neonatal exposure to the ERα agonist PPT, bisphenol-A or phytoestrogens on hypothalamic kisspeptin fiber density in male and female rats. *Neurotoxicology* **2009**, *30*. [CrossRef]
61. Patisaul, H.B. Effects of environmental endocrine disruptors and phytoestrogens on the kisspeptin system. *Adv. Exp. Med. Biol.* **2013**, *784*. [CrossRef]
62. Konigsmark, B.W. Methods for the counting of neurons. In *Contemporary Research Methods in Neuroanatomy*; Springer: New York, NY, USA, 1970.

 metabolites

MDPI

Article

Early Postnatal Genistein Administration Affects Mice Metabolism and Reproduction in a Sexually Dimorphic Way

Marilena Marraudino [1], Giovanna Ponti [1], Chantal Moussu [2], Alice Farinetti [1,3], Elisabetta Macchi [4], Paolo Accornero [4], Stefano Gotti [1,3], Paloma Collado [5], Matthieu Keller [2,†] and Giancarlo Panzica [1,3,*,†]

1 Neuroscience Institute Cavalieri Ottolenghi (NICO), Regione Gonzole 10, Orbassano, 10043 Torino, Italy; marilena.marraudino@unito.it (M.M.); gponti2@gmail.com (G.P.); alice.farinetti@unito.it (A.F.); stefano.gotti@unito.it (S.G.)

2 UMR Physiologie de la Reproduction et des Comportements, Institut National de Recherche pour l'agriculture, l'Alimentation et l'Environnement (INRAE), Centre National de la Recherche Scientifique (CNRS), Institut Français du Cheval et de l'Equitation (IFCE), Université de Tours, 37380 Nouzilly, France; chantal.porte@inrae.fr (C.M.); matthieu.keller@inrae.fr (M.K.)

3 Laboratory of Neuroendocrinology, Department of Neuroscience Rita Levi Montalcini, University of Torino, Via Cherasco 15, 10125 Torino, Italy

4 Department of Veterinary Sciences, University of Torino, Largo Braccini 2, Grugliasco, 10095 Torino, Italy; elisabetta.macchi@unito.it (E.M.); paolo.accornero@unito.it (P.A.)

5 Department of Psychobiology, Universidad Nacional de Educación a Distancia (UNED), C/Juan del Rosal 10, 28040 Madrid, Spain; pcollado@psi.uned.es

* Correspondence: giancarlo.panzica@unito.it; Tel.: +39-011-6706607; Fax: +39-011-2367054

† Equally contributed and should be considered as joint senior authors.

Citation: Marraudino, M.; Ponti, G.; Moussu, C.; Farinetti, A.; Macchi, E.; Accornero, P.; Gotti, S.; Collado, P.; Keller, M.; Panzica, G. Early Postnatal Genistein Administration Affects Mice Metabolism and Reproduction in a Sexually Dimorphic Way. *Metabolites* **2021**, *11*, 449. https://doi.org/10.3390/metabo11070449

Academic Editor: Amedeo Lonardo

Received: 1 June 2021
Accepted: 7 July 2021
Published: 10 July 2021

Abstract: The phytoestrogen genistein (GEN) may interfere with permanent morphological changes in the brain circuits sensitive to estrogen. Due to the frequent use of soy milk in the neonatal diet, we aimed to study the effects of early GEN exposure on some physiological and reproductive parameters. Mice of both sexes from PND1 to PND8 were treated with GEN (50 mg/kg body weight, comparable to the exposure level in babies fed with soy-based formulas). When adult, we observed, in GEN-treated females, an advanced pubertal onset and an altered estrous cycle, and, in males, a decrease of testicle weight and fecal testosterone concentration. Furthermore, we observed an increase in body weight and altered plasma concentrations of metabolic hormones (leptin, ghrelin, triiodothyronine) limited to adult females. Exposure to GEN significantly altered kisspeptin and POMC immunoreactivity only in females and orexin immunoreactivity in both sexes. In conclusion, early postnatal exposure of mice to GEN determines long-term sex-specific organizational effects. It impairs the reproductive system and has an obesogenic effect only in females, which is probably due to the alterations of neuroendocrine circuits controlling metabolism; thus GEN, should be classified as a metabolism disrupting chemical.

Keywords: phytoestrogens; endocrine disruptor; dimorphism; obesity; kisspeptin; POMC; orexin

1. Introduction

Genistein (GEN; 4′,5,7-trihydroxyisoflavone) is an isoflavonoid compound. Its chemical structure shares features with 17β-estradiol, enabling it to bind to estrogen receptors. GEN is produced by many plants, is highly present in Leguminosae species, and, due to its estrogenic activity, is considered a phytoestrogen [1]. The main sources of GEN, in our diet, are soybeans and soy-based foods. Most foods contain a small quantity of isoflavones, but when consumed regularly and from various sources, they can reach a cumulative dose that can contribute to long-term effects [2]. In adults, phytoestrogens, including GEN, have been generally associated with beneficial effects, i.e., obesity and diabetes [3], menopause [4], cancer [5], and hypertension associated to metabolic syndrome [6]; however, many studies suggest that phytoestrogens are harmful to human health [7,8]. Above all, babies may be subjected to higher levels of phytoestrogens (6–9 mg/kg/day) than typical adult exposures (approximately 1 mg/kg/day) [9]. In fact, in addition to soy-based infant formulas also many soy-based foods are specifically prepared for babies [9,10]. In rodents, sex-specific alterations were reported in several neuronal populations, including hypothalamic and amygdaloid circuits containing many gonadal hormone-sensitive neurons [11–16].

The neuroendocrine control of food intake and energy expenditure is based on many circuits including both orexinergic and anorexinergic elements that are targets for a series of chemical signals coming from the periphery (for a review see [17]). The main centers of this control are the arcuate nucleus (ARC), the dorsomedial hypothalamic nucleus (DMH), the paraventricular hypothalamic nucleus (PVN), and the lateral hypothalamus (LH). The ARC contains orexinergic elements (NPY/AgRP neurons) and anorexinergic ones (the pro-opiomelanocortin/cocaine- and amphetamine-regulated transcript neurons (POMC/CART)) while the LH encompasses orexin-A/hypocretin-1 (OX) neurons. Unlike NPY/AgRP neurons (stimulating food intake), OX stimulates both feeding and energy expenditure, in response to physiological variation in glucose blood levels between meals [18]. Among the peripheral signals, leptin, a satiety hormone produced by adipocytes, stimulates POMC neurons in the ARC [19] and depresses the OX system. In the ARC, OX fibers contact POMC cells decreasing their synthesis and promoting hyperphagia and weight gain [20]. Interestingly, POMC and OX fibers directly project to the PVN, the most important center of metabolic control. The PVN expresses the receptor for OX [21] and the two peptides released from POMC neurons: melanocortin and α-melanocyte-stimulating hormone (α-MSH) [22]. Moreover, the PVN modulates metabolism through its action on the hypothalamus-pituitary-adrenal and on the hypothalamus-pituitary-thyroids axis [23]. The medial part of the PVN, where corticotropin-releasing hormone (CRH) and thyrotropin-releasing hormone (TRH) neurons are located, is strongly innervated by kisspeptin (kiss) fibers [24]. The kiss system is formed by a neuronal population more numerous in females than in males [25], and, in rodents, kiss cells are clustered in the ARC and the rostral periventricular area of the third ventricle (RP3V). This peptide was, at first, identified as one of the key controllers of GnRH neurons (for a review see [26]), but today, kiss neurons seem to have a major nodal role in integrating the different signals transmitting metabolic information, from both the periphery and central nervous system, onto reproductive centers [27]. The connection between energy balance, puberty, and reproduction is evident, particularly in females, in conditions of anorexia (energy insufficiency) or obesity (excess of energy) [27]. Kiss neurons at the time of puberty are directly modulated by leptin [28]. Hypothalamic colocalization of kiss and leptin receptors is controversial and limited only to the ARC group; in particular, the estimated degree of colocalization varies between 40% [29], 15% [30], and 6–8% [31] of the kiss neurons in the mouse ARC. In addition, the signaling system becomes fully mature after the completion of puberty [29,30]. Moreover, leptin deficiency reduces kiss mRNA expression in the ARC [29,32] and the number of positive neurons in the RP3V [32]. Finally, kiss has strong effects on both POMC and NPY systems in the ARC: kiss has an excitatory effect on POMC neurons via the kiss receptor [33], which is expressed in about 63% of POMC-immunoreactive neurons in female rats [34] and inhibits NPY cells via the GABA system [33]. However, in vitro, kiss induces an increase of NPY and a decrease of BDNF expression [35]. In addition to peripheral stimuli such as leptin and other hormones produced by the gastrointestinal system (i.e., Ghrelin and others), estrogen also plays an important role in controlling food intake and metabolism [36].

Hypothalamic circuits involved in the regulation of food intake and energy metabolism are therefore potential targets for xenoestrogens, including phytoestrogens and GEN [37]. Much evidence suggests that they could have an obesogenic effect [36] or might counteract aspects of the metabolic syndrome [38]. Treatment with GEN affects body weight in rats, but the effect depends on sex, age, and hormonal status [38,39]. In the young, in ovariectomized [40,41] and intact adult female mice [42] GEN shows an anti-obesogenic effect, but this appears to depend on the dose: adipogenesis is inhibited at low doses of

GEN [43]. However, in males, the low concentration has an obesogenic effect [44]. In rats, the perinatal treatment of GEN has different effects: obesogenic in females [45] and anti-obesogenic in males [46].

The ability of GEN to bind the estrogen receptors (ERs) [12] makes the hypothalamic circuits highly sensitive to this molecule. A large body of literature is present on the effect of GEN on the kiss system with in vitro [47] and in vivo studies (for a review see [48]). No significant alterations of the kiss system were found in adult male rats postnatally exposed to GEN [49], while in females this system was mostly affected with a reduction of fiber density [11,50]. Yet, little is known about the GEN effects on the circuits that control food intake and energetic metabolism. A single study [51] shows that high phytoestrogens (daidzein and genistein) concentrations decrease, in male mice hypothalamus, AgRP and increase MCH, orexin A, and TRH mRNA levels in postnatal life, but it does not influence NPY, POMC, and CART expression. However, in rats, the postnatal administration of two doses of GEN (10 or 50 µg/g body weight) from PND6 to PND13 decreased the number of POMC-ir cells in the ARC only in females, while the OX system did not appear to be affected by the treatment in either sex [52].

Our study aimed to enclose in a single experiment the potential for sexually dimorphic effects of early postnatal administration of GEN in mice (at a dose comparable to that of babies fed with soy-based formulas) on: (I) female reproductive peripheral parameters (puberty onset, estrus cycle, mammary gland development); (II) male reproductive peripheral parameters (testosterone level, testicle size); (III) control of food intake and metabolic regulation through hormonal (leptin, triiodothyronine, ghrelin) and physiological (body weight, food consumption, and feed efficiency) parameters; (IV) hypothalamic circuits involved in both reproductive and metabolic control (POMC, orexin, kisspeptin systems).

2. Results

2.1. Female Reproductive Parameters

2.1.1. Vaginal Opening and Estrous Cycle

Vaginal opening (VO) is one of the parameters employed to evaluate the potential effects of endocrine disrupting chemicals on puberty onset in rodents [53]. GEN treatment stimulated an anticipation of the VO (F-CON = 27.8 ± 0.29; F-GEN = 26.5 ± 0.30, $p = 0.005$) (Figure 1a).

GEN treatment altered the estrous cycle: F-GEN spent more time in estrus and diestrus phases compared to F-CON (Estrus = 30% and Diestrus = 43% vs. 35% and 51% respectively in F-GEN) with a significant reduction of proestrus phase (t-Test, $p = 0.03$; F-CON = 16.67 vs. F-GEN = 6.67) (Figure 1c).

2.1.2. Uterus Weight

The uterus weight of F-GEN was higher than F-CON starting from PND22, however, the difference was significant only at PND22 (Student's t-test $p = 0.035$), but not after puberty (PND30, $p = 0.120$ and PND60, $p = 0.123$) (Figure 1b). Two-way ANOVA for age and treatment as independent variables confirmed this result (respectively F = 30.489, $p = 0.001$; F = 6.797, $p = 0.014$), but no difference was present in their interaction (F = 0.951, $p = 0.398$).

2.1.3. Progesterone Level

Progesterone serum levels in control females did not change significantly in young (PND30 6.84 ± 1.19) and adult (PND60 6.73 ± 0.90; $p = 0.565$) animals. GEN treatment did not affect progesterone levels in young animals (PND30 F-CON vs. F-GEN; $p = 0.41$), while it significantly increased it in adults (PND60 F-CON vs. F-GEN; $p = 0.007$) (Figure 1d).

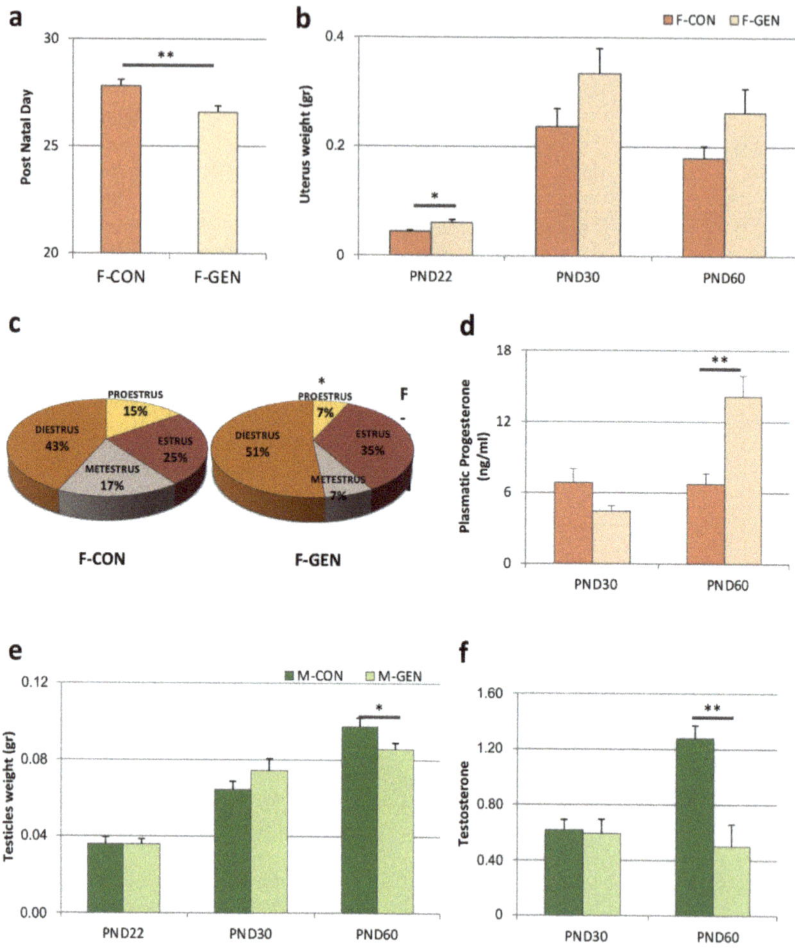

Figure 1. Parameters related to reproduction. (**a**) Histograms representing the evaluation of the day of vaginal opening (VO) in control (F-CON) and treated (F-GEN) female CD1 mice (mean $\pm$ SEM). (**b**) Histograms representing the variations of the uterus weight (expressed in gram; mean $\pm$ SEM) measured during development at postnatal day (PND) 22, PND30, and PND60 of control (F-CON) and treated (F-GEN) females. (**c**) Pie charts illustrating the time spent (expressed as percentage) in each phase of estrus cycle (estrus, metestrus, diestrus, and proestrus) in control (F-CON) and treated (F-GEN) females. (**d**) Histograms representing circulating variations of plasma concentration of progesterone (expressed in ng/mL, mean $\pm$ SEM) in control (F-CON) and treated (F-GEN) females, during development at PND30 and PND60. (**e**) Histogram representing the increase of testicle weight (expressed in gram; mean $\pm$ SEM) measured during development at postnatal day (PND) 22, PND30, and PND60 of control (M-CON) and treated (M-GEN) males. (**f**) Histogram representing variations of fecal testosterone levels (expressed in ng/mL) in control (M-CON) and treated (F-GEN) males, during development at PND30 and PND60. * $p < 0.05$; ** $p \leq 0.01$. VO, uterus and testicles weight, and time spent in each phase of estrus cycle were compared using Student's *t*-test, while the progesterone was analyzed using Tukey's test.

2.1.4. Mammary Gland Analysis

Mammary gland length gradually increased with development with no significant differences between the experimental groups (Table S1, Supplementary Materials) (Figure 2).

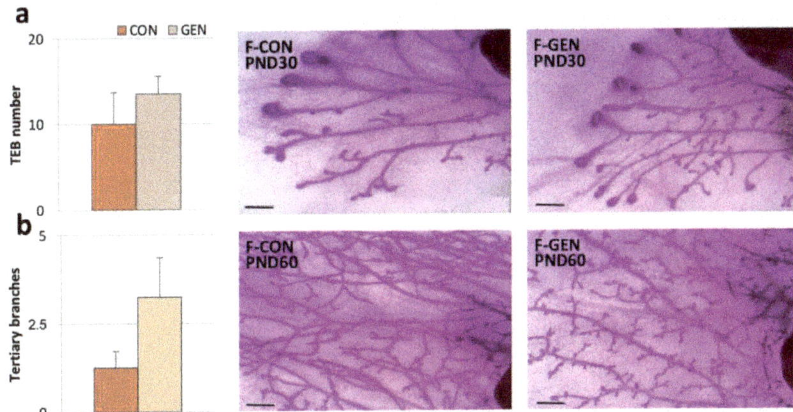

Figure 2. Mammary gland. (**a**) Histograms (**left**) and photomicrograph (**right**) showing the number of terminal end buds (TEBs) of mammary glands in control (CON) and treated (GEN) female CD1 mice at PND30. (**b**) Histograms (**left**) and photomicrograph (**right**) representing the number of tertiary branches of mammary glands in control (CON) and treated (GEN) female CD1 mice at PND60 (expressed as mean ± SEM). Scale bar = 0.5 mm.

Terminal end buds (TEBs) in virgin mice are present only at pubertal age when they are stimulated by endogenous estrogens and other factors [54]. Indeed, TEBs were present in control animals at PND22 and PND30, while, as expected [55], control mammary glands (at PND60) did not have TEBs. GEN treatment did not affect TEBs at any of the developmental stages considered (Table S1, Supplementary Materials) (Figure 2a).

Moreover, the treatment did not affect the overall architecture of the adult mammary gland. In fact, at PND60, the number of branches was similar in F-CON (6.67 ± 0.22) and F-GEN (7.17 ± 0.51; $p = 0.408$). However, GEN-treated females tended to have mammary glands with a higher score of tertiary branches (F-CON = 1.25 ± 0.48; F-GEN = 3.25 ± 1.11, $p = 0.17$), which develop cyclically during diestrus (Figure 2b).

2.2. *Male Reproductive Parameters*

2.2.1. Testicle Weight

Testicle weight was similar at PND22 (*t*-test; $p = 1$) and PND30 ($p = 0.207$), but in adults, it was significantly lower in M-GEN than M-CON ($p = 0.05$) (Figure 1e). The analyses using two-way ANOVA for age and treatment as independent variables demonstrated a significant difference only for age (F = 67.472; $p = 0.001$), but not for treatment (F = 0.015; $p = 0.904$) and their interaction (F = 2.611, $p = 0.090$).

2.2.2. Testosterone Level

Control males had higher testosterone levels at PND60 (CON vs. PND30 CON; $p < 0.001$) while GEN-treated mice did not show this increase, resulting in significantly lower levels than in control PND60 ($p = 0.003$) (Figure 1f, Table S2).

2.3. *Leptin*

In control animals of both sexes, plasma concentration of leptin was low and with no sex differences until PND22 (Figure 3). In females, leptin levels significantly increased at PND60 (PND60 vs. PND30; $p < 0.001$) (Figure 3b). On the contrary, in control males, leptin levels significantly increased at PND30 (PND30 vs. PND22; $p = 0.001$) then decreased at PND60 (PND60 vs. PND30; $p < 0.001$) (Figure 3a). An interesting dimorphism was present at PND30 when males showed a strong peak of plasma leptin (M-CON vs. F-CON;

$p = 0.014$), while at PND60 the concentration was significantly higher in females compared to males (M-CON vs. F-CON; $p = 0.008$).

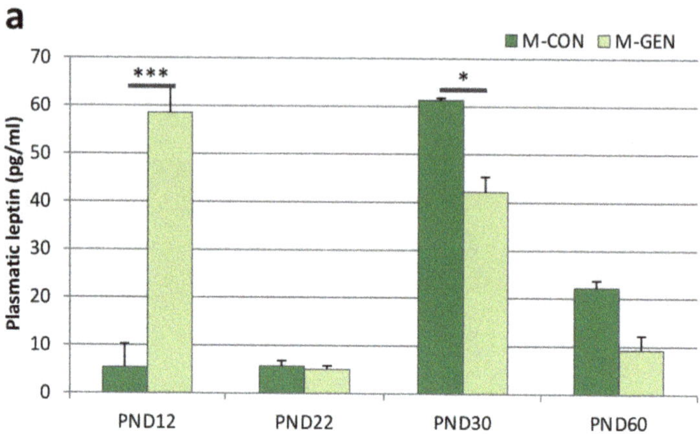

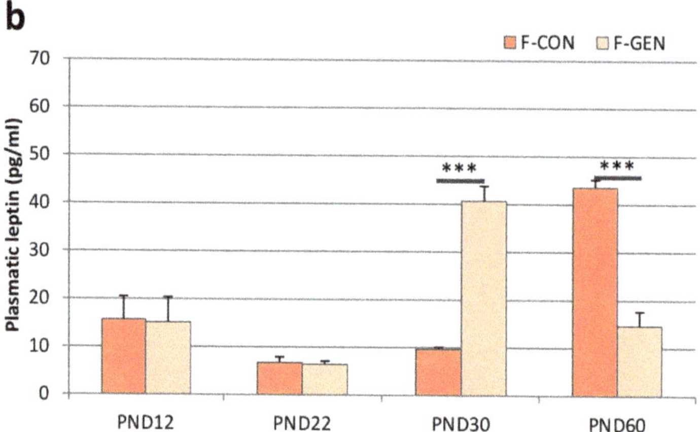

Figure 3. Leptin levels. Histograms representing variations of circulating leptin levels (expressed in pg/mL; mean ± SEM) in males (**a**) and females (**b**), control (CON) and treated (GEN), during the development at postnatal day (PND) 12, PND22, PND30, and PND60. * $p < 0.05$; *** $p \leq 0.001$ (Tukey test).

This dimorphism was abolished by GEN treatment. In fact, in males, GEN treatment induces an early significant increase in leptin level at PND12 in comparison to control animals (M-GEN vs. M-CON; $p < 0.001$). Plasma concentration of leptin reached a peak at PND30 in both GEN experimental groups, although it was significantly lower in GEN treated males than in CON (M-GEN vs. M-CON; $p = 0.032$). No significant difference was observed at PND60 (M-GEN vs. M-CON; $p = 0.456$; Figure 3a). A similar trend was observed in the GEN-treated females, where changes in circulating leptin were similar to M-GEN and M-CON, with a peak at PND30 and a decrease at PND60, both significant in comparison to F-CON ($p < 0.001$; Figure 3b). Plasma concentrations of leptin in treated males were significantly higher than F-GEN only at PND12 (M-GEN vs. F-GEN; $p < 0.001$).

2.4. Kisspeptin System

The pattern of distribution, development, and dimorphism of kiss-ir structures observed within the hypothalamic region of control animals was consistent with previous observations in rodents [24,25]. The quantitative analysis of CON and GEN groups suggested that postnatal exposure to GEN significantly altered the development of kisspeptin expression in the hypothalamic nuclei under study in a dimorphic manner (all data are reported in Supplementary Materials, Table S3).

In the ARC, the high density of kiss-ir precluded our ability to distinguish cell bodies (Figure 4a); therefore, we quantified the fractional area (FA). Early postnatal exposure to GEN deeply influenced the expression of the kiss in the ARC. In males, GEN influenced kiss expression only at PND12, when M-GEN showed a significantly higher expression of kiss, which was also higher than in females (M-GEN vs. F-GEN; $p < 0.001$), reversing the sex dimorphism of the system (Figure 4b). No difference was detected in the comparison M-CON versus M-GEN at PND22, PND30, and PND60. In females, the situation was different: at PND12 ($p < 0.001$) and at PND22 the FA in F-GEN remained significantly lower in comparison to F-CON ($p = 0.004$), but after puberty, at PND30, the kiss-ir signal increased in treated females ($p < 0.001$). Then, at PND60 FA decreased in F-GEN animals (Figure 4a). GEN treatment seems to cause anticipation of the development of the system in females at PND30 (Figure 4b).

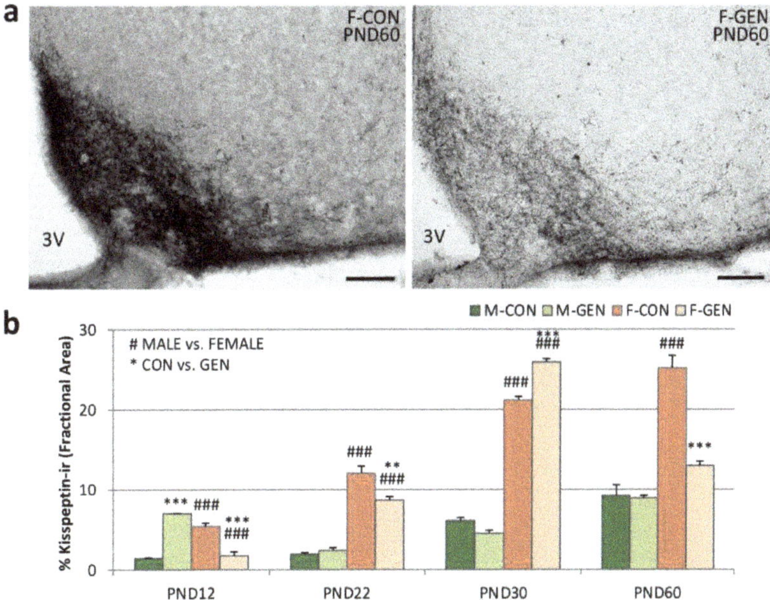

Figure 4. Kisspeptin immunohistochemistry. (**a**) Photomicrographs showing kisspeptin immunostaining in the ARC in adult (PND60) CD1 control (F-CON) and treated (F-GEN) female mice at PND60. 3V, third ventricle. Scale bar = 100 µm. (**b**) Histograms representing the variations of the percentage of area (FA; mean ± SEM) covered by kisspeptin immunopositive elements in the ARC of both sexes (M-CON, M-GEN, F-CON, and F-GEN) at different ages (PND2, PND22, PND30, and PND60). ** $p \leq 0.01$; *** $p \leq 0.001$; ### $p \leq 0.001$ (Bonferroni test).

Within RP3V we analyzed the number of kiss-ir cells. In control animals, the system was dimorphic at every age, with a higher number of cells in females. In addition, while the number of cells was stable for all the considered ages in males, there was a significant increase in females, from PND12 to PND60. Early GEN postnatal treatment had no significant effect on males at any of the considered ages (Figure 5b), while females were significantly affected. The number of cells was significantly higher in F-GEN at PND12 ($p = 0.002$), PND22 ($p = 0.002$), and PND30 ($p = 0.003$) in comparison to F-CON. At PND60 the number of cells in the F-GEN group was close to PND30. Thus, at PND60 the F-CON showed a higher number of positive cells than F-GEN ($p < 0.001$) (Figure 5a,b). Therefore, it appears that kiss-ir cells reached a plateau at PND30 in F-GEN, whereas this peak is reached only at PND60 in F-CON.

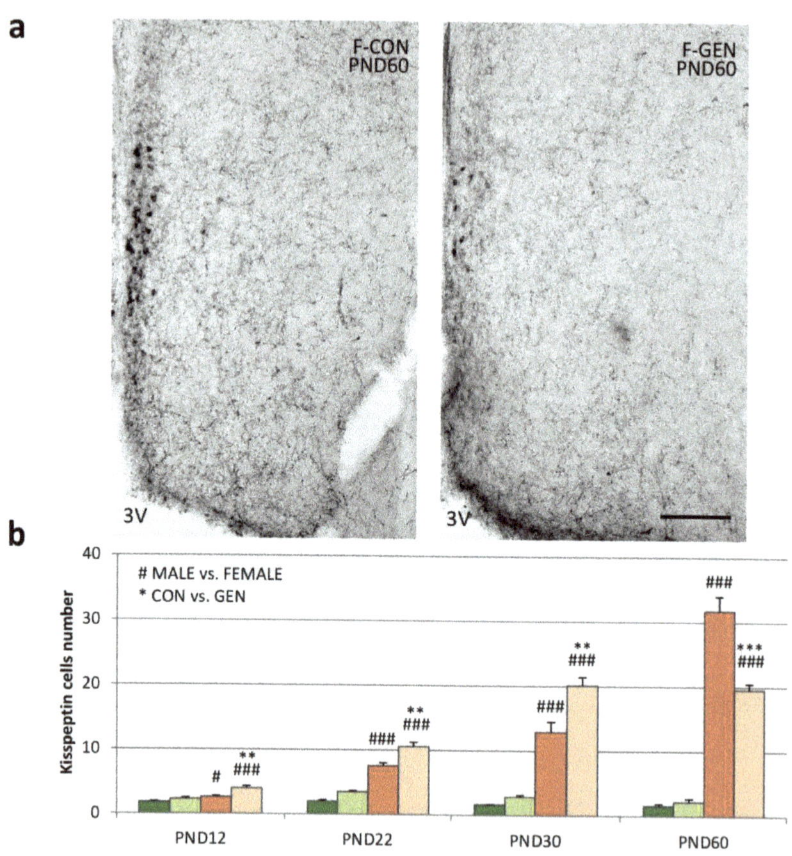

Figure 5. Kisspeptin immunoreactivity. (**a**) Photomicrographs showing kisspeptin immunostaining in the RP3V of adult (PND60) control (F-CON) and treated (F-GEN) CD1 female mice. 3V, third ventricle. Scale bar = 100 µm. (**b**) Number of kisspeptin-positive cells (mean ± SEM) in the RP3V of both sexes (M-CON, M-GEN, F-CON, and F-GEN) at different ages (PND12, PND22, PND30, and PND60). ** $p \leq 0.01$; *** $p \leq 0.001$; # $p \leq 0.05$; ### $p \leq 0.001$ (Bonferroni test).

In the PVN (Figure 6a), the effect of GEN treatment followed more that observed in RP3V than in the ARC. We observed no effects in males at all the considered ages, whereas in females we observed a precocious increase of kiss-ir in F-GEN animals at PND22 and PND30, with a sharp decrease at PND60. As for the ARC and RP3V in F-CON, the kiss-ir increased in the PVN, reaching the highest value at PND60 (Figure 6b).

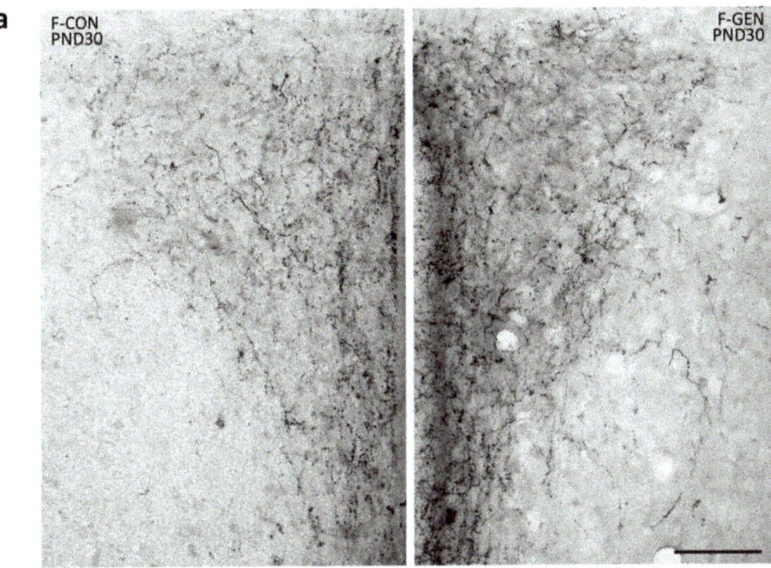

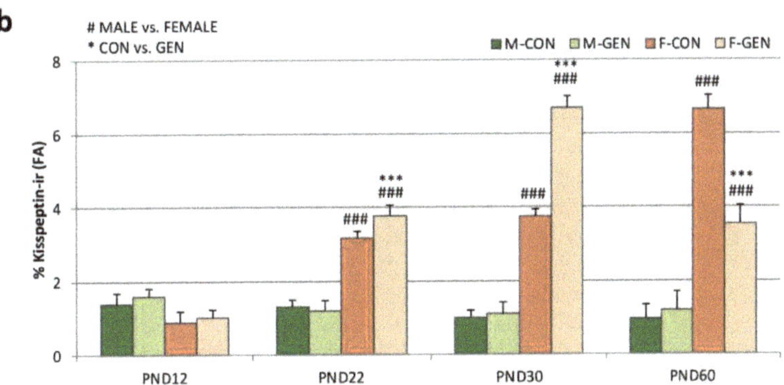

Figure 6. Kisspeptin innervation of the PVN. (**a**) Comparison of the PVN innervation at PND30, when the F-GEN group has its peak. (**b**) Variations of the fractional area (FA; mean ± SEM) covered by kisspeptin immunoreactive fibers in the PVN of both sexes (M-CON, M-GEN, F-CON, and F-GEN) at different ages (PND12, PND22, PND30, and PND60). Scale bar = 100 µm. *** $p \leq 0.001$; ### $p \leq 0.001$ (Bonferroni test).

The distribution of kiss-ir fibers within the PVN, as also reported in previous studies [24,56], was not homogeneous but was denser in the medial versus lateral PVN (Figure 7a). In M-CON, kiss-ir was similar in the different parts of the PVN, although it tended to be higher in the ventro-medial part. No significant differences were observed at any age or after GEN treatment. On the other hand, in females (Figure 7b), the GEN effect was evident within dorso- and ventral-medial parts, where the kiss-ir fibers were denser. F-GEN reached the highest value of FA in the DM and VM part of the PVN at PND30, with a decrease at PND60, whereas the F-CON reached the highest value at PND60.

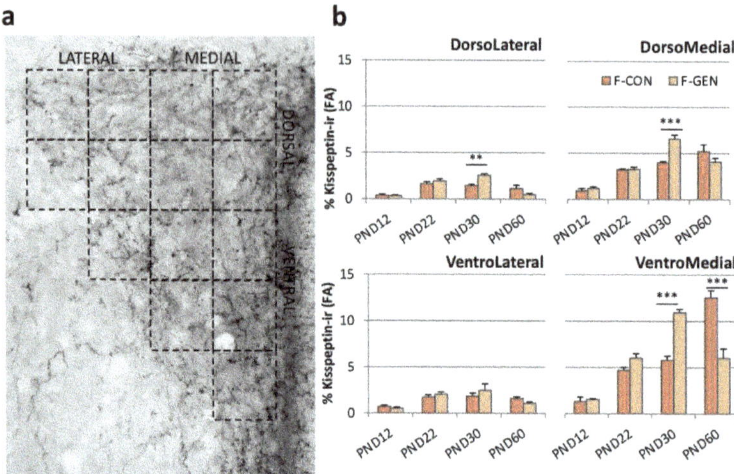

Figure 7. (**a**) Photomicrograph of kisspeptin immunostaining in the PVN of adult control females, showing the subdivision of the PVN in fourteen quadrants to identify the four parts of the nucleus (DM, dorso-medial; DL, dorsolateral; VM, ventro-medial; VL, ventro-lateral). (**b**) Variations of the FA covered by kisspeptin immunoreactive fibers in the different parts of the PVN (DM, DL, VM, VL) in control (F-CON) and treated (F-GEN) female CD1 mice at different ages of sacrifice (PND12, PND22, PND30, and PND60). Scale bar = 50 µm. ** $p \leq 0.01$; *** $p \leq 0.001$ (Bonferroni test).

2.5. Metabolic Parameters

2.5.1. Body Weight, Food Consumption, and Feed Efficiency (FE)

The analysis using three-way ANOVA of body weight measured, with age, sex, and treatment considered as independent variables, showed a significant effect both of the interaction sex and treatment (F = 8.90; $p = 0.003$) and sex and age (F = 34.735; $p = 0.001$), but not for the interaction treatment and age (F = 1.806; $p = 0.079$) or among all independent variables (sex * age * treatment; F = 0.698; $p = 0.693$). We did not observe any effect of GEN on the metabolic parameters considered during and after the treatment until day 30. Starting from PND30, the body weight presented strong differences among experimental groups (Table S4, Supplementary Materials). The analysis via Tukey's test for M-CON vs. F-CON reported a significantly higher weight in males (PND30, PND40, PND50, and PND60, $p < 0.001$).

Differences were observed also in the percentage of body weight gained since PND30 (Figure 6a). Two-way ANOVA for sex and treatment confirmed this result (respectively F = 15.39; $p = 0.001$ and F = 5.901; $p = 0.025$). No differences were present between controls and treated males, while a significant increase was observed starting from PND40 in F-GEN compared to control females (Tukey test; F-CON vs. F-GEN at PND40 $p = 0.007$; at PND50 $p = 0.001$; at PND60 $p = 0.005$) (Figure 8a).

Daily food consumption per animal during the weeks after weaning was calculated as reported in Materials and Methods. Generally, we observed lower food consumption in F-CON in comparison to M-CON. This trend was observed in GEN-treated animals with males treated with GEN eating significantly more food than F-GEN only during the second ($p = 0.038$) and fourth ($p = 0.011$) week (Table S5, Supplementary Materials) (Figure 8b). However, no differences were found between control and treated animals either in males or in females. The two-way ANOVA for repeated measures (independent variables: sex and treatment, repeated measure: weekly food consumption) showed, in fact, only an effect of sex (F = 21.015 and $p = 0.001$), with no effect of treatment (F = 0.165 and $p = 0.689$) and no significant interaction between sex and treatment (F = 0.056 and $p = 0.815$).

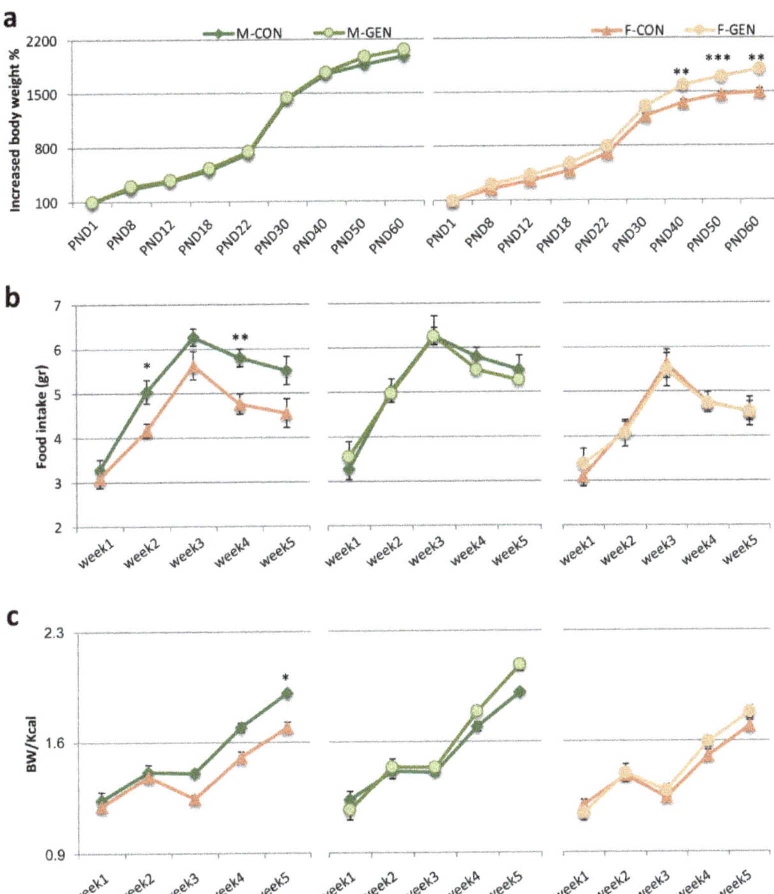

Figure 8. Metabolic parameters. (**a**) Lines representing the increase of body weight in percentage, with the weight at PND1 = 100 (mean ± SEM) during the development of male control (M-CON) vs. male-treated (M-GEN), on the left, and female control (F-CON) vs. female-treated (F-GEN), on the right. (**b**) Variations of the amount of food intake expressed in grams (mean ± SEM) during five weeks after weaning in both sexes (M-CON, M-GEN, F-CON, F-GEN). On the left a comparison among control males and females; the other two diagrams illustrate the comparison within each sex among control and genistein-treated mice. (**c**) Variations of the daily feed efficiency (body weight/kcal introduced; mean ± SEM) calculated during five weeks after weaning in both sexes (M-CON, M-GEN, F-CON, F-GEN). On the left a comparison among control males and females; the other two diagrams illustrate the comparison within each sex among control and genistein-treated mice. * $p < 0.05$; ** $p \leq 0.01$; *** $p \leq 0.001$; (Bonferroni test).

Feed efficiency (body weight gain/Kcal) was analyzed using two-way ANOVA for repeated measures (independent variables: sex and treatment, repeated measure: feed efficiency) showing a global effect of sex (F = 6.722 and $p = 0.0017$), but no effect of treatment (F = 0.704 and $p = 0.411$). The multiple comparisons between groups displayed a sexual dimorphism in controls, with a significant increase in M-CON vs. F-CON in the last week of the experiment (from PND53 to PND60; Tukey test, $p = 0.038$). This difference was present also between M-GEN and F-GEN at week 3 ($p = 0.05$) and week 4 ($p = 0.03$; Figure 8c). GEN treatment did not alter feed efficiency either in males or females (Table S6).

2.5.2. Triiodothyronine (T3) and Ghrelin Levels

Plasma concentration of T3 at PND30 showed no significant differences. In the adult animals, T3 levels were significantly higher in females than in males (F-CON vs. M-CON; $p = 0.001$). GEN treatment did not affect T3 levels in males (M-CON vs. M-GEN; $p = 1$), but it reduced the plasma concentration of T3 levels in adult females at PND60 close to significant (F-CON vs. F-GEN; $p = 0.05$), abolishing sex dimorphism (F-GEN vs. M-GEN; $p = 1$; Table 1).

Table 1. Plasma concentration of T3 and ghrelin. The concentration of T3 (expressed as ng/mL, mean ± SEM) and concentration of ghrelin in the plasma (expressed as ng/mL, mean ± SEM) at PND30 and PND60 for different groups of mice.

	Plasma Concentration of T3 (ng/mL)			
	M-CON	**M-GEN**	**F-CON**	**F-GEN**
	(mean ± SEM)	(mean ± SEM)	(mean ± SEM)	(mean ± SEM)
PND30	5.3 ± 0.32	5.8 ± 0.28	5.6 ± 0.19	6.2 ± 0.34
PND60	6.4 ± 0.06	5.7 ± 0.11	7.5 ± 0.19	6.3 ± 0.09
	Plasma Concentration of ghrelin (ng/mL)			
	M-CON	**M-GEN**	**F-CON**	**F-GEN**
	(mean ± SEM)	(mean ± SEM)	(mean ± SEM)	(mean ± SEM)
PND30	0.30 ± 0.04	0.40 ± 0.06	0.34 ± 0.07	0.7 ± 0.11
PND60	0.87 ± 0.18	4.23 ± 0.69	4.33 ± 0.73	2.76 ± 0.26

The analysis of plasma concentration of ghrelin at PND30 did not present significant differences among the experimental groups. In adults, at PND60, F-CON animals showed a higher plasma concentration of hormone in comparison to males (F-CON vs. M-CON; $p = 0.001$). The postnatal treatment with genistein completely reversed this situation, the sex difference disappeared (F-GEN vs. M-GEN; $p = 0.146$): the plasma concentration of ghrelin significantly increased in GEN males (M-CON vs. M-GEN; $p = 0.001$) and decreased in GEN females, although not significantly (F-CON vs. F-GEN; $p = 0.06$), compared to control animals (Table 1).

2.6. *Hypothalamic Systems Controlling Food-Intake*

2.6.1. POMC System

According to our previous study [57], in the present experiment, the POMC-ir in the ARC was very similar in adult mice of both sexes (Table S7). In fact, the Bonferroni test did not show any significant difference in POMC-ir cell number ($p = 0.896$) or in FA covered by ir structures ($p = 0.918$) in control animals. GEN treatment significantly affects the adult profile of POMC expression in the ARC in a sex-specific manner. Two-way ANOVA reported a significant effect on the sex and treatment interaction (cells number, $F = 13.710$; $p = 0.002$; FA, $F = 7.371$; $p = 0.015$). Indeed, the alterations of the POMC-ir elements were limited to females (Figure 9a). The Bonferroni post hoc test showed statistically significant higher values in F-GEN than F-CON for the number of POMC cells ($p = 0.047$), but not for FA ($p = 0.081$) in the ARC, while no difference was present in males ($p = 0.657$ and $p = 0.134$, respectively for FA and cellular number) (Figure 9b; full quantitative data are in Table S7 for adults and in Table S8 for the development).

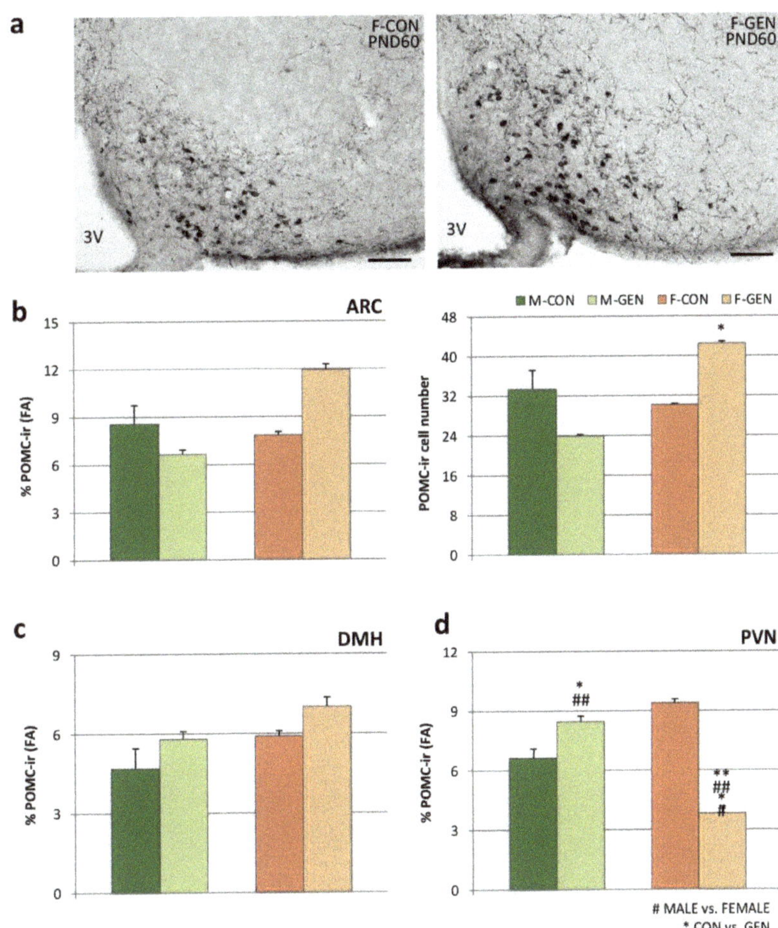

Figure 9. Pro-opiomelanocortin (POMC) immunoreactivity. (**a**) Photomicrographs of the arcuate nucleus (ARC) immunostained for POMC at PND60 in control (F-CON) and treated (F-GEN) female mice. 3V, third ventricle. Scale bar = 100 μm. (**b**) Variations of both the percentage of area (FA: mean ± SEM) covered by POMC immunopositive structures (left) and the number of positive cell bodies (right) in the ARC of both sexes (M-CON, M-GEN, F-CON, F-GEN) for adult mice. (**c**,**d**) Histograms representing the FA (mean ± SEM) covered by POMC immunoreactivity within (**c**) the dorsomedial hypothalamic nucleus (DMH) and (**d**) paraventricular nucleus (PVN). * $p < 0.05$; ## $p \leq 0.01$ (Bonferroni test).

The immunoreactivity of POMC fibers within DMH did not result in significant differences both in controls and in GEN-treated animals (Figure 9c).

POMC-ir was dimorphic in the PVN of control animals: Bonferroni test showed that the FA covered by ir structures (in this case only fibers) was higher in females ($p = 0.002$). The two-way ANOVA demonstrated a significant effect of sex (F = 5.455; $p = 0.036$), of treatment (F = 21.667; $p < 0.001$), and of the interaction sex/treatment (F = 84.642; $p < 0.001$). The early postnatal GEN exposure decreased the immunoreactivity in the female PVN ($p < 0.001$) and increased it in males ($p = 0.048$), with a completely dimorphic effect. In fact, in GEN animals the dimorphism was inverted, higher in males than in females ($p < 0.001$; Figure 9d). This effect was predominant in the medial part of the PVN, where the FA strongly decreased in treated females (DM, $p < 0.001$; VM, $p < 0.001$).

2.6.2. Orexin System

As reported in previous studies (for a review see [58]), in both sexes of adult CD1 mice, the lateral hypothalamus (LH), in its full rostro-caudal extension (Figure 10a), contains a large number of orexin-ir cells. Considering the total number of positive cells, the two-way ANOVA for sex and treatment showed a significant effect of treatment (F = 17.502; $p = 0.001$). In controls, the system is sexually dimorphic with males having a significantly higher number of cells (313.7 ± 18.21) than females (189.9 ± 19.29, Bonferroni test M-CON vs. F-CON; $p < 0.001$). Post hoc Bonferroni showed a statistically significant decrease of cell number in GEN-treated male mice (253.4 ± 14.70) in comparison to M-CON ($p = 0.041$), while in females (260.7 ± 10.90) a significant increase of cell number was present in F-GEN vs. F-CON ($p < 0.05$). Because of the different effects in males and females, the dimorphism completely disappeared in GEN-treated mice (Bonferroni test, M-GEN vs. F-GEN; $p = 1$) (adult data are in Table S9A and changes during the development are reported in Table S8).

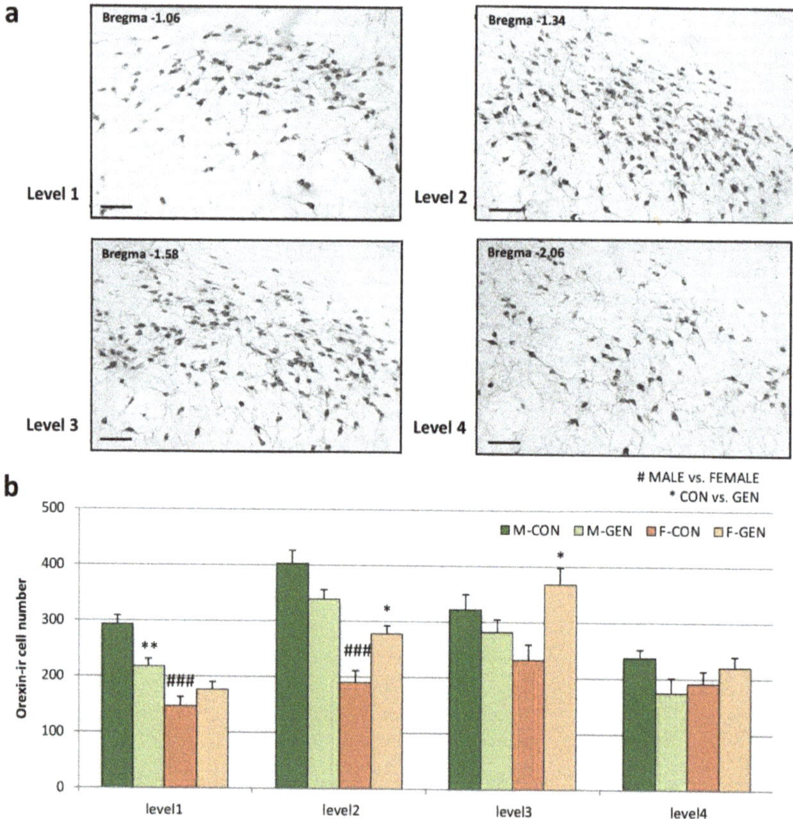

Figure 10. Distribution of orexin cells within the lateral hypothalamic nucleus. (**a**) Photomicrographs showing the distribution of orexin-positive cells in four different rostro-caudal levels of the lateral hypothalamus (LH), corresponding to Bregma −1.06, −1.34, −1.58, and −2.06 of the Mouse Brain Atlas [59] in male control CD1 mice at postnatal day (PND) 60. Scale bar = 100 μm. (**b**) Variations in the number of orexin-positive cells (mean ± SEM) in different levels of the LH of adult CD1 mice of both sexes (M-CON, M-GEN, F-CON, F-GEN). * $p < 0.05$; ** $p \leq 0.01$; ### $p \leq 0.001$ (Bonferroni test).

Considering the distribution of orexin-positive cells along the rostro-caudal axis (Figure 10a), the two-way ANOVA revealed a significant difference for level and a significant interaction for level and sex (respectively $F = 43.843$ $p < 0.001$; $F = 17.727$ $p < 0.001$). The two-by-two comparison (Bonferroni test) showed a significant decrease in the anterior region of M-GEN (M-GEN vs. M-CON; level1 $p = 0.013$), while in females the cell number increased in the more caudal region of the LH (F-GEN vs. F-CON level2 $p = 0.04$ and level3 $p = 0.025$) (Figure 10b, full data are in Table S9A).

In the PVN of control animals, the OX system was not dimorphic via comparison with the Bonferroni test ($p = 0.8$). Two-way ANOVA showed a significant effect of the interaction sex/treatment in adult animals within the PVN ($F = 16.09$; $p = 0.002$). The early postnatal GEN treatment significantly decreased the OX-ir fibers in males ($p = 0.045$), while they slightly increased in females (not significantly; $p = 0.088$). Interestingly, this sexual dimorphism was more pronounced in treated animals, in which the FA was higher in GEN females than in GEN males ($p = 0.009$), especially in the ventromedial part of the nucleus ($p = 0.002$) (Table S9B).

3. Discussion

Genistein, an isoflavone contained in soy, has an estrogen-like structure and exerts its effects by binding to the estrogen receptors: *in vitro* studies demonstrated that ERα, ERβ, and the estrogen membrane receptor (GPR30) are important for the neuritogenic effect of GEN [12]. The impact of GEN as an endocrine disrupting chemical (EDC) on health is still debated [37]. GEN alters estrous cyclicity, fecundity, ovulation, and female reproductive behavior [60,61], and when the exposure occurs at neonatal age, even at environmentally relevant doses, it affects female development and persists into adulthood [62]. In the present experiment, we observed precocious vaginal opening (an index of precocious female puberty [53]), an increase of uterus weight at pre-puberal age, irregular estrous cycles, with a reduction of the proestrus phase, probably correlated to an increase of plasma concentration of progesterone, and elongation of the diestrus phase.

In addition, we also analyzed the mammary gland, another important female reproductive target of EDCs. The analysis reported that the mammary glands were not directly affected by postnatal GEN treatment. This was not surprising since early development and sexual differentiation of the mammary gland occur before birth, thus perturbations of mammary gland development by EDCs have been observed only when the exposure was during the gestational period (i.e., GEN [63] or BPA [64]). The allometric postnatal growth of the mammary gland is independent of sexual steroids [55], while it responds to endocrine stimulation from puberty [55]. We observed that the number of animals displaying TEB at PND22 was higher in the GEN-treated group than in the controls, even if the difference was not significant. Since branching morphogenesis initiates at puberty [55], these data confirm that puberty is premature in GEN-treated animals. In addition, post-weaning GEN treatment induced advanced puberty, and this treatment had a more apparent effect on the mammary gland anatomy [65].

All these peripheral effects could reflect the alterations of the kiss system described in the same animals. Many studies demonstrated that the kiss is an important target for EDCs action (i.e., bisphenol A [66]) but very limited data are available on the impact of phytoestrogens on kisspeptin circuits in mice. In female rats, neonatal exposure to GEN (10 mg/kg) induced a persistent masculinization of kiss-ir fibers in RP3V (lower density), but not in the ARC [11], while developmental estrogen exposure significantly decreased kiss immunostaining in the ARC [11]. Moreover, postnatal GEN exposure does not significantly affect kiss-ir levels in adult males [49]. Consistently, we observed that early postnatal exposure to GEN induced sexually dimorphic effects on the kiss system. In males, GEN induced only a transient increase in the kiss-ir FA in the ARC at PND12. Interestingly, this increase was concomitant to a significant peak of plasma concentration of leptin, which is considered a positive modulator of the kiss system [29,67]; in fact, the energy balance is undoubtedly strictly correlated to the reproductive function. In our animals, the same correlation was also present in females, but at a post-pubertal age. The observed early increase of kiss immunoreactivity in the ARC, RP3V, and PVN correlated with a peak of plasma concentration of leptin measured in PND30 females treated with GEN.

Not many studies have focused on the possibility that a soy-based diet could alter circulating leptin levels. However, from the literature, the leptin concentrations in serum were affected by GEN, but the effect depends on sex, age, and the hormonal status in which the treatment was carried out (as reviewed in [68]). Unsurprisingly, we also observed here a strong variability in serum leptin levels during the development of treated mice based on both sex and age. Nevertheless, a decrease in plasma levels of leptin levels occurred in both sexes in adulthood, but especially in GEN females. The hipoleptinemic action of GEN was documented in male adult rats, showing a reduction in leptin levels after only three days of treatment with a dose of 5 mg/kg body weight [69]. In female rats, GEN has no effects on the plasma levels of normal females, but it induces a reduction of serum leptin in pregnant females [70], as well as in obesity models [71,72].

Negative effects of EDCs on reproduction are well described, but, in recent years, some EDCs have been considered among the multiple environmental factors that have been linked to the increase in obesity and metabolic syndrome, and they were named metabolic disrupting chemicals (MDCs) [73]. These MDCs can act indirectly to promote adipogenesis and cause weight gain by shifting energy balance to promote calories accumulation, altering basal metabolism [74], and altering hormonal control of appetite and satiety [8]. More recently, GEN has also been included in the list of MDCs [37], and these effects emerged strongly in our work. Note that susceptibility to obesity begins during development (in utero and early life), critical periods in which MDCs can influence developmental planning and thus disrupt the set point for weight gain later in life [75]. Contradictory data exist, underlining the importance of exposure times, dose/concentration, and sex to establish safety recommendations for the intake of GEN in the diet, especially if in early life. The same treatment may have a different outcome depending on the age and the sex of the animal. Compared to dams on a soy-free diet, rats on a soy-enriched diet gain less weight during pregnancy, and although they consume more food, they do not become heavier during lactation. Their offspring, however, are significantly heavier (both sexes, but more pronounced in males), show higher food intake, and females have an earlier pubertal onset [39]. GEN postnatal oral administration (PND1 to PND22) of 50mg/kg GEN (the same dose as our study) in rat pups resulted in a similar effect in females and increases adipocyte size and number, fat mass, and fat/lean mass ratio and decreased the size of muscle fiber [45]. Consistently, in a previous study [14] we demonstrated an obesogenic effect of postnatal GEN (administrated from PND1 to PND8) in adult female CD1 mice only. Here, we confirmed this effect, from puberty until adulthood. The increase in body weight was not correlated to alterations of food intake and daily feed efficiency, indicating a probable metabolic disruption. Concurrently, only in GEN-treated females, plasma concentration of two important metabolic hormones, leptin and T3, were significantly decreased, while the serum ghrelin showed a strong increase in GEN males only.

Indeed, GEN induces similar metabolic changes as well as alterations in the T3, ghrelin, and leptin in other models [68]. In fact, this increased plasma concentration of T3 only in female mice has also been shown in NIH/S female mice postnatally exposed to GEN (8 mg/kg body weight/day) [76] and in the golden Syrian hamster exposed to a soy protein diet for 28 days [77]. GEN also affected in a sexually dimorphic way the ghrelin concentration, with a decrease in females, as was observed in previous studies in mice [76]. In addition, in other experimental models (i.e., Mustela family) in adult females, there was a reduction in levels of plasma concentration of ghrelin, although not fully significant [78], while a strong increase was observed in males [79]. The reported alterations in plasma concentrations of the analyzed metabolites, associated with increase of weight gain only in female mice treated with GEN, without any alteration in the amount of ingested food, suggest the control of energy expenditure is susceptible to postnatal treatment with genistein in a sexually dimorphic way.

In addition to the previously discussed direct (or indirect) effects on the kiss system, leptin decrease in the adult may, in turn, impair the activation of the hypothalamic POMC system. POMC neurons in the ARC have an anorexigenic action on food-intake control. These neurons co-express different neuropeptides and a wide variety of receptors, including leptin receptors [80]. Moreover, the POMC system is sensitive to gonadal steroids [81], and we recently demonstrated that this system could be a good target for MDCs action by using a chronic exposure to tributyltin [57], as well as long-term exposure to BPA, DES, and TBT [82]. Here, we confirm that also early postnatal exposure to GEN induces long-term sex-specific organizational effects on the POMC system in the ARC and PVN.

Another hypothalamic system controlling metabolism and food intake is the OX system. This peptide modulates energy balance based on food intake by discriminating the physiological variation of glucose levels between meals [18]. Moreover, OX enhances spontaneous physical activity and regulates energy expenditure thus promoting obesity resistance [83]. In adult mice, the number of OX neurons in the LH is higher in males than in females [84]. This dimorphism could be associated with sexual male maturation, since in adult male rodents, OX neurons are markedly activated during copulation [85]. Interestingly, we observed that this dimorphism, observed in our control groups, was totally reverted in GEN mice: GEN increased the cell number in females while decreased it in males. OX neurons co-express a few estrogen (ER-alpha) receptors and no androgen (AR) receptors [85]; therefore, its sex dimorphism may be due to indirect control by gonadal steroids, and early exposure to GEN may, thus, permanently interfere with OX system differentiation. Moreover, we demonstrated that different rostrocaudal levels of the LH harbor OX subpopulations with specific features. In fact, males had more OX-ir cells than females in the most rostral levels, while females presented a higher number of OX cells in the most caudal levels. Furthermore, those subpopulations displayed a sexually dimorphic response to GEN postnatal treatment, with a decrease in the number of OX cells in adult males in rostral levels of the LH and an increase in more caudal levels of females. Future studies should investigate if these different OX subpopulations, located in specific rostrocaudal domains of the LH, have different targets and roles.

Our hypothesis that postnatal treatment with GEN may act on the control of energy expenditure is supported by the observed alterations of all the analyzed systems projecting to the PVN. In fact, the PVN is the most important hypothalamic center of metabolic control, modulating feeding behavior through the action of CRH and TRH, both indirectly via effects on energy expenditure (hypothalamus-pituitary-thyroid axis, HPT) and directly through the HPA axis [23,86].

As demonstrated in previous studies [24,56], the PVN is rich in kiss fibers, especially in the medial part of the nucleus, where CRH and TRH neurons are located, suggesting a strong correlation between reproductive and metabolic control [24]. In fact, PVN kiss fibers are part of a sexual network essential for the control of the HPG axis through the control of the GnRH system. The PVN is also rich in NPY and POMC fibers [82], as well as in orexin fibers (present study).

Cell numbers in the ARC and their projections may respond differently to treatment with EDCs, which is particularly evident for the POMC system [57,82]. Present results show a reduction of POMC innervation of the PVN in females and an increase in males, while cell numbers in the ARC have an opposite trend. This could be due to the presence of subpopulations of POMC cells in the ARC expressing different receptors [34,80,81,87] and that potentially send their projections to different targets. In addition, a potential direct effect of kisspeptin on POMC neurons also cannot be ruled out. In fact, kisspeptin in the ARC directly excites POMC neurons [33] that express kiss1 receptors [34]. Both systems are affected by GEN treatments in our animals, and this could also have a differential impact on their projections.

As previously discussed, OX regulates food intake and energy homeostasis, but also reproduction. Orexin has in fact a direct impact on GnRH [88] and kiss [89] neurons. Moreover, OX receptors have also been found in rat testes, and testosterone secretion is directly stimulated by OX [90]. In a previous experiment, we have shown, in males postnatally exposed to GEN, a decrease in the ratio prostate/BW and to a lesser extent in the ratio testis/BW [14]. However, in this experiment, we have observed a decrease of fecal testosterone and of the weight of testicles, and we also observed a decrease in the number of OX neurons in LH, suggesting that OX may act at the testicular level also in mice. Furthermore, a previous study has shown that exposure to GEN (8 mg/kg body weight/day) early in life leads to changes in the reproductive organs of males, with relative weights of the prostate and seminal vesicles being greater than in control males. [76].

4. Materials and Methods

4.1. Animals

We purchased from Charles River, France, 26 female and 13 male adult virgin CD-1 mice. All experiments were performed according to the EU directive on animal experimentation 2010/63 and approved by the local ethical committee (Comité d'Ethique en Expérimentation Animale Centre-Val de Loire) under approval 426-201504031706655.

Animals were housed in monosexual groups of 3 mice in conventional polycarbonate cages (45 × 25 × 15 cm) with water and food (standard diet 150 low phytoestrogen certificate, SAFE, France) ad libitum and exposed to a 12-h light/dark cycle. After 2 weeks of the adaptation period, females were housed with males in groups of 2 females and 1 male for one night, beginning at 18:00 h (at the end of the light phase); after the mating, verified by the presence of a vaginal plug (generally 3–5 days), the females were placed in single cages.

4.2. Genistein Treatment

The day after the birth, postnatal day one (PND1), litters were reduced to 8 pups, 4 males and 4 females, sexed via anogenital distance [91]. Pups were then allocated randomly to two groups and subjected from PDN1 to PND8 to oral administration of vehicle (10 μL/g sesame oil) or genistein (GEN 50 mg/kg body weight; cat. Number G6649, Sigma-Aldrich, St. Quentin Fallavier Cedex, France) diluted in sesame oil. This protocol mimics the exposure of babies fed with soy-based formulas [92]. Moreover, as previously shown in a pharmacokinetic study, the dose we used produces serum levels of GEN in neonatal mice, within the human range [93]. Mice spontaneously drank the solution through a micropipette directly into the mouth [14,15]. Three weeks after birth (PND21), the pups were weaned and housed in single-sex groups of 3–5 animals, differentiated by treatment, in polypropylene mouse cages.

Animals were divided into 4 groups: control males (M-CON), control females (F-CON), genistein-treated males (M-GEN), and genistein-treated females (F-GEN), and sacrificed at PND12, PND22, PND30, and PND60. Six mice per group were perfused for immunohistochemical studies of the neuronal circuits, while the others were killed by decapitation (N = 6 per group) to study the peripheral parameters.

4.3. Reproductive and Metabolic Parameters

4.3.1. Vaginal Opening and Estrous Cycle

After weaning, GEN-treated and control females (N = 24 per group of treatment) were checked daily, from PND19 to PND29, for vaginal opening (VO) to detect the time of puberty [53].

From PND40 to PND55 daily microscopic inspection of vaginal smears flushed with physiological saline solution was performed in F-CON (N = 6) and F-GEN (N = 6) to determine the phase of the estrous cycle. The percentage of the days in each phase was calculated.

4.3.2. Uterus and Testicle Weight

The weight of reproductive organs, uterus, and testicles, obtained from female (N = 6) and male (N = 6) mice, were manually dissected and measured in each group at PND22,

PN30, and PND60 after decapitation. Testicle weight was calculated by summing the weights of the two testicles for each male.

4.3.3. Mammary Gland Analysis

Mammary glands were collected at PND22, PND30, and PND60 from females in the diestrus phase of the cycle (determined by vaginal smear). Briefly, the fourth mammary gland (inguinal) was dissected from the skin, stretched on a glass slide, and fixed in Carnoy's fixative at 4 °C for 2 h and then stored in the same solution until processing. Whole mounts were gradually re-hydrated, stained with Carmine Alum (Stem Cell Technologies) overnight, disdained for 2 h in 70% EtOH with 2% HCl, progressively dehydrated, clarified in methylsalicilate overnight, and photographed.

Whole mounts were photographed at 1× and 4× on a Leica S8AP0 stereomicroscope equipped with a Leica EC3 digital camera. Mammary gland length was measured with Image J software (version 1.47v; Wayne Rasband, NIH, Bethesda, MD, USA), as the distance between the beginnings of the duct arising from the nipple and the end of the more distal ducts of the glands. The total number of branches was counted at PND30 and PND60 as the number of branches 3.5 mm before and 3.5 mm after the lymph node after it. The number of tertiary branches was scored on a scale ranging from no branches (0) to high density (5).

4.3.4. Body Weight, Food Consumption, and Feed Efficiency (FE)

Body weight was recorded daily during the treatment and every two days from PND8 to sacrifice with an electronic precision balance (Mod. Kern-440-47N). To eliminate differences due to the variability between animals, we normalized the absolute body weight into a percentage of the body weight of the first day of the treatment (PND1), conventionally considered equal to 100.

Animals were fed with a standard diet 150 low phytoestrogen certificate (SAFE, Augy, France) containing 3264 Kcal/g of metabolizable energy with 21% as protein, 12.6% as lipid, and 66.4% as carbohydrate. Mean food consumption (mean grams per mouse per day) was determined every two days at 10.00 a.m. All animals from each group were housed in standard cages (each containing 3 animals). The daily food consumption per animal was estimated by dividing the total food consumption (total amount of food supplied per cage minus the weight of the residual food in each cage) by the number of mice in the cage and the number of days after the last measurement. After the measurement, fresh food was given to the mice. Daily energy intake was calculated by multiplying daily food intake by the caloric value of the chow (3264 Kcal/gr), and daily feed efficiency was expressed as body weight (gr)/Kcal eaten [94].

4.3.5. Hormonal Levels

Blood samples were collected at PND12, PND22, PND30, and PND60 from animals killed by decapitation, which always took place in the morning between 9 AM and 12 AM. PND30 and PND60 females were killed in diestrus (assessed by vaginal smears).

Blood samples were collected in the morning, in EDTA-treated tubes, centrifuged at 3500 g for 20 min, and then the plasma was stored frozen at −80 °C. Samples were processed using standard procedures provided by manufacturers with the following kits: progesterone EIA-1561 (intra-assay variation (CV) is 5,4%, while the analytical sensitivity of this assay is 0.045 ng/mL), leptin ELI-4564 (intra-assay is 1,64% and inter-assay 3,96%, while the limit of sensitivity of this assay is 0.05 ng/mL (~3.13 pM) using a 10 μL sample size), total triiodothyronine (T3) EIA-4569 (intra-assay variation (CV) is 6.54% and inter-assay 5.23%, the analytical sensitivity of this assay is 0.1 ng/mL), ghrelin EZRGRA-90K (intra-assay variation is 1,60% and inter-assay 3.41%, while the limit of sensitivity of this assay is 8 pg/mL when using a 20 μL sample size)(DRG Instruments GmbH).

To minimize the stress for the animals, we measured testosterone levels in feces in young (PND30) and adult (PND60) male mice. Animals were isolated in a clean cage in the late morning. After 2 h 1.7 ± 0.3 mL of fecal pellet were collected, and animals returned to their home cage. Pellets were stored at −80 °C. Extraction of fecal testosterone was carried out on pulverized dried feces by using diethyl ether as previously reported [95]. Fecal testosterone level was determined using an enzyme immunoassay kit (K032; Arbor Assays, Ann Arbor, MI, USA) validated for multi-species dried fecal extracts.

4.4. *Immunohistochemistry*

4.4.1. Fixation and Sampling

Mice were perfused at PND12, PND22, PND30, and PND60. Females at PND30 and PND60 were in diestrus (assessed by vaginal smear). Animals were deeply anesthetized with pentobarbital, monitored until the pedal reflex was abolished, and killed via intracardiac perfusion with saline solution (NaCl 0.9%) followed by fixative (4% paraformaldehyde, PAF, in 0.1 M phosphate buffer, pH 7.3). Dissected brains were stored in a freshly prepared PAF solution for 2 h at 4 °C, followed by washing in a 30% sucrose solution at 4 °C overnight. Finally, brains were frozen in liquid isopentane pre-cooled in dry ice at −35 °C and stored in a deep freezer at −80 °C until sectioning.

Three series of adjacent 40-µm-thick coronal sections were obtained with a cryostat (Leica CM 1900), collected in a cryoprotectant solution [96], and kept at −20 °C. We stained these three series respectively for kiss, POMC, and OX immunohistochemistry. To avoid between-assays variance due to systematic group differences, sections were processed into groups containing samples from each treatment and sex. Sections were washed overnight in PBS at pH 7.3 before immunohistochemical processing. The following day, after washing in PBS containing 0.2%Triton X-100 for 30 min the endogenous peroxidase activity was blocked with methanol/hydrogen peroxide solution (1:1) in PBS for 20 min at room temperature. Sections were then pre-incubated with normal goat serum (Vector Laboratories, Burlingame, CA, USA) for 30 min before the use of the specific antibodies. After the immunohistochemical reaction, the sections were collected on chrome alum pretreated slides, air-dried, washed in xylene, and cover slipped with Entellan mounting medium (Merck, Milano, Italy).

4.4.2. Kisspeptin Immunohistochemistry

Kisspeptin immunostaining was performed according to our previous studies [24,56], by using an overnight incubation at 4 °C of floating sections with a rabbit polyclonal antibody (AC#566, Drs A. Caraty and I. Franceschini, Tours, France) at a dilution of 1:10,000 in PBS-Triton X-100 0.2%. Sections were then incubated, at room temperature, in biotinylated goat anti-rabbit IgG (Vector Laboratories) for 60 min at a dilution of 1:200. The antigen–antibody reaction was revealed with a biotin–avidin system (Vectastain ABC Kit Elite, Vector Laboratories, Burlingame, CA, USA) with an incubation of 60 min at room temperature. The peroxidase activity was visualized with 0.400 mg/mL of 3.30-diamino-benzidine (SIGMA-Aldrich, Milan, Italy) and 0.004% hydrogen peroxide in 0.05M Tris–HCl buffer pH 7.6. The specificity of the AC566 antibody for immunohistochemistry was previously reported [97].

4.4.3. POMC Immunohistochemistry

POMC immunostaining was performed according to our previous studies [57,82], with an overnight incubation of floating sections at 4 °C with a rabbit polyclonal antibody against POMC (POMC precursor 27–52, H029-30, Phoenix Pharmaceuticals, Inc., Burlingame, CA, USA) diluted 1:5000 in PBS-Triton X-100 0.2%, pH 7.3–7.4. Sections were then incubated in biotinylated goat anti-rabbit IgG (Vector Laboratories, 1:250) for 60 min. The antigen-antibody reaction was revealed as described for kisspeptin immunohistochemistry. The POMC antibody specificity for immunohistochemistry was tested by the factory and in previous studies [57,82].

4.4.4. Orexin Immunohistochemistry

Floating sections were incubated overnight in a rabbit anti-orexin A antibody (PC345, Calbiochem, Merck KGaA, Darmstadt, Germany) diluted 1:2000 in PBS at 4 °C. The sections were then incubated with biotinylated anti-rabbit IgG serum (Pierce, Vector, CA, USA, 1:200, 90 min). The antigen-antibody reaction was revealed as described for kisspeptin immunohistochemistry. The specificity of the orexin antibody for immunohistochemistry was tested by the factory and in previous studies [98].

In addition to the reported specificity, for each antibody we performed additional controls, omitting the primary antibody (negative control) or the secondary antibody. In these control sections, cells and fibers were completely unstained.

4.5. Quantitative Analysis

4.5.1. Cell Counting and Fractional Area Evaluation

Based on the different nuclei and immunochemical markers, we evaluated the number of positive cells that were present in a nucleus or the extent of immunoreactivity (fractional area, FA), including cell bodies, dendrites, and fibers, within a nucleus. Digital images were acquired using a NIKON Eclipse 80i microscope (Nikon Italia SpA, Firenze, Italy) connected to a NIKON Digital Sight DS-Fi1 video camera. Images were then processed and analyzed with ImageJ (version 1.47v; Wayne Rasband, NIH, Bethesda, MD, USA). To have a better resolution, for each microscopic field, we took several pictures at different levels of focus collecting them in a stack of 4–7 pictures. These stacks were processed using the Z Project-Maximum intensity function of Image J, which created an output image, each of whose pixels contained the maximum value overall of the images in the stack, determining an optimal focus for all the structures contained in the stack.

Cell counting was performed only in those regions where cell bodies were clearly labeled with the specific antibody, in predetermined fields (region of interest, ROI). If cell bodies could be easily extracted from the background using the threshold function, the digitized images were processed and analyzed using the Analyze Particles automatic function of ImageJ (OX cells in LH nucleus); in the other cases we used the manual Image J Cell counter plugin (POMC cells in ARC nucleus).

The fractional area (FA) was evaluated, according to the general principles described by Mize et al. [99], by calculating the percentage of pixels covered by the immunopositive structures highlighted using the threshold function of Image J in a predetermined ROI. Due to differences in the immunostaining, the range of the threshold was individually adjusted for each section. By using the Analyze-Measure function of Image J the percentage of area covered by threshold within the ROI was automatically measured. The results were grouped to provide mean (±S.E.M.) values.

4.5.2. Kisspeptin

For the immunostaining of the kiss system, three standardized sections were selected for each of the 3 analyzed nuclei, matching the Mouse Brain Atlas [59]: ARC (Bregma −1.58 mm, −1.70 mm, and −1.82 mm), PVN (Bregma −0.58 mm, −0.82 mm, and −0.94 mm), and RP3V (Bregma 0.26 mm, 0.02 mm, and −0.22 mm). Digital microphotographs were acquired with x40 objective (PVN) or x20 objective (ARC and RP3V) and were processed and analyzed with ImageJ (see above). Measurements were performed within predetermined ROIs. The PVN, in each selected section, was divided, as in our previous study [24], into fourteen squares (each of 31,100 μm^2 at PND12; 37,050 μm^2 at PND22; 40,150 μm^2 at PND30 and PND60) to cover its full extension (see Figure 7a) and grouped in four regions: dorso-medial, dorso-lateral, ventro-medial, and ventro-lateral. The measure of total PVN was a mean of the FA measured in each of the four regions. The ROI for the ARC changed during development (at PND12 550,000 μm^2, at PND22 600,000 μm^2, and in adults at 890,000 μm^2) to include the immunopositive region, and it was placed using the third ventricle as a reference to always have the same orientation. We easily identified and counted positive kiss neurons (characterized by the presence of a clearly labeled cell body) only within RP3V, while in the ARC and PVN, we quantified the FA covered by immunoreactive material.

4.5.3. POMC

The number of POMC positive cells and the FA covered by immunoreactivity were analyzed in three selected standardized sections of comparable levels of ARC adjacent to the ones stained for the kiss, as well as the measurement of FA within PVN [59]. For ARC and PVN we used the same ROIs and analysis as for the kiss immunostained sections. For the dorsomedial hypothalamic nucleus (DMH), three standardized serial sections of comparable level were selected to analyze POMC-ir fibers (Bregma −1.46 mm; −1.82 mm; −1.94 mm); we acquired images using a 20× objective and used an ROI (1250 μm^2) located within borders of the nucleus as evidenced by the immunoreactivity.

4.5.4. Orexin

We measured the number of OX cells within four coronal sections through the region of the lateral hypothalamic area (LH; Bregma −1.06 mm; −1.34 mm; −1.58 mm; −2.06 mm) according to the Mouse Brain Atlas [59], which is where most OX-ir cells were found in the caudal hypothalamus [58]. The sections were acquired using a 10× objective. The OX-ir cells were counted within a ROI (1,580,000 μm^2) that covered the entire extension of the nucleus. The results were grouped to provide mean (±S.E.M.) values. Furthermore, the FA was analyzed in three selected standardized sections of comparable levels of the PVN adjacent to the ones stained for POMC [59], and to measure OX-ir fibers, we used the same ROIs and analysis used for the kiss immunostained sections.

4.6. Statistical Analysis

Quantitative data were examined with SPSS statistic software (SPSS Inc, Chicago, IL, USA) via three-way ANOVA, where age, sex, and treatment were considered independent variables, or/and two-way ANOVA (considering sex and treatment or age and treatment as independent variables), and one-way ANOVA. When appropriated, we performed the Bonferroni or Tukey multivariate test to compare groups or Student's *t*-test. The data are presented as mean ± SEM and the differences between groups are considered significant for values of $p \leq 0.05$.

5. Conclusions

In conclusion, early postnatal exposure to GEN determines long-term sex-specific organizational effects on neural circuits controlling food intake, energy metabolism, and reproduction in CD1 mice, which are more pronounced in females. At the same time, other parameters related to reproduction are also altered (i.e., puberty and estrous cycle), as well as those related to metabolism (i.e., body weight, food consumption, and feed efficiency). Our hypothesis is therefore that these effects are strictly linked to alterations of neuroendocrine circuits controlling both reproduction and energy expenditure.

According to this view, GEN may be classified not only as an EDC with strong effects on reproduction but also as a metabolism-disrupting chemical (MDC). The danger to human health of synthetic contaminants in food, such as pesticides, is widely known and a subject of debate even among non-specialists. Much less known are the dangers associated with some molecules of natural origin, such as phytoestrogens, including GEN, that are present in many foods. Genistein, by binding estrogen receptors, can alter the functional processes that depend on them (for example, reproduction and energy metabolism) and the development of the neuroendocrine circuits that regulate these activities. The alteration of these nervous circuits could be at the root of some problems (constantly growing in our society) that are found in the human field, such as the predisposition to obesity in children fed with soy milk. Furthermore, the effects of GEN on development may be due to epigenetic modifications in the offspring. It is therefore important, for food safety and human health, to better investigate the effects of phytoestrogens on the central nervous system and the repercussions they can have in the organization of many nervous circuits regulated by hormones.

Supplementary Materials: The following are available online at https://www.mdpi.com/article/10.3390/metabo11070449/s1, Table S1: Mammary gland length and terminal end bud (TEB) numbers, Table S2: Fecal testosterone concentration, Table S3: Kisspeptin system: quantitative data, Table S4: Body weight gain, Table S5: Daily food eaten, Table S6: Daily feed efficiency, Table S7: POMC system: quantitative data, Table S8: POMC and orexin system: developmental data, Table S9: Orexin system: quantitative data.

Author Contributions: M.M. performed experiments, analyzed data, and wrote the paper; G.P. (Giovanna Ponti), A.F., C.M., E.M., and P.A. performed experiments and analyzed data; S.G., P.C., and G.P. (Giancarlo Panzica) supervised the paper; G.P. (Giancarlo Panzica) and M.K. designed experiments and supervised the paper. All authors have read and agreed to the published version of the manuscript.

Funding: This research was funded by COST (European Cooperation in Science and Technology) Action BM1105, Fondazione Cavalieri Ottolenghi, Ministero dell'Istruzione, dell'Università e della Ricerca–MIUR project Dipartimenti di Eccellenza 2018–2022 to the Department of Neuroscience "Rita Levi Montalcini", University of Torino, by local research grants of the University of Torino to G.P., S.G., and G.C.P, and by PSI2014-57362-P and PSI2017-86396-P to PC.

Institutional Review Board Statement: The study was performed in accordance with the EU directive on animal experimentation 2010/63 and approved by the local ethical committee (*Comité d'Ethique en Expérimentation Animale Centre-Val de Loire*) under approval 426-201504031706655.

Informed Consent Statement: Not applicable.

Data Availability Statement: All the data are available from the authors upon reasonable request. There are no restrictions on the data availability.

Acknowledgments: We are very grateful to Chiara Violino for her technical help. The fellowships of M.M. were granted by G.C. Bergui (2018–2020) and by the Fondazione Umberto Veronesi (2021).

Conflicts of Interest: The authors declare no conflict of interest.

References

1. Dixon, R. Genistein. *Phytochemistry* **2002**, *60*, 205–211. [CrossRef]
2. Liggins, J.; Grimwood, R.; Bingham, S.A. Extraction and Quantification of Lignan Phytoestrogens in Food and Human Samples. *Anal. Biochem.* **2000**, *287*, 102–109. [CrossRef] [PubMed]
3. Bhathena, S.J.; Velasquez, M.T. Beneficial role of dietary phytoestrogens in obesity and diabetes. *Am. J. Clin. Nutr.* **2002**, *76*, 1191–1201. [CrossRef]
4. Aidelsburger, P.; Schauer, S.; Grabein, K.; Wasem, J. Alternative methods for the treatment of post-menopausal troubles. *GMS Health Technol. Assess.* **2012**, *8*, Doc 03. [CrossRef]
5. Khan, S.; Zhao, J.; Khan, I.; A Walker, L.; DasMahapatra, A.K. Potential utility of natural products as regulators of breast cancer-associated aromatase promoters. *Reprod. Biol. Endocrinol.* **2011**, *9*, 91. [CrossRef]
6. Hemati, N.; Asis, M.; Moradi, S.; Mollica, A.; Stefanucci, A.; Nikfar, S.; Mohammadi, E.; Farzaei, M.H.; Abdollahi, M. Effects of genistein on blood pressure: A systematic review and meta-analysis. *Food Res. Int.* **2020**, *128*, 108764. [CrossRef]
7. Patisaul, H.B.; Jefferson, W. The pros and cons of phytoestrogens. *Front. Neuroendocr.* **2010**, *31*, 400–419. [CrossRef]
8. Petrine, J.C.P.; Del Bianco-Borges, B. The influence of phytoestrogens on different physiological and pathological processes: An overview. *Phytother. Res.* **2021**, *35*, 180–197. [CrossRef] [PubMed]
9. Setchell, K.D. Phytoestrogens: The biochemistry, physiology, and implications for human health of soy isoflavones. *Am. J. Clin. Nutr.* **1998**, *68*, 1333S–1346S. [CrossRef] [PubMed]
10. Franke, A.; Custer, L.J.; Tanaka, Y. Isoflavones in human breast milk and other biological fluids. *Am. J. Clin. Nutr.* **1998**, *68*, 1466S–1473S. [CrossRef] [PubMed]
11. Losa, S.M.; Todd, K.L.; Sullivan, A.W.; Cao, J.; Mickens, J.A.; Patisaul, H.B. Neonatal exposure to genistein adversely impacts the ontogeny of hypothalamic kisspeptin signaling pathways and ovarian development in the peripubertal female rat. *Reprod. Toxicol.* **2011**, *31*, 280–289. [CrossRef]
12. Marraudino, M.; Farinetti, A.; Arevalo, M.-A.; Gotti, S.; Panzica, G.; Garcia-Segura, L.-M. Sexually Dimorphic Effect of Genistein on Hypothalamic Neuronal Differentiation in Vitro. *Int. J. Mol. Sci.* **2019**, *20*, 2465. [CrossRef]
13. Patisaul, H.B.; Fortino, A.E.; Polston, E.K. Neonatal genistein or bisphenol-A exposure alters sexual differentiation of the AVPV. *Neurotoxicol. Teratol.* **2006**, *28*, 111–118. [CrossRef]

14. Ponti, G.; Farinetti, A.; Marraudino, M.; Panzica, G.; Gotti, S. Postnatal genistein administration selectively abolishes sexual dimorphism in specific hypothalamic dopaminergic system in mice. *Brain Res.* **2019**, *1724*, 146434. [CrossRef] [PubMed]
15. Ponti, G.; Rodriguez-Gomez, A.; Farinetti, A.; Marraudino, M.; Filice, F.; Foglio, B.; Sciacca, G.; Panzica, G.; Gotti, S. Early postnatal genistein administration permanently affects nitrergic and vasopressinergic systems in a sex-specific way. *Neuroscience* **2017**, *346*, 203–215. [CrossRef] [PubMed]
16. Rodriguez-Gomez, A.; Filice, F.; Gotti, S.; Panzica, G. Perinatal exposure to genistein affects the normal development of anxiety and aggressive behaviors and nitric oxide system in CD1 male mice. *Physiol. Behav.* **2014**, *133*, 107–114. [CrossRef] [PubMed]
17. Timper, K.; Brüning, J.C. Hypothalamic circuits regulating appetite and energy homeostasis: Pathways to obesity. *Dis. Model. Mech.* **2017**, *10*, 679–689. [CrossRef]
18. Messina, G.; Dalia, C.; Tafuri, D.; Monda, V.; Palmieri, F.; Dato, A.; Russo, A.; De Blasio, S.; Messina, A.; De Luca, V.; et al. Orexin-A controls sympathetic activity and eating behavior. *Front. Psychol.* **2014**, *5*, 997. [CrossRef]
19. Cowley, M.; Smart, J.L.; Rubinstein, M.; Cerdán, M.G.; Diano, S.; Horvath, T.L.; Cone, R.D.; Low, M.J. Leptin activates anorexigenic POMC neurons through a neural network in the arcuate nucleus. *Nat. Cell Biol.* **2001**, *411*, 480–484. [CrossRef]
20. Morello, G.; Imperatore, R.; Palomba, L.; Finelli, C.; Labruna, G.; Pasanisi, F.; Sacchetti, L.; Buono, L.; Piscitelli, F.; Orlando, P.; et al. Orexin-A represses satiety-inducing POMC neurons and contributes to obesity via stimulation of endocannabinoid signaling. *Proc. Natl. Acad. Sci. USA* **2016**, *113*, 4759–4764. [CrossRef]
21. Bäckberg, M.; Hervieu, G.; Wilson, S.; Meister, B. Orexin receptor-1 (OX-R1) immunoreactivity in chemically identified neurons of the hypothalamus: Focus on orexin targets involved in control of food and water intake. *Eur. J. Neurosci.* **2002**, *15*, 315–328. [CrossRef]
22. Kishi, T.; Aschkenasi, C.J.; Choi, B.J.; Lopez, M.E.; Lee, C.E.; Liu, H.; Hollenberg, A.N.; Friedman, J.M.; Elmquist, J.K. Neuropeptide Y Y1 receptor mRNA in rodent brain: Distribution and colocalization with melanocortin-4 receptor. *J. Comp. Neurol.* **2004**, *482*, 217–243. [CrossRef]
23. van Swieten, M.; Pandit, R.; Adan, R.; van der Plasse, G. The neuroanatomical function of leptin in the hypothalamus. *J. Chem. Neuroanat.* **2014**, *61-62*, 207–220. [CrossRef] [PubMed]
24. Marraudino, M.; Miceli, D.; Farinetti, A.; Ponti, G.; Panzica, G.; Gotti, S. Kisspeptin innervation of the hypothalamic paraventricular nucleus: Sexual dimorphism and effect of estrous cycle in female mice. *J. Anat.* **2017**, *230*, 775–786. [CrossRef] [PubMed]
25. Clarkson, J.; Herbison, A.E. Postnatal Development of Kisspeptin Neurons in Mouse Hypothalamus; Sexual Dimorphism and Projections to Gonadotropin-Releasing Hormone Neurons. *Endocrinology* **2006**, *147*, 5817–5825. [CrossRef] [PubMed]
26. Herbison, A.E. Control of puberty onset and fertility by gonadotropin-releasing hormone neurons. *Nat. Rev. Endocrinol.* **2016**, *12*, 452–466. [CrossRef] [PubMed]
27. Castellano, J.M.; Tena-Sempere, M. Metabolic control of female puberty: Potential therapeutic targets. *Expert Opin. Ther. Targets* **2016**, *20*, 1181–1193. [CrossRef]
28. Qiu, X.; Dao, H.; Wang, M.; Heston, A.; Garcia, K.M.; Sangal, A.; Dowling, A.R.; Faulkner, L.D.; Molitor, S.; Elias, C.F.; et al. Insulin and Leptin Signaling Interact in the Mouse Kiss1 Neuron during the Peripubertal Period. *PLoS ONE* **2015**, *10*, e0121974. [CrossRef]
29. Smith, J.; Acohido, B.V.; Clifton, D.K.; Steiner, R.A. KiSS-1 Neurones Are Direct Targets for Leptin in the ob/ob Mouse. *J. Neuroendocr.* **2006**, *18*, 298–303. [CrossRef]
30. Cravo, R.; Margatho, L.; Osborne-Lawrence, S.; Donato, J.; Atkin, S.; Bookout, A.; Rovinsky, S.; Frazao, R.; Lee, C.; Gautron, L.; et al. Characterization of Kiss1 neurons using transgenic mouse models. *Neuroscience* **2011**, *173*, 37–56. [CrossRef]
31. Louis, G.W.; Greenwald-Yarnell, M.; Phillips, R.; Coolen, L.; Lehman, M.; Myers, M.G. Molecular Mapping of the Neural Pathways Linking Leptin to the Neuroendocrine Reproductive Axis. *Endocrinology* **2011**, *152*, 2302–2310. [CrossRef]
32. Quennell, J.H.; Mulligan, A.C.; Tups, A.; Liu, X.; Phipps, S.J.; Kemp, C.J.; Herbison, A.; Grattan, D.; Anderson, G.M. Leptin Indirectly Regulates Gonadotropin-Releasing Hormone Neuronal Function. *Endocrinology* **2009**, *150*, 2805–2812. [CrossRef]
33. Fu, L.-Y.; Pol, A.N.V.D. Kisspeptin Directly Excites Anorexigenic Proopiomelanocortin Neurons but Inhibits Orexigenic Neuropeptide Y Cells by an Indirect Synaptic Mechanism. *J. Neurosci.* **2010**, *30*, 10205–10219. [CrossRef]
34. Higo, S.; Iijima, N.; Ozawa, H. Characterisation ofKiss1r(Gpr54)-Expressing Neurones in the Arcuate Nucleus of the Female Rat Hypothalamus. *J. Neuroendocr.* **2017**, *29*, 29. [CrossRef] [PubMed]
35. Orlando, G.; Leone, S.; Ferrante, C.; Chiavaroli, A.; Mollica, A.; Stefanucci, A.; Macedonio, G.; Dimmito, M.P.; Leporini, L.; Menghini, L.; et al. Effects of Kisspeptin-10 on Hypothalamic Neuropeptides and Neurotransmitters Involved in Appetite Control. *Molecules* **2018**, *23*, 3071. [CrossRef]
36. Carrillo, B.; Collado, P.; Díaz, F.; Chowen, J.A.; Pérez-Izquierdo, M. Ángeles; Pinos, H. Physiological and brain alterations produced by high-fat diet in male and female rats can be modulated by increased levels of estradiol during critical periods of development. *Nutr. Neurosci.* **2017**, *22*, 29–39. [CrossRef] [PubMed]
37. Marraudino, M.; Bonaldo, B.; Farinetti, A.; Panzica, G.; Ponti, G.; Gotti, S. Metabolism Disrupting Chemicals and Alteration of Neuroendocrine Circuits Controlling Food Intake and Energy Metabolism. *Front. Endocrinol.* **2019**, *9*, 766. [CrossRef]
38. Jungbauer, A.; Medjakovic, S. Phytoestrogens and the metabolic syndrome. *J. Steroid Biochem. Mol. Biol.* **2014**, *139*, 277–289. [CrossRef] [PubMed]

39. Cao, J.; Echelberger, R.; Liu, M.; Sluzas, E.; McCaffrey, K.; Buckley, B.; Patisaul, H.B. Soy but not bisphenol A (BPA) or the phytoestrogen genistin alters developmental weight gain and food intake in pregnant rats and their offspring. *Reprod. Toxicol.* **2015**, *58*, 282–294. [CrossRef] [PubMed]
40. Kim, H.-K.; Nelson-Dooley, C.; Della-Fera, M.A.; Yang, J.-Y.; Zhang, W.; Duan, J.; Hartzell, D.L.; Hamrick, M.W.; Baile, C.A. Genistein Decreases Food Intake, Body Weight, and Fat Pad Weight and Causes Adipose Tissue Apoptosis in Ovariectomized Female Mice. *J. Nutr.* **2006**, *136*, 409–414. [CrossRef] [PubMed]
41. Naaz, A.; Yellayi, S.; Zakroczymski, M.A.; Bunick, D.; Doerge, D.R.; Lubahn, D.B.; Helferich, W.G.; Cooke, P.S. The Soy Isoflavone Genistein Decreases Adipose Deposition in Mice. *Endocrinology* **2003**, *144*, 3315–3320. [CrossRef]
42. Penza, M.; Montani, C.; Romani, A.; Vignolini, P.; Pampaloni, B.; Tanini, A.; Brandi, M.L.; Magdalena, P.A.; Nadal, A.; Ottobrini, L.; et al. Genistein Affects Adipose Tissue Deposition in a Dose-Dependent and Gender-Specific Manner. *Endocrinology* **2006**, *147*, 5740–5751. [CrossRef] [PubMed]
43. Harmon, A.W.; Harp, J.B. Differential effects of flavonoids on 3T3-L1 adipogenesis and lipolysis. *Am. J. Physiol. Physiol.* **2001**, *280*, C807–C813. [CrossRef] [PubMed]
44. Zanella, I.; Marrazzo, E.; Biasiotto, G.; Penza, M.; Romani, A.; Vignolini, P.; Caimi, L.; Di Lorenzo, D. Soy and the soy isoflavone genistein promote adipose tissue development in male mice on a low-fat diet. *Eur. J. Nutr.* **2015**, *54*, 1095–1107. [CrossRef] [PubMed]
45. Strakovsky, R.S.; Lezmi, S.; Flaws, J.A.; Schantz, S.L.; Pan, Y.-X.; Helferich, W.G. Genistein Exposure During the Early Postnatal Period Favors the Development of Obesity in Female, But Not Male Rats. *Toxicol. Sci.* **2013**, *138*, 161–174. [CrossRef] [PubMed]
46. Zhang, Y.B.; Yan, J.D.; Yang, S.Q.; Guo, J.P.; Zhang, X.; Sun, X.X.; Na, X.L.; Dai, S.C. Maternal Genistein Intake Can Reduce Body Weight in Male Offspring. *Biomed. Environ. Sci.* **2015**, *28*, 769–772.
47. Mueller, J.K.; Heger, S. Endocrine disrupting chemicals affect the Gonadotropin releasing hormone neuronal network. *Reprod. Toxicol.* **2014**, *44*, 73–84. [CrossRef]
48. Patisaul, H.B. Effects of Environmental Endocrine Disruptors and Phytoestrogens on the Kisspeptin System. *Adv. Exp. Med. Biol.* **2013**, *784*, 455–479. [CrossRef] [PubMed]
49. Patisaul, H.B.; Todd, K.L.; Mickens, J.A.; Adewale, H.B. Impact of neonatal exposure to the ERα agonist PPT, bisphenol-A or phytoestrogens on hypothalamic kisspeptin fiber density in male and female rats. *NeuroToxicology* **2009**, *30*, 350–357. [CrossRef]
50. Bateman, H.L.; Patisaul, H.B. Disrupted female reproductive physiology following neonatal exposure to phytoestrogens or estrogen specific ligands is associated with decreased GnRH activation and kisspeptin fiber density in the hypothalamus. *NeuroToxicology* **2008**, *29*, 988–997. [CrossRef]
51. Cederroth, C.R.; Vinciguerra, M.; Kühne, F.; Madani, R.; Doerge, D.R.; Visser, T.J.; Foti, M.; Rohner-Jeanrenaud, F.; Vassalli, J.-D.; Nef, S. A Phytoestrogen-Rich Diet Increases Energy Expenditure and Decreases Adiposity in Mice. *Environ. Health Perspect.* **2007**, *115*, 1467–1473. [CrossRef]
52. Fernandez-Garcia, J.; Carrillo, B.; Tezanos, P.; Collado, P.; Pinos, H. Genistein during Development Alters Differentially the Expression of POMC in Male and Female Rats. *Metabolism* **2021**, *11*, 293. [CrossRef]
53. Laffan, S.B.; Posobiec, L.M.; Uhl, J.E.; Vidal, J.D. Species Comparison of Postnatal Development of the Female Reproductive System. *Birth Defects Res.* **2018**, *110*, 163–189. [CrossRef]
54. Paine, I.S.; Lewis, M.T. The Terminal End Bud: The Little Engine that Could. *J. Mammary Gland. Biol. Neoplasia* **2017**, *22*, 93–108. [CrossRef] [PubMed]
55. Macias, H.; Hinck, L. Mammary gland development. *Wiley Interdiscip. Rev. Dev. Biol.* **2012**, *1*, 533–557. [CrossRef] [PubMed]
56. Marraudino, M.; Martini, M.; Trova, S.; Farinetti, A.; Ponti, G.; Gotti, S.; Panzica, G. Kisspeptin system in ovariectomized mice: Estradiol and progesterone regulation. *Brain Res.* **2018**, *1688*, 8–14. [CrossRef] [PubMed]
57. Farinetti, A.; Marraudino, M.; Ponti, G.; Panzica, G.; Gotti, S. Chronic treatment with tributyltin induces sexually dimorphic alterations in the hypothalamic POMC system of adult mice. *Cell Tissue Res.* **2018**, *374*, 587–594. [CrossRef]
58. Ferguson, A.V.; Samson, W.K. The orexin/hypocretin system: A critical regulator of neuroendocrine and autonomic function. *Front. Neuroendocr.* **2003**, *24*, 141–150. [CrossRef]
59. Paxinos, G.; Franklin, K.B.J.; Franklin, K.B.J. *The Mouse Brain in Stereotaxic Coordinates*, 2nd ed.; Academic Press: San Diego, CA, USA, 2001.
60. Crain, S.M.; Shen, K.-F. Low doses of cyclic AMP-phosphodiesterase inhibitors rapidly evoke opioid receptor-mediated thermal hyperalgesia in naïve mice which is converted to prominent analgesia by cotreatment with ultra-low-dose naltrexone. *Brain Res.* **2008**, *1231*, 16–24. [CrossRef]
61. Jefferson, W.N.; Padilla-Banks, E.; Newbold, R.R. Studies of the Effects of Neonatal Exposure to Genistein on the Developing Female Reproductive System. *J. AOAC Int.* **2006**, *89*, 1189–1196. [CrossRef]
62. Jefferson, W.N.; Padilla-Banks, E.; Newbold, R.R. Disruption of the developing female reproductive system by phytoestrogens: Genistein as an example. *Mol. Nutr. Food Res.* **2007**, *51*, 832–844. [CrossRef]
63. Rudel, R.A.; Fenton, S.E.; Ackerman, J.M.; Euling, S.Y.; Makris, S.L. Environmental Exposures and Mammary Gland Development: State of the Science, Public Health Implications, and Research Recommendations. *Environ. Health Perspect.* **2011**, *119*, 1053–1061. [CrossRef] [PubMed]
64. Muñoz-De-Toro, M.; Markey, C.M.; Wadia, P.R.; Luque, E.H.; Rubin, B.S.; Sonnenschein, C.; Soto, A.M. Perinatal Exposure to Bisphenol-A Alters Peripubertal Mammary Gland Development in Mice. *Endocrinology* **2005**, *146*, 4138–4147. [CrossRef]

65. Li, R.; El Zowalaty, A.E.; Chen, W.; Dudley, E.A.; Ye, X. Segregated responses of mammary gland development and vaginal opening to prepubertal genistein exposure in Bscl2(-/-) female mice with lipodystrophy. *Reprod. Toxicol.* **2015**, *54*, 76–83. [CrossRef] [PubMed]
66. Ruiz-Pino, F.; Miceli, D.; Franssen, D.; Vazquez, M.J.; Farinetti, A.; Castellano, J.M.; Panzica, G.; Tena-Sempere, M. Environmentally Relevant Perinatal Exposures to Bisphenol A Disrupt Postnatal Kiss1/NKB Neuronal Maturation and Puberty Onset in Female Mice. *Environ. Health Perspect.* **2019**, *127*, 107011. [CrossRef]
67. Luque, R.M.; Kineman, R.; Tena-Sempere, M. Regulation of Hypothalamic Expression of KiSS-1 and GPR54 Genes by Metabolic Factors: Analyses Using Mouse Models and a Cell Line. *Endocrinology* **2007**, *148*, 4601–4611. [CrossRef] [PubMed]
68. Szkudelska, K.; Nogowski, L. Genistein—A dietary compound inducing hormonal and metabolic changes. *J. Steroid Biochem. Mol. Biol.* **2007**, *105*, 37–45. [CrossRef] [PubMed]
69. Szkudelski, T.; Nogowski, L.; Pruszyńska-Oszmałek, E.; Kaczmarek, P.; Szkudelska, K. Genistein restricts leptin secretion from rat adipocytes. *J. Steroid Biochem. Mol. Biol.* **2005**, *96*, 301–307. [CrossRef]
70. Nogowski, L.; Szkudelska, K.; Szkudelski, T.; Pruszyńska-Oszmałek, E. The effect of a phytoestrogen, genistein, on the hormonal and metabolic status of pregnant rats. *J. Anim. Feed. Sci.* **2006**, *15*, 275–286. [CrossRef]
71. Li, R.-Z.; Ding, X.-W.; Geetha, T.; Al-Nakkash, L.; Broderick, T.L.; Babu, J.R. Beneficial Effect of Genistein on Diabetes-Induced Brain Damage in the ob/ob Mouse Model. *Drug Des. Dev. Ther.* **2020**, *ume 14*, 3325–3336. [CrossRef]
72. Weigt, C.; Hertrampf, T.; Zoth, N.; Fritzemeier, K.H.; Diel, P. Impact of estradiol, ER subtype specific agonists and genistein on energy homeostasis in a rat model of nutrition induced obesity. *Mol. Cell. Endocrinol.* **2012**, *351*, 227–238. [CrossRef]
73. Heindel, J.J.; Saal, F.S.V.; Blumberg, B.; Bovolin, P.; Calamandrei, G.; Ceresini, G.; Cohn, B.A.; Fabbri, E.; Gioiosa, L.; Kassotis, C.; et al. Parma consensus statement on metabolic disruptors. *Environ. Health* **2015**, *14*, 1–7. [CrossRef]
74. Snedeker, S.M.; Hay, A. Do Interactions Between Gut Ecology and Environmental Chemicals Contribute to Obesity and Diabetes? *Environ. Health Perspect.* **2012**, *120*, 332–339. [CrossRef]
75. Heindel, J.J.; Blumberg, B.; Cave, M.; Machtinger, R.; Mantovani, A.; Mendez, M.A.; Nadal, A.; Palanza, P.; Panzica, G.; Sargis, R.; et al. Metabolism disrupting chemicals and metabolic disorders. *Reprod. Toxicol.* **2017**, *68*, 3–33. [CrossRef] [PubMed]
76. Ryökkynen, A.; Kukkonen, J.; Nieminen, P. Effects of dietary genistein on mouse reproduction, postnatal development and weight-regulation. *Anim. Reprod. Sci.* **2006**, *93*, 337–348. [CrossRef]
77. Wright, S.M.; Salter, A.M. Effects of Soy Protein on Plasma Cholesterol and Bile Acid Excretion in Hamsters. *Comp. Biochem. Physiol. Part B Biochem. Mol. Biol.* **1998**, *119*, 247–254. [CrossRef]
78. Nieminen, P.; Mustonen, A.-M.; Lindström-Seppä, P.; Asikainen, J.; Mussalo-Rauhamaa, H.; Kukkonen, J. Phytosterols Act as Endocrine and Metabolic Disruptors in the European Polecat (Mustela putorius). *Toxicol. Appl. Pharmacol.* **2002**, *178*, 22–28. [CrossRef]
79. Ryökkynen, A.; Mustonen, A.-M.; Pyykönen, T.; Nieminen, P. Endocrine and metabolic alterations in the mink (Mustela vison) due to chronic phytoestrogen exposure. *Chemosphere* **2006**, *64*, 1753–1760. [CrossRef] [PubMed]
80. Cheung, C.C.; Clifton, D.K.; Steiner, R.A. Proopiomelanocortin Neurons Are Direct Targets for Leptin in the Hypothalamus. *Endocrinology* **1997**, *138*, 4489–4492. [CrossRef] [PubMed]
81. Xu, Y.; Faulkner, L.D.; Hill, J.W. Cross-talk between metabolism and reproduction: The role of POMC and SF1 neurons. *Front. Endocrinol.* **2012**, *2*, 98. [CrossRef] [PubMed]
82. Marraudino, M.; Bo, E.; Carlini, E.; Farinetti, A.; Ponti, G.; Zanella, I.; Di Lorenzo, D.; Panzica, G.C.; Gotti, S. Hypothalamic Expression of Neuropeptide Y (NPY)and Pro-OpioMelanoCortin (POMC) in Adult Male Mice is Affected by Chronic Exposure to Endocrine Disruptors. *Metabolites* **2021**, *11*, 368. [CrossRef]
83. Chieffi, S.; Carotenuto, M.; Monda, V.; Valenzano, A.; Villano, I.; Precenzano, F.; Tafuri, D.; Salerno, M.; Filippi, N.; Nuccio, F.; et al. Orexin System: The Key for a Healthy Life. *Front. Physiol.* **2017**, *8*, 357. [CrossRef]
84. Brownell, S.E.; Conti, B. Age- and gender-specific changes of hypocretin immunopositive neurons in C57Bl/6 mice. *Neurosci. Lett.* **2010**, *472*, 29–32. [CrossRef]
85. Muschamp, J.W.; Dominguez, J.M.; Sato, S.M.; Shen, R.-Y.; Hull, E. A Role for Hypocretin (Orexin) in Male Sexual Behavior. *J. Neurosci.* **2007**, *27*, 2837–2845. [CrossRef]
86. Kalra, S.P.; Kalra, P.S. Neuroendocrine Control of Energy Homeostasis: Update on New Insights. *Prog. Brain Res.* **2010**, *181*, 17–33. [CrossRef] [PubMed]
87. Biglari, N.; Gaziano, I.; Schumacher, J.; Radermacher, J.; Paeger, L.; Klemm, P.; Chen, W.; Corneliussen, S.; Wunderlich, C.M.; Sue, M.; et al. Functionally distinct POMC-expressing neuron subpopulations in hypothalamus revealed by intersectional targeting. *Nat. Neurosci.* **2021**, *24*, 913–929. [CrossRef] [PubMed]
88. Gaskins, G.T.; Moenter, S.M. Orexin A Suppresses Gonadotropin-Releasing Hormone (GnRH) Neuron Activity in the Mouse. *Endocrinology* **2012**, *153*, 3850–3860. [CrossRef]
89. Hosseini, A.; Khazali, H. Central Orexin A Affects Reproductive Axis by Modulation of Hypothalamic Kisspeptin/Neurokinin B/Dynorphin Secreting Neurons in the Male Wistar Rats. *Neuromolecular Med.* **2018**, *20*, 525–536. [CrossRef]
90. Barreiro, M.L.; Pineda, R.; Navarro, V.M.; López, M.; Suominen, J.S.; Pinilla, L.; Señaris, R.; Toppari, J.; Aguilar, E.; Diéguez, C.; et al. Orexin 1 Receptor Messenger Ribonucleic Acid Expression and Stimulation of Testosterone Secretion by Orexin-A in Rat Testis. *Endocrinology* **2004**, *145*, 2297–2306. [CrossRef] [PubMed]

91. Manno, F.A.M. Measurement of the digit lengths and the anogenital distance in mice. *Physiol. Behav.* **2008**, *93*, 364–368. [CrossRef] [PubMed]
92. Cimafranca, M.A.; Davila, J.; Ekman, G.C.; Andrews, R.N.; Neese, S.L.; Peretz, J.; Woodling, K.A.; Helferich, W.G.; Sarkar, J.; Flaws, J.; et al. Acute and Chronic Effects of Oral Genistein Administration in Neonatal Mice1. *Biol. Reprod.* **2010**, *83*, 114–121. [CrossRef]
93. Doerge, D.R.; Twaddle, N.C.; Banks, E.P.; Jefferson, W.N.; Newbold, R.R. Pharmacokinetic analysis in serum of genistein administered subcutaneously to neonatal mice. *Cancer Lett.* **2002**, *184*, 21–27. [CrossRef]
94. Bo, E.; Farinetti, A.; Marraudino, M.; Sterchele, D.; Eva, C.; Gotti, S.; Panzica, G. Adult exposure to tributyltin affects hypothalamic neuropeptide Y, Y1 receptor distribution, and circulating leptin in mice. *Andrology* **2016**, *4*, 723–734. [CrossRef]
95. Macchi, E.; Cucuzza, A.S.; Badino, P.; Odore, R.; Re, F.; Bevilacqua, L.; Malfatti, A. Seasonality of reproduction in wild boar (Sus scrofa) assessed by fecal and plasmatic steroids. *Theriogenology* **2010**, *73*, 1230–1237. [CrossRef] [PubMed]
96. Watson, R.E.; Wiegand, S.J.; Clough, R.W.; Hoffman, G.E. Use of cryoprotectant to maintain long-term peptide immunoreactivity and tissue morphology. *Peptides* **1986**, *7*, 155–159. [CrossRef]
97. Franceschini, I.; Lomet, D.; Cateau, M.; Delsol, G.; Tillet, Y.; Caraty, A. Kisspeptin immunoreactive cells of the ovine preoptic area and arcuate nucleus co-express estrogen receptor alpha. *Neurosci. Lett.* **2006**, *401*, 225–230. [CrossRef] [PubMed]
98. Pinos, H.; Pérez-Izquierdo, M.; Carrillo, B.; Collado, P. Effects of undernourishment on the hypothalamic orexinergic system. *Physiol. Behav.* **2011**, *102*, 17–21. [CrossRef]
99. Mize, R.; Holdefer, R.N.; Nabors, L. Quantitative immunocytochemistry using an image analyzer. I. Hardware evaluation, image processing, and data analysis. *J. Neurosci. Methods* **1988**, *26*, 1–23. [CrossRef]

Article

The Crowded Uterine Horn Mouse Model for Examining Postnatal Metabolic Consequences of Intrauterine Growth Restriction vs. Macrosomia in Siblings

Julia A. Taylor [1,*], Benjamin L. Coe [1], Toshi Shioda [2] and Frederick S. vom Saal [1]

[1] Division of Biological Sciences, University of Missouri-Columbia, Columbia, MO 65211, USA; coeb@health.missouri.edu (B.L.C.); vomsaalf@missouri.edu (F.S.v.S.)

[2] Center for Cancer Research, Massachusetts General Hospital and Harvard Medical School, Building 149, 13th Street, Charlestown, MA 02129, USA; shioda@helix.mgh.harvard.edu

* Correspondence: taylorja@missouri.edu

Abstract: Differential placental blood flow and nutrient transport can lead to both intrauterine growth restriction (IUGR) and macrosomia. Both conditions can lead to adult obesity and other conditions clustered as metabolic syndrome. We previously showed that pregnant hemi-ovariectomized mice have a crowded uterine horn, resulting in siblings whose birth weights differ by over 100% due to differential blood flow based on uterine position. We used this crowded uterus model to compare IUGR and macrosomic male mice and also identified IUGR males with rapid (IUGR-R) and low (IUGR-L) postweaning weight gain. At week 12 IUGR-R males were heavier than IUGR-L males and did not differ from macrosomic males. Rapid growth in IUGR-R males led to glucose intolerance compared to IUGR-L males and down-regulation of adipocyte signaling pathways for fat digestion and absorption and type II diabetes. Macrosomia led to increased fat mass and altered adipocyte size distribution compared to IUGR males, and down-regulation of signaling pathways for carbohydrate and fat digestion and absorption relative to IUGR-R. Clustering analysis of gonadal fat transcriptomes indicated more similarities than differences between IUGR-R and macrosomic males compared to IUGR-L males. Our findings suggest two pathways to adult metabolic disease: macrosomia and IUGR with rapid postweaning growth rate.

Keywords: intrauterine growth restriction; macrosomia; glucose tolerance; abdominal adipocyte gene expression; thrifty phenotype hypothesis

Citation: Taylor, J.A.; Coe, B.L.; Shioda, T.; vom Saal, F.S. The Crowded Uterine Horn Mouse Model for Examining Postnatal Metabolic Consequences of Intrauterine Growth Restriction vs. Macrosomia in Siblings. *Metabolites* **2022**, *12*, 102. https://doi.org/10.3390/metabo12020102

Academic Editor: Silvia Ravera

Received: 18 December 2021
Accepted: 19 January 2022
Published: 22 January 2022

1. Introduction

In the United States and many developed countries, the incidence of obesity and related diseases, collectively referred to as metabolic syndrome, are increasing at a rapid rate [1]. Evidence from epidemiological studies links the rate of fetal growth, body weight at birth, rate of growth during early postnatal life, and adult metabolic diseases [2]. Namely, when there is reduced fetal growth and body weight at birth, but the subsequent postnatal growth rate is markedly higher than the median, this results in body weight centile crossing, and significant metabolic abnormalities occur. This sequence of events is a central feature of the developmental basis of health and disease (DOHaD) hypothesis for metabolic diseases [3].

Many studies have investigated the effects of maternal nutrition during pregnancy. Decreased maternal nutrition during pregnancy leads to intrauterine growth-restriction (IUGR), and the small for gestational age babies in the bottom 10th percentile for birth weight are at increased risk of being overweight in adulthood [4,5]. Similarly, babies that are in the top fifth percentile for body weight at birth (macrosomic), a condition often associated with maternal obesity and/or diabetes, are also at risk for adult obesity [6]. Regardless of the basis for their obesity, the outcome is that most obese individuals are at increased risk for developing other disorders associated with metabolic syndrome. Among these co-morbidities are increased blood pressure, cardiovascular disease, insulin insensitivity, glucose intolerance, diabetes, fatty liver disease, and elevated triglycerides and cholesterol [1,2,7].

A large number of animal studies investigating these conditions have manipulated maternal nutrition to induce IUGR or macrosomia in the offspring [8–10]. These models alter fetal growth trajectories via maternal protein restriction, caloric restriction, use of a feed with a very high percentage of fat, or streptozocin-induced type 1 diabetes mellitus to increase fetal glucose uptake. However, although maternal malnutrition is a cause of IUGR in non-developed countries, war zones, and in cases of eating disorders [4,11], dietary restriction paradigms do not adequately represent the human condition associated with IUGR in developed countries, where the cause of IUGR is not typically severe caloric or protein restriction.

Many factors can contribute to reduced fetal nutrition without a decrease in maternal nutrition, such as insufficient blood flow to the placenta or deficits in placental transport of specific nutrients [12]. Another approach to creating fetal undernutrition has been to experimentally create insufficient transport of nutrients to fetuses from the maternal circulation as a result of ligation of the uterine blood vessels [13]. However, a problem with this approach is that there are physiological responses to trauma in addition to surgically restricting blood flow to specific placentae [14].

While it was once not uncommon for parents to be advised to over-feed IUGR babies, there is now extensive epidemiological evidence showing that IUGR babies who experience a rapid "catch-up" growth spurt during infancy or early childhood are at high risk for adult obesity, type 2 diabetes and other co-morbidities of metabolic syndrome, consistent with the "thrifty phenotype" hypothesis [15]. The hypothesis is that the physiological "program" of IUGR babies is one that makes them adapted for a lifetime of reduced nutrition, and these babies are thus at high risk for becoming overweight when they are exposed to a typical highly processed Western diet that is higher in calories than their physiological program's "set point" [16,17]. Thus, in humans, fetal growth rate interacts with the growth rate during early postnatal life to determine whether IUGR leads to adult obesity and other metabolic diseases.

We examined here the consequences for male mice of developing in a crowded uterine horn that results in differential placental blood flow based on the random implantation of fetuses in the middle vs. either end of the uterine horn. This is an IUGR and macrosomia model that results in animals that match the clinical description of IUGR in the bottom 10th percentile for human fetuses and of macrosomia at the other end of the birth weight spectrum [18].

In more detail, previous studies in mice suggested that crowding a uterus with more fetuses than is normal resulted in the production of offspring that varied in the rate of fetal growth as a result of differences in the amount of placental blood flow [19]. In a series of studies, we demonstrated [20,21] that blood flows from two directions into the artery supplying each independent uterine horn in rats and mice, which have a duplex uterus. Specifically, blood flows into each uterine artery from both the cranial end, branching off of the descending aorta, and the caudal end, which branches off of the ipsilateral iliac artery (Figure 1). The unusual "loop" vascular structure results in a greater flow of blood to the placentae located at the cranial and caudal ends of each uterine horn relative to placentae located in the middle of each horn in both rats and mice. The magnitude of the effect on blood flow to each placenta and on fetal growth as a result of being positioned at the ends or middle of a uterine horn in mice is greatly exaggerated if one ovary is removed since a change in follicular dynamics in the remaining ovary causes hyper-ovulation, referred to as compensatory ovarian hypertrophy [19,22,23]. A mouse that becomes pregnant after hemi-ovariectomy will thus have a crowded uterine horn since embryos cannot migrate from one uterine horn to the other in mice. This crowded uterine horn results in dramatic differences between siblings in fetal nutrient availability based on implantation site, which is a random event [24].

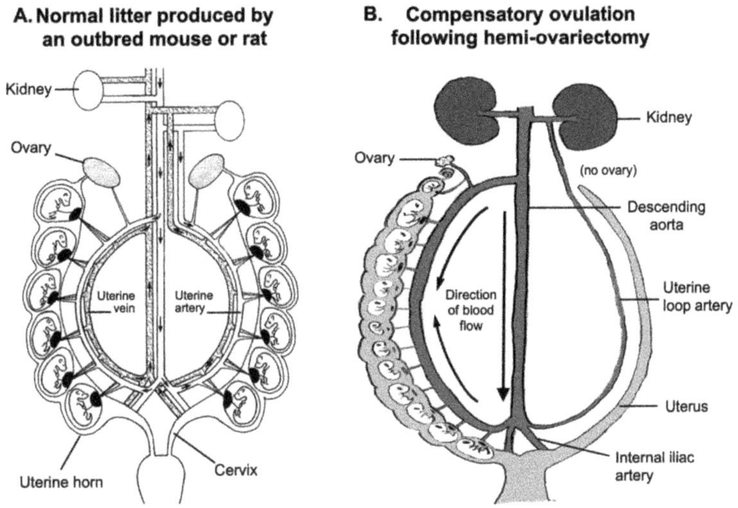

Figure 1. The pregnant mouse uterus. (**A**) The intact duplex uterus of the mouse or rat, with each uterine horn having an independent cervix. The bi-direction blood flow in each uterine artery and vein is indicated by arrows. Modified from Even et al. [21], with permission. The left kidney is caudal to the right kidney, which leads to variability in the cranial vascular anatomy of the left uterine horn. (**B**) The consequence of hemi-ovariectomy on ovulation in litter-bearing species that normally ovulate from both ovaries, which is referred to as compensatory ovarian hypertrophy due to the remaining ovary ovulating the normal number of oocytes that would have been produced by both ovaries. Due to the variability in the vascular anatomy of the left uterine horn, all mice had the left ovary removed, which resulted in compensatory ovulation by the remaining right ovary and crowding of fetuses in the right uterine horn. Fetuses that end up randomly implanted in the middle portion of the uterus [24] have reduced blood flow and nutrient transport across the placenta relative to siblings at the ends of the horn, due to the bi-directional uterine arterial blood flow; however, if the fetus at the cranial end of the crowded uterus has a placental artery that branches off of the ovarian artery (as shown in Panel **B**), this fetus will be IUGR rather than macrosomic [19,22]. Modified from Coe et al. [22], with permission.

In the present study, we used the crowded uterus phenomenon as a model system to increase differences between siblings in the rate of fetal growth and thus body weight at birth. This allowed us to examine the consequences for postnatal growth rate, and for adipocyte number, size and gene expression in the largest abdominal fat pad, which in mice is associated with the gonads. We also weighed the other major fat pads in mice: the abdominal renal fat pads and the subcutaneous inguinal fat pads. In addition, we examined glucose tolerance in male mice that were classified as IUGR (<10% of birth weight range), median (~50%) or macrosomic (>93%) at birth due to differential placental blood flow based purely on their random position within the uterus.

We report here results from the crowded uterus model, which provides the opportunity to examine the etiology of differences caused by differential fetal and postnatal growth without nutrient or surgical intervention. Our hypothesis was that IUGR males that experienced a rapid period of postnatal growth (referred to as IUGR-R males) would in adulthood appear similar to males identified as macrosomic at birth. However, we expected both IUGR-R and macrosomic males to show significant differences from IUGR males that exhibited a low rate of postnatal growth (referred to as IUGR-L males). We identified significant differences in adult phenotype between IUGR male mice based on their rate of growth (low vs. rapid), providing support in this animal model for the "thrifty phenotype" hypothesis. Interestingly, there were also differences in phenotype between IUGR-R males and macrosomic males, suggesting that these two pathways to adult metabolic diseases need to be considered when seeking approaches to mitigate the metabolic abnormalities of these individuals.

2. Results

2.1. Placental Blood Flow and Fetal Growth

The placement of fetuses in the uterus impacts fetal growth [19,22]. Here we chose to only examine offspring that developed in one crowded uterine horn. A schematic of a normal pregnancy is shown in Figure 1A, and a crowded uterine horn is shown in Figure 1B.

2.2. Birth Weight Criteria and Postnatal Growth Rate

CD-1 mouse litters were produced in two blocks. In Block 1, 52 litters were produced by hemi-ovariectomized postpartum females resulting in a total of 605 offspring, of which 585 survived until weaning. The overall pre-weaning mortality rate was 3.3%, with 25 percent of all pre-weaning deaths occurring in IUGR animals, and only 0.05% in macrosomic animals. The total number of animals at birth consisted of 297 females and 308 males; the distribution of all male offspring body weights from birth until 12 weeks old is presented in Figure 2A; body weight at birth data were normally distributed. The mean weight for all male mice on the day of birth was 1.64 ± 0.24 g (mean ± SEM).

Of the animals that survived until weaning, the weight range for males with birth weights in the bottom 5th percentile (designated as IUGR; n = 14) was 0.95–1.23 g (mean ± SEM: 1.12 ± 0.03 g) (Figure 2A). Males categorized as being at the "median" (~1.64 g) body weight range were in the 47.4–51.2 percentile of all birth weights (n = 16 saved for follow-up experiments). The weight range for males with birth weights in the top 5th percentile (designated as macrosomic) was 2.02–2.40 g (n = 15); the mean (±SEM) was 2.20 ± 0.04 g.

Body weight between birth and weaning (week 3) increased as a function of body weight at birth (Figure 2A). However, the percent weight gain in the first week after weaning (between weeks 3–4) showed a dramatic decrease as a function of body weight category at birth (Figure 2B). Namely, the IUGR males (bottom 5th percentile) exhibited a 95% increase in their body weight between week 3 (weaning) and week 4 of life, while macrosomic males (top 95th percentile) showed only a 37% increase during the same time. Thus, by week 4, the IUGR males had reached a body weight that was not significantly different from males identified as macrosomic at birth. In contrast, the males in the 5–50% body weight at birth categories were significantly lighter than IUGR males by postnatal week 4 and into adulthood, with the last measurement being made when all males were 12 weeks old (Figure 2A).

2.3. Different Sub-Groups of IUGR Males

A sub-group of the IUGR males did not go through a rapid post-weaning growth phase, although in Figure 2 they are included in the mean growth rate for all IUGR males. Of the animals that were below the 10th percentile for birth weight, 45% (n = 13/29) did not experience a post-weaning rapid weight gain. These males, who gained less than 65% of their weight during weeks 3–4, referred to here as IUGR-low postweaning growth rate (IUGR-L) males, were compared with IUGR-rapid post-weaning growth rate males (IUGR-R) that gained over 65% body weight during this period. IUGR-L males achieved a lower adult body weight by week 12 relative to IUGR-R or macrosomic males, while IUGR-L did not differ from median body weight at birth males (Figure 3A). The percent weight gain between weeks 3–4 is shown in Figure 3B. The IUGR-R males had a lower birth weight than IUGR-L males (1.166 ± 0.030 g vs.1.253 ± 0.021 g, p = 0.029) and were also in a lower birth weight percentile than IUGR-L males (3.925 ± 0.618% vs. 6.248 ± 0.748%, $p = 0.023$).

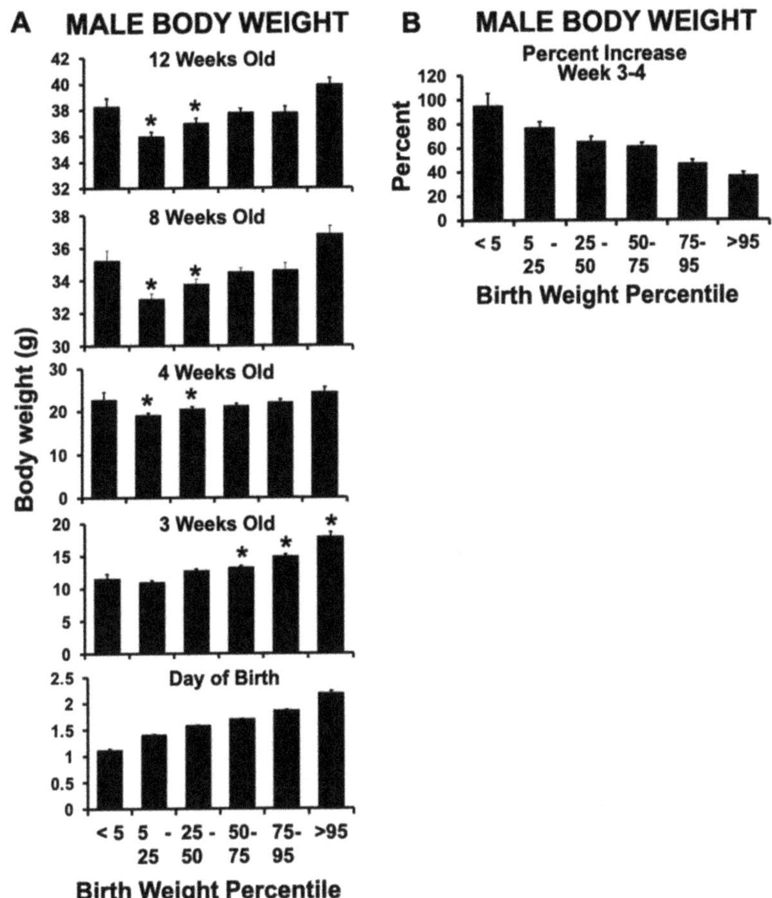

Figure 2. Body weight and growth of Block 1 male mice. (**A**) Body weights of male mice produced in 52 litters by Block 1 hemi-ovariectomized females based on birth weight categories, from the ≤5th percentile to ≥95th percentile. Beginning at postnatal week 3 (weaning), body weights were measured for all surviving male offspring until postnatal week 12. (**B**) the percent increase in body weight for these males during the first week of free-feeding after weaning (week 3–4). A subset of these Block 1 males (from IUGR, median and macrosomic birth weight categories) were retained after week 12 for additional studies. (See Table S1). * $p < 0.05$ compared to males from the bottom 5th percentile (IUGR males). Values are mean ± SEM.

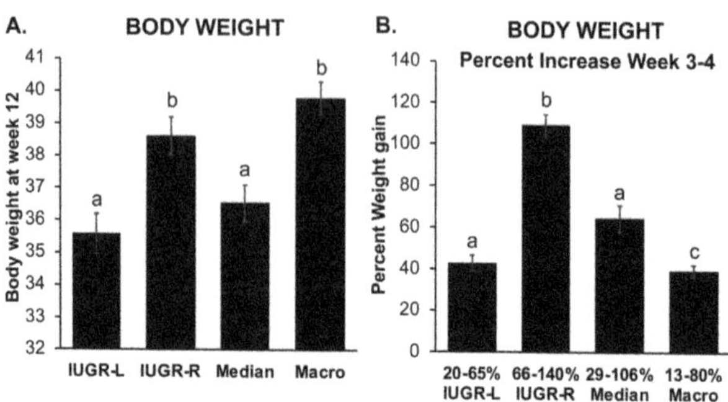

Figure 3. Body weight and percent growth of IUGR-L (n = 13), IUGR-R (n = 16), median (n = 17) and macrosomic (n = 21) males. (**A**) body weights at week 12, when IUGR-L and median males weighed less than IUGR-R and macrosomic males. (**B**) the percent postweaning (week 3–4) weight gain, showing that IUGR-R males exhibited the most rapid post-weaning weight gain, while macrosomic males showed the lowest week 3–4 percent weight gain. Groups with different letters above the error bar are significantly different from each other; groups that share a letter in common are not statistically different. Values are mean ± SEM.

Within the macrosomic group of males, the range of body weight gain during the week after weaning was smaller, and there was no significant difference in adult body weight based on the rate of growth during the week after weaning (data not shown).

2.4. Experiments on Block 1 and Block 2 Animals When 6 Months Old

After week 12, selected IUGR, median and macrosomic males from Block 1, here defined as below the 10th percentile at birth, at the median, and above the 93rd percentile respectively, were saved for further studies. IUGR males were subdivided into IUGR-L and IUGR-R animals based on differential growth rate during postnatal weeks 3–4, as described above. Block 1 males included the first set of animals that were subjected to a 14-h fast prior to a glucose tolerance test (GTT) and a second set that did not experience either a fast or a GTT (Sets 1 and 2, respectively, see Table S1).

We generated a second cohort of animals (Block 2) from an additional 21 hemi-ovariectomized females (see Table S1), IUGR, median and macrosomic at birth animals were identified using the same criteria as for Block 1. IUGR males were again separated into rapid- and slow-growing based on the differential postweaning growth rate, although for Block 2 we selected a slightly higher cutoff, at less than or above 76% postweaning weight gain.

We conducted experiments on males from the three different birth weight percentiles (IUGR, median and macrosomic), as well as differing post-weaning growth percentiles for IUGR animals (IUGR-L and IUGR-R) when they were six months old.

First, we conducted a glucose tolerance test (GTT) on selected males (Set 1 in Table S1) from Block 1. IUGR-L, IUGR-R, median and macrosomic males were fasted for 14 h overnight (during the dark phase of the light:dark cycle prior to the GTT and were sacrificed immediately after the GTT; organ weights were measured, and the fat pads were collected for analysis of gonadal adipocyte cell number and size, and analysis of expression of selected genes by quantitative reverse transcription polymerase chain reaction (qPCR) in gonadal fat.

Second, we also conducted a GTT on the Block 2 males. These animals were only fasted during the first 4 h of the light phase, for comparison with the effects of the overnight 14-h fast, and were not sacrificed for at least 1–2 weeks post-test. Body weights and gonadal fat pad weights were also measured, but no further analyses were conducted.

Finally, to rule out any effect of fasting associated with the GTT on fat, we also collected fat from the second set of males from Block 1 (Set 2 in Table S1), without fasting the animals or conducting a GTT prior to collection of fat. We collected gonadal, renal and inguinal fat pads from these males, and then conducted an analysis of gonadal adipocyte number and size, as well as analysis of gonadal fat gene expression by both qPCR and microarray.

2.5. Glucose Tolerance Test (GTT)

2.5.1. GTT in 14-h Fasted Males

We observed a significant difference in response to the glucose challenge between median and IUGR-R compared to macrosomic and IUGR-L males. Specifically, blood glucose concentrations after the glucose challenge were similar in animals for the median and IUGR-R birth weight males, which showed impaired glucose tolerance in comparison to macrosomic and IUGR-L males based on the area under the blood glucose concentration-time curve (AUC; Figure 4A). Importantly, males in the slower post-weaning growth sub-group of IUGR males (IUGR-L) had better glucose tolerance than the heavier IUGR sub-group (IUGR-R) males (Figure 4A). Specifically, blood glucose concentrations tended to be lower in IUGR-L males compared to the IUGR-R males based on AUC ($p = 0.06$), and at both 30 and 60 min after glucose challenge ($p = 0.08$ for each comparison). What was surprising was that the median males had impaired glucose tolerance relative to the macrosomic males ($p < 0.05$). Macrosomic males' body weight and gonadal fat pad weight were greater than all other groups (Figure 4B,C).

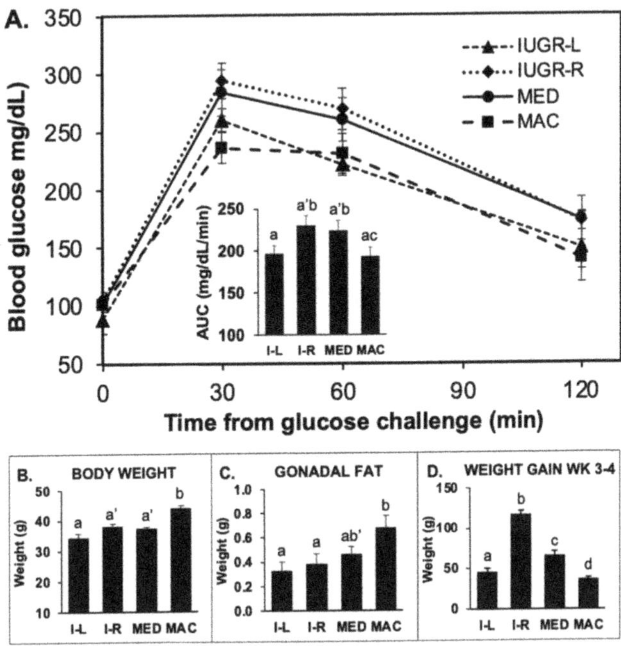

Figure 4. Glucose tolerance test in 14-h fasted Block 1 males. (**A**) glucose tolerance test data for IUGR-L (I-L; $n = 6$), IUGR-R (I-R; $n = 9$; one animal died in adulthood), median (MED, $n = 16$) and macrosomic (MAC, $n = 16$) males that underwent a 14 hr-fast prior to glucose challenge. Inset: Area under the curve (AUC) for this test (a' $p < 0.08$ vs. IUGR-L). (**B**) body weight at the time of tissue collection following the GTT (a' $p < 0.08$ compared to IUGR-L). (**C**) gonadal fat weight (b' $p = 0.06$ compared to macrosomic). (**D**) the week 3–4 percent weight gain for these males. Groups with different letters above the error bar are significantly different from each other; groups that share a letter in common are not statistically different; letters with a dash indicate p values between 0.05 and 0.01 as indicated above. Values are mean $\pm$ SEM.

2.5.2. GTT in 4-h Fasted (Block 2) Males

The results from the 4-h fasted males in Block 2 were different from those observed for the 14-h fasted males. First, the blood glucose levels were markedly higher for the 4-h fasted males relative to 14-h fasted males both at baseline and at 30 and 60 min after glucose injection. Second, based on AUC for 4-h fasted males, macrosomic males had reduced glucose tolerance relative to the median males ($p = 0.055$; Figure 5A), while both IUGR-L and IUGR-R showed somewhat lower glucose tolerance but did not differ significantly from median males.

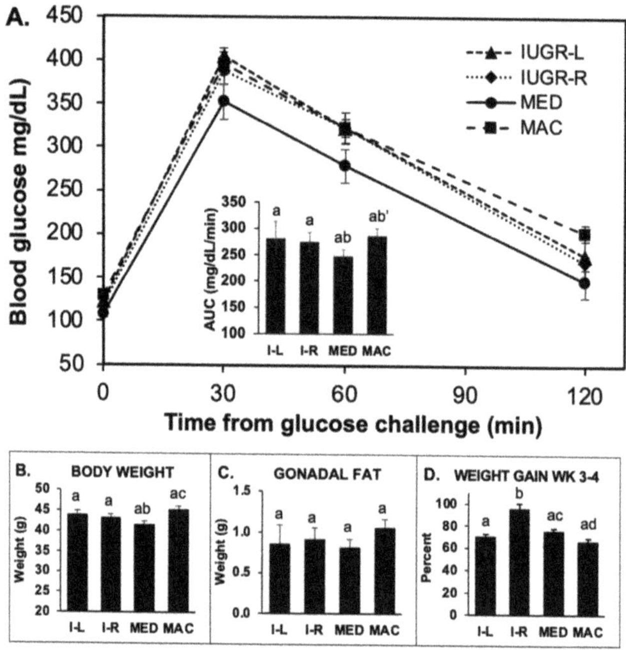

Figure 5. Glucose tolerance test in 4-h fasted Block 2 males. (**A**) glucose tolerance test for male IUGR-L (I-L; $n = 7$), IUGR-R (I-R; $n = 19$), median (MED; $n = 25$) and macrosomic (MAC; $n = 25$) mice that underwent a 4-h fast prior to glucose challenge. Inset: Area under the curve (AUC) for this test. The macrosomic males had reduced glucose tolerance relative to median males (b′ $p = 0.055$). (**B**) body weight at the time of tissue collection following the GTT. Macrosomic males were significantly heavier than IUGR-R and median males ($p < 0.05$). (**C**) gonadal fat weight. While gonadal fat pad weight was slightly greater in macrosomic males, it did not differ significantly between the groups. (**D**) the week 3–4 percent weight gain for these males. Groups with different letters above the error bar are significantly different from each other; groups that share a letter in common are not statistically different; letters with a dash indicate p values between 0.05 and 0.01 as indicated above. Values are mean ± SEM.

2.6. *Body Weight, Organ Weights, Fat Pad Weights, and Gonadal Fat Pad Adipocyte Number and Size: Comparison of 14-h Fasted and Non-Fasted Block 1 Males*

2.6.1. Data from 14-h Fasted Males

An important issue identified in this experiment was that the 14-h fasting procedure prior to the glucose tolerance test reduced body weight in all groups of males (IUGR-L: 12.5%, IUGR-R 9.2%, median: 11%, macrosomic: 6.3%), although less so in macrosomic males compared with IUGR-L males (Figure 6A). Weight loss in mice due to fasting is consistent with prior findings [25]. This weight loss resulted in the non-fasted IUGR-L and IUGR-R males being heavier at the time of fat pad collection than fasted IUGR animals. The

14-h fast altered the adipocyte size distribution (Figure 7) and also affected gene expression in gonadal fat collected after the GTT experiment relative to non-fasted males from the same prenatal growth category groups (see Section 2.7).

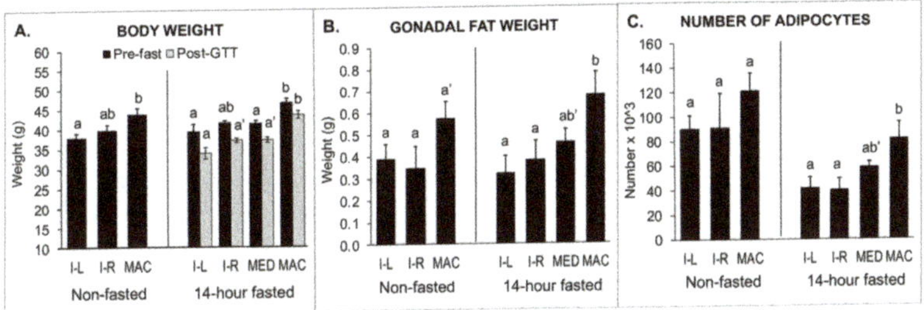

Figure 6. Comparison of body weight, fat pad weights, and adipocyte number in non-fasted vs. 14-h fasted animals. Data are from non-fasted males (IUGR-L $n = 4$, IUGR-R $n = 5$ and macrosomic $n = 5$) that did not undergo a GTT, and males fasted for 14 h prior to GTT and fat pad collection (IUGR-L $n = 5$, IUGR-R $n = 8$, median $n = 14$, macrosomic $n = 13$). (**A**) body weights, (**B**) gonadal fat pad weights, and (**C**) the number of adipocytes per total mass of fat. Groups with different letters are significantly different from each other; groups that share a letter in common are not statistically different; a′ and b′ $p < 0.1$. Values are mean ± SEM.

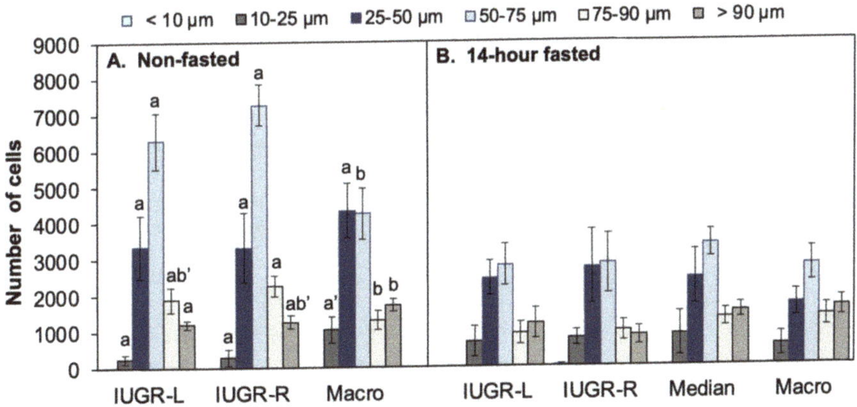

Figure 7. Gonadal adipocyte size distribution. Distribution of size categories for gonadal adipocytes collected from (**A**) non-fasted and (**B**) 14-h fasted males. For non-fasted males (**A**), the IUGR-L and IUGR-R males had significantly more mid-size adipocytes relative to macrosomic males, while macrosomic males had more of the largest adipocytes. a′ $p < 0.1$; b′ $p < 0.08$. For the 14-h fasted males, there were no significant differences between males from the different birth weight categories in any of the size categories (**B**). Different letters indicate statistically significant differences between birth weight groups within a size category ($p < 0.05$); groups that share a letter in common are not statistically different; letters with a dash indicate p values between 0.05 and 0.01 as indicated above. Values are mean ± SEM.

In 14-h fasted males, there were significant effects of birth weight category on body weight both prior to the GTT and also at the time of fat pad collection. Both prior to testing and after the 14-h fast and GTT, macrosomic animals were significantly heavier than IUGR-L and median animals (Figure 6A). IUGR-R animals were not statistically different from IUGR-L and median animals prior to testing, but at the time of collection, following the GTT, IUGR-L animals tended to be lighter than IUGR-R and median animals ($p < 0.1$, Figure 6A).

There were significant differences between IUGR-R and macrosomic males in heart, kidney and spleen weights, with macrosomic males having significantly heavier organ weights (Table S2). There were also significant differences between the different groups of 14-h fasted males in the weights of the gonadal fat pad weight collected immediately after the GTT test that indicated an effect of birth weight category. The gonadal fat pads from the macrosomic males were significantly heavier than for IUGR-L and IUGR-R males and tended ($p = 0.07$) to be heavier than the median male fat pads (Figure 6B; Table S3). The total amount of fat (summed renal, inguinal and gonadal fat pad weights) was significantly greater in macrosomic males than either IUGR or median animals (Table S3).

The total number of adipocytes in the gonadal fat pads is shown in Figure 6C. Macrosomic males had the highest mean adipocyte count and a significantly greater number of gonadal adipocytes than either IUGR-L males or IUGR-R males. Males in the median group tended to have fewer gonadal adipocytes than macrosomic males ($p < 0.1$). For these 14-h fasted males, there were no differences in the size distribution of gonadal adipocytes from IUGR-L, IUGR-R, median or macrosomic animals (Figure 7B).

2.6.2. Data from Non-Fasted Males

In non-fasted males, there were significant effects of birth weight category on body weight at the time of fat pad collection, with macrosomic animals being heavier than either IUGR group (Figure 6A).

Gonadal fat pad weights did not differ significantly between IUGR-L, IUGR-R and macrosomic animals (Figure 6B) although gonadal fat pads from macrosomic males tended ($p = 0.07$) to be heavier than those of IUGR-R males. Total fat weight was significantly higher in macrosomic animals compared to either IUGR group (Table S3).

Total adipocyte number did not differ significantly for non-fasted males (Figure 6C). However, analysis of adipocyte size distribution in gonadal fat pads of non-fasted males indicated that the size distribution was similar for IUGR-L and IUGR-R males and that both were significantly different from the macrosomic males, that had fewer adipocytes in the mid-size range ($p < 0.05$; Figure 7A). In contrast, macrosomic males had significantly more large adipocytes compared to IUGR-L and IUGR-R males. Thus, in non-fasted males, macrosomic males had a significantly different gonadal adipocyte size distribution relative to either IUGR-L or IUGR-R males.

2.7. *Gonadal Adipose Tissue Gene Expression by qPCR*

Six target genes were selected for analysis by qPCR based on their known roles in adipose tissue function. Gene expression was first measured in samples prepared from the gonadal fat from non-fasted males not previously administered a GTT, and these results are shown in Figure 8. Due to the limited number of IUGR and macrosomic males examined, we did not attempt to separate the IUGR males into IUGR-L and IUGR-R categories.

In non-fasted animals, the expression of Pparg2, Cebpa, Glut4 and Lpl was significantly higher in IUGR animals than in macrosomic animals ($p < 0.05$ for Pparg2, Cebpa and Lpl; $p < 0.01$ for Glut4). There were no significant differences between IUGR and macrosomic animals in the expression of Hsd11b1 and Cyp19, which appeared to be due to the high variance in samples from IUGR males.

We also measured the expression of these genes in fat samples from selected 14-h fasted male mice. In fat collected after the long fasting conditions and GTT, no differences were seen between the IUGR and macrosomic animals in the expression of these same six genes (Figure S3). Details of postnatal growth and body fat of the non-fasted and 14-h fasted animals whose tissues were used in the qPCR analyses are presented in Table S4.

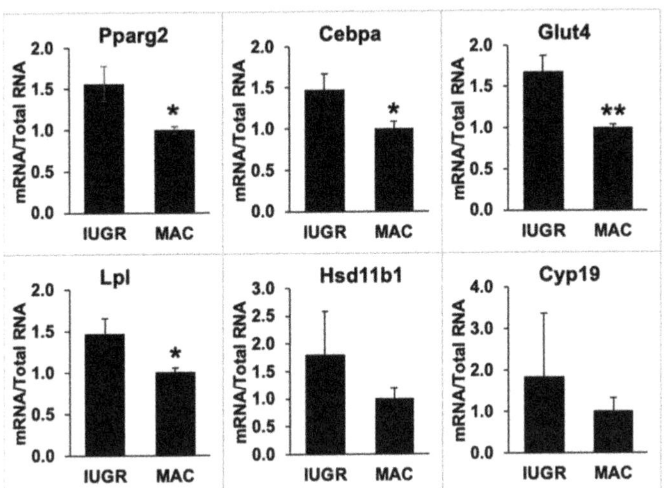

Figure 8. qPCR quantification of gene expression in gonadal fat of non-fasted IUGR and macrosomic male mice (n = 4–5 per group.) Values for IUGR animals are expressed relative to values for the macrosomic animals. Values are mean ± SEM. * $p < 0.05$, ** $p < 0.01$ compared to IUGR.

2.8. Gonadal Adipose Tissue Gene Expression by Microarray Analysis

We conducted a microarray analysis of gene expression in gonadal fat from non-fasted males that had not experienced a GTT. We examined 4 macrosomic males, 3 IUGR-L males and 3 IUGR-R males. Postnatal growth, body fat and other characteristics of these animals are presented in Table S5. A list of differentially expressed genes is given in Table S7.

Analysis of the microarray data was specifically aimed at identifying differences between the IUGR-L and IUGR-R groups, and between the IUGR-R and macrosomic animals (median body weight at birth males were not examined). Initial analysis of the array data was performed using a two-tailed t-test with Benjamini-Hochberg correction and a two-fold difference cutoff, and further analysis was conducted by t-test for the comparisons of interest. Impacted signaling pathways and gene ontologies were also identified. In addition, we used ANOVA to identify genes that were differentially expressed among the pooled microarray data from all three treatment groups.

2.8.1. Direct Comparison of Gene Expression in IUGR-L, IUGR-R and Macrosomic Males

Examining the two IUGR groups, from the t-test with Benjamini-Hochberg correction, only one gene was identified as being differentially expressed between the two IUGR groups: Fnip1 (folliculin-interacting protein 1) was down-regulated 2.1-fold in the IUGR-R animals compared to the IUGR-L animals (Figure 9). When the data were analyzed without the multiple hypothesis correction, 1197 genes were identified as being differentially expressed between the two groups. KEGG pathways impacted (Table 1) included Type II Diabetes Mellitus (down-regulated in IUGR-R samples, z-score = 3.86) and Fat Digestion and Absorption (down-regulated in IUGR-R samples, z-score = 2.12).

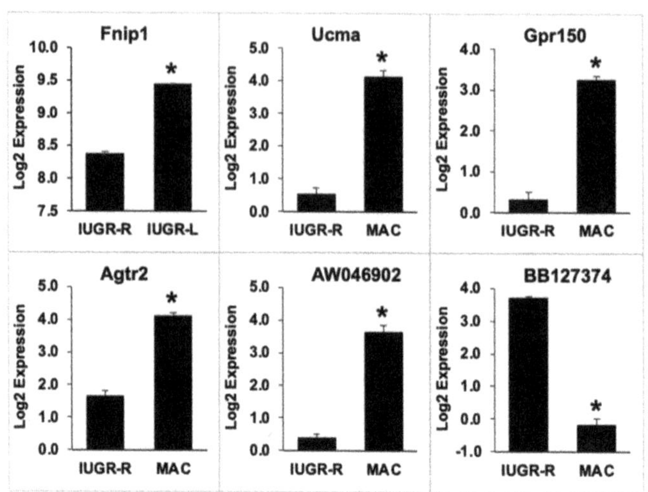

Figure 9. Differentially expressed genes. Statistical comparisons were between either IUGR-L and IUGR-R animals or between IUGR-R and Macrosomic animals. * $p < 0.05$ compared to IUGR-R. p values are Benjamini-Hochberg-corrected; uncorrected p values are all < 0.001. Values are mean ± SEM.

Table 1. KEGG Pathways differentially impacted by growth. Statistical comparisons were between either IUGR-L and IUGR-R animals or between IUGR-R and Macrosomic animals.

Groups Compared	KEGG Pathway	Ratio	Direction	*p*-Value	Gene Identifier	Gene ID
IUGR-L Vs. IUGR-R	**Type 2 Diabetes Mellitus**	9.5	Up	0.002	NM_008840	Pik3cd
	down-regulated in IUGR-R	4.31	Down	0.047	BE943756	Cacna1a
	z score = 3.86	3.45	Up	0.019	NM_010438	Hk1
		2.92	Down	0.045	BB184171	Mapk8
		2.11	Down	0.004	BB205102	Pik3cg
		2.11	Down	0.002	BB048682	Cacna1d
		2.03	Down	0.038	BE947490	Mapk10
	Fat digestion and absorption	3.04	Up	0.025	BG070618	-
	down-regulated in IUGR-R	3.01	Down	0.047	NM_011128	Pnliprp2
	z-score = 2.12	2.56	Down	0.023	NM_025469	Clps
		2.05	Down	0.006	AI326372	Pnlip
Macrosomic vs. IUGR-R	**Carbohydrate digestion and absorption**	5.67	Up	0.019	NM_008840	Pik3cd
		2.89	Down	0.011	BI696040	Atp1a1
	up-regulated in IUGR-R	2.42	Up	0.028	NM_019741	Slc2a5
	z score = 4.67	2.22	Up	0.037	BC027319	Atp1b1
		2.2	Up	0.043	BC027319	Atp1b1
	Fat digestion and absorption	6.49	Up	0.048	AI194999	Apoa1
	up-regulated in IUGR-R	4.96	Up	0.015	NM_007980	Fabp2
	z-score = 4.09	4.29	Up	0.006	AI527359	Apoa1
		3.41	Up	0.019	NM_025469	Clps

For the IUGR-R vs. macrosomic comparison, analysis with Benjamini-Hochberg correction identified 5 differentially expressed genes, which included Ucma (unique cartilage matrix-associated protein, down-regulated 12.2-fold in macrosomic animals), Gpr150 (G protein-coupled receptor 150, down-regulated 7.6-fold) and Agtr2 (angiotensin II receptor, type 2, down-regulated 5.6-fold). The relative expression of these and other genes is shown in Figure 9. Analysis without the multiple hypothesis correction yielded a list of 659 genes. Impacted KEGG pathways identified (Table 1) included Carbohydrate Digestion and Absorption (up-regulated in IUGR-R, z-score = 4.67) and Fat Digestion and Absorption (up-regulated in IUGR-R, z-score = 4.09).

2.8.2. Clustering Analysis of Differentially Expressed Genes

Due to the small sample size, ANOVA was conducted without multiple hypothesis correction, yielding 855 genes that were differentially expressed among all groups at $p < 0.01$. These genes were subjected to hierarchical clustering, yielding the heatmap in Figure 10. The heatmap shows the separation of six clusters of differentially expressed genes identified as A–F, which segregate by birth weight, post-weaning growth rate, or adult weight.

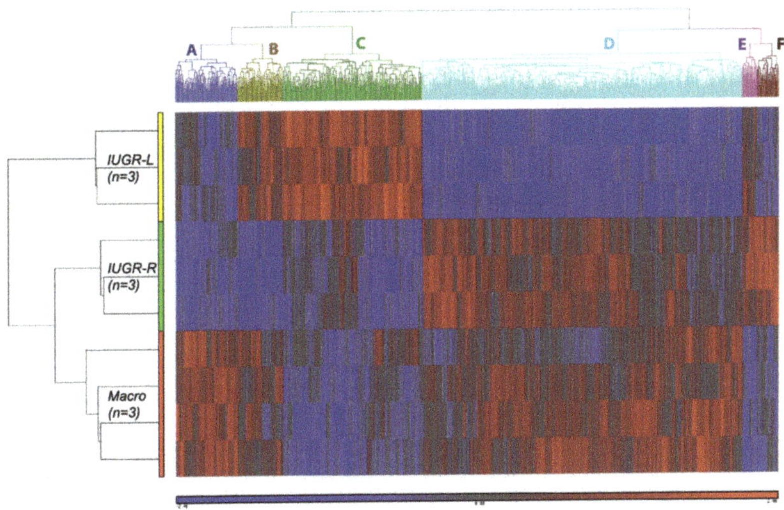

Figure 10. Unsupervised hierarchical clustering analysis of transcriptomes of gonadal fat. The heatmap shows differential expression of gene clusters in male adipose tissue from IUGR-L (first three rows; yellow bar) IUGR-R (center three rows; green bar) and macrosomic ("macro", last four rows; red bar) groups. Each column represents a gene and each row represents a different animal. Red indicates up-regulation and blue indicates down-regulation; grey indicates unchanged expression. Six clusters were identified (A–F at top). Genes were identified by ANOVA and were significant at $p < 0.01$.

An unsupervised hierarchical clustering analysis revealed that the macrosomic group was linked closely to the IUGR-R group (red and green bars on the left) and separated from the IUGR-L group (yellow bar). This was largely due to the gene Clusters C and D, which together (651 genes) represent 76% of the entire set of differentially expressed genes. Cluster D represents the largest gene node consisting of 454 genes (53%) that were more strongly expressed in the macrosomic and IUGR-R groups compared to the IUGR-L group, and Cluster C represents the second-largest node consisting of 197 genes (23%) that were more strongly expressed in the IUGR-L group compared to the Macrosomic or IUGR-R groups. These data show a close similarity of abdominal (gonadal) adipose tissue gene expression in IUGR-R and macrosomic males compared to IUGR-L males.

The clusters were subjected to gene ontology (GO) analysis using DAVID, but enrichment results were limited. No significant GO results were obtained for Clusters A, B, E or F. Cluster C genes were enriched in phosphoprotein, acetylation, nucleus, mRNA splicing, mRNA processing, Spliceosome, cell division, and ubl conjugation. Cluster D genes were enriched in phosphorylation, acetylation, cytoskeleton, phosphoprotein, and cell division. The biological significance of these enrichments was not clear. Combining Clusters C and D drove enrichment toward cell division, especially spindle and microtubule formation, but overall these GO results did not provide ready explanations of the observed phenotypic differences between the three groups of animals. Similar results were obtained using DAVID to identify impacted pathways. No significant functional pathways were obtained for Clusters A, B, E or F when analyzed singly. Submitting Cluster C indicated effects within

Spliceosome, transcriptional misregulation in cancer, and oocyte meiosis. More pathways were indicated for Cluster D, and included B cell receptor signaling, MAPK signaling and Ras signaling, and pathways in cancer. Combining Clusters B and F pointed toward MAPK signaling and combining Clusters C and D pointed toward B cell receptor signaling, MAPK signaling and Ras signaling, among others.

It is likely that the small sample size limited the sensitivity of the formal analyses. In both the results from the *t*-test comparisons and those of the clustering analysis, we identified changes in the expression of several genes of interest that were not assigned to pathways by the software, several of which were directly relevant to adipocyte function. Accordingly, as an additional/alternative approach, we drew from both sets of data and identified groupings of genes that suggested potential impacts within specific signaling pathways. Examples (Table S6) were effects on the renin-angiotensin system (eight genes), PPAR signaling (five genes), adipocytokine signaling (5 genes) and glycolysis and gluconeogenesis (five genes), all of which generally distinguished between heavier compared to lighter-weight adult animals.

We also found changes in the expression of a number of genes in the Krüppel-Like Factor (Klf) family. *t*-test results identified a 2.23-fold up-regulation of Klf9 in IUGR-R vs. IUGR-L animals; expression was similar in macrosomic and IUGR-R animals. Klf2, Klf4 and Klf13 were present in Cluster D and Klf6 was present in Cluster A. Thus, for four of these five genes (Klf2, Klf4, Klf9 and Klf13) expression was higher in the heavier-in-adulthood macrosomic and IUGR-R males, while Klf6 expression was higher as a function of birth weight category (Figure S4A).

In addition, we saw standalone effects on three genes of interest, all of which were downregulated in IUGR-R compared to IUGR-L: Tbx15 (T-box transcription factor 15, an early patterning gene recently identified as a master regulator of obesity genes [26]) and Repin1 (replication initiator 1, strongly associated with adipogenesis) were down-regulated 13.16-fold and 2.36-fold respectively in IUGR-R, and in both cases expression was similar in IUGR-R and macrosomic samples. Dlk1 (delta-like 1 homolog) was down-regulated 2.12-fold in IUGR-R compared to IUGR-L (Figure S4B).

3. Discussion

Using a novel crowded uterus model, we evaluated the consequences of IUGR and macrosomia on outcomes associated with metabolic syndrome. An important part of this work was the comparison of IUGR males that experienced a rapid post-weaning increase in body weight (IUGR-R males) and those that did not (IUGR-L males). At this time, we do not have an explanation for why IUGR males segregated after weaning into rapid and low growth groups. IUGR-R males showed about a 2.5-fold greater weight gain than IUGR-L males during the first week after weaning, and by 12 weeks of age, IUGR-R males had reached the same body weight as macrosomic animals and were significantly heavier than slow-growing IUGR males. At six months of age, IUGR-R males showed impaired response to a glucose challenge compared to IUGR-L males and altered expression of many genes in abdominal adipose tissue was revealed by microarray analysis. Clustering analysis of gonadal fat transcriptomes (discussed further below) showed that rapidly-growing IUGR males (IUGR-R) were aligned with macrosomic males in outcomes that are related to metabolic abnormalities throughout life [6]. However, at that age IUGR-R males differed from macrosomic males in terms of body weight, fat weight, adipocyte counts and size, as well as heart, kidney and spleen weights. There were also differences in abdominal fat gene expression between IUGR-R and macrosomic males revealed by microarray analysis.

Thus, these findings provide evidence for two distinct pathways to abdominal fat gene regulation and glucose regulation, and reveal that IUGR that is followed by a rapid period of weight gain during postnatal life results in changes in adipose tissue gene activity that in many respects are similar to those seen due to macrosomia at birth. However, there are also significant differences between the IUGR-R phenotype and the macrosomic phenotype, and both IURG-L and IUGR-R males were in some respects more similar to each other than to macrosomic males. These results are complex, and it is not possible at this time to predict disease or metabolic outcomes in these animals. Clearly, more research is needed to unravel the complex interaction of fetal under- or over-growth with postnatal growth trajectories in order to understand the factors that lead to adult overweight, which, itself, is a predictor of dysregulation of abdominal adipocyte gene expression.

A consequence of IUGR (often associated with a period of rapid postnatal weight increase), as well as macrosomia, is the development of glucose intolerance and other co-morbidities of obesity [2,15]. Using the 14-h fasting procedure, the faster-growing IUGR-R males showed reduced glucose tolerance compared to the slower-growing IUGR-L and macrosomic males. We did not anticipate that macrosomic animals would have superior glucose tolerance; however, since the median birth weight animal also showed reduced glucose tolerance compared to these groups; this test did seem to distinguish between animals with slow and more rapid post-weaning weight gain. Using the shorter 4-h fast, both of the IUGR groups and the macrosomic males showed reduced glucose tolerance relative to males in the median birth weight category, although the comparisons did not reach statistical significance; it is likely that the fasting time was too short to allow for sensitive testing. Thus, under the 14-h fasting regime, there was a tendency for faster-growing animals to have reduced glucose tolerance, and under the 4-h regime, both the IUGR and macrosomic animals tended toward impaired glucose tolerance. It was beyond the scope of this study to include insulin measurements in these animals, but this would provide very valuable information on glucose regulation in these animals and should be included in future work.

The 14-h fasting procedure is a common approach for a mouse GTT, but we show here that the consequences to the animal may preclude further useful analyses. Although a decrease in body weight with fasting is consistent with other findings [25], the degree of weight loss observed in these animals was suggestive of physiological stress and indicated that alternative testing protocols should be considered in future work, as discussed elsewhere [27]. In this study, macrosomic animals appeared to be more resilient in the face of this fast; the degree of body weight loss was less, fat mass and adipocyte counts were less impacted, and glucose tolerance was less impacted compared to IUGR-L males that lost a greater percentage of their body weight during an overnight fast than IUGR-R or macrosomic males. Taken together with other findings, these results suggest that IUGR creates a phenotype that is sensitive to environmental stressors, such as starvation, relative to non-IUGR animals. We also showed elsewhere that another source of environmental stress, namely, exposure during fetal life to a very low dose of the manmade toxic chemical bisphenol A (BPA), exacerbated the rate of postweaning growth in low birth and low weaning weight male mice, and also significantly impaired glucose tolerance [28]. BPA is one of many manmade chemicals referred to as metabolic disrupting chemicals [1,29].

We did not monitor weights between week 12 (three months) and six months of age, but by six months body the weight differences seen at 12 weeks had shifted, and the two IUGR groups were more similar and significantly lighter than macrosomic animals. This may reflect slowed growth during this period in IUGR-R animals, but the reason for this shift is not clear and further investigation is needed. Of note, in the Goto-Kakizaki diabetic rat, growth slows, and adipose tissue accumulation stops after week 12 [30], and it is interesting to speculate on possible mechanistic similarities although these are very different models. Overall, for body weight, there is a birth weight effect but also a post-weaning growth rate effect that is age-dependent.

Birth weight category appeared to strongly influence both fat weight and adipocyte number and size. In both 14-h fasted and non-fasted animals, macrosomic animals had higher amounts of total fat; gonadal fat pads tended to be heavier in macrosomic animals compared to either IUGR group, which did not differ, although this did not reach significance in non-fasted animals. Analysis of adipocyte size distribution in non-fasted animals clearly showed a difference between the two IUGR groups and macrosomic animals. After the 14-h fast this difference was eliminated, owing to a marked decrease in mid-sized adipocytes in IUGR-L and IUGR-R males, and a smaller decrease in these adipocytes in macrosomic males. However, the Coulter counter would not have detected any particle than ~8 µm diameter, so the apparent loss of adipocytes due to the 14-h fast might also reflect such a significant decrease in adipocyte size that they were not counted.

Analysis of gene expression revealed further differences and similarities between IUGR and macrosomic animals. The up-regulation of Pparg2, Cebpa, Glut4 and Lpl expression in IUGR males (or down-regulation in macrosomic males) is consistent with findings of clustering effects on PPAR signaling identified in the microarray analysis. This indicates the effects of IUGR on genes associated with an increased predisposition to store fat in adipocytes and the potential development of obesity-related abnormalities. This finding may be reflected in the increased number of mid-sized adipocytes observed in IUGR compared to macrosomic animals, although gonadal fat weights do not differ between these two groups of non-fasted animals. To confirm these results, future experiments should consider confirmation of effects on gene expression through Western blot and other approaches.

The objective of the microarray analysis was to provide preliminary information regarding the potential pathways in gonadal adipocytes that differed based on the rate of fetal growth as well as the rate of postnatal growth. Importantly, clustering analysis demonstrated that in terms of gene expression the association between the heavier-in-adulthood (week 12) IUGR-R and macrosomic animals was closer than the association between low birthweight IUGR-L and IUGR-R animals; the similarity of IUGR-R and macrosomic adipocyte gene expression was in spite of the very different birth weights, and also in spite of the more similar adipose tissue weights and adipocyte counts in the two groups of IUGR animals. In the mouse, adipocyte number may be determined prenatally but adipogenesis is initiated after birth [31]. Thus, although the amount of fat and adipocyte size in the adult may be influenced by the relatively low prenatal and pre-weaning nutrient availability, the accelerated growth during the immediate postweaning period occurred during the ongoing development of adipose tissue and permanently impacted overall gene expression in this tissue. Our results are limited by the small sample size, but we identified differences in metabolic pathways related to fat digestion and absorption, carbohydrate digestion, and type II diabetes mellitus signaling. We showed using clustering analysis that the expression of many genes in adipose tissue was influenced by either fetal growth or post-weaning growth or both.

Two of the differentially expressed genes identified using the more stringent *t*-test were particularly interesting. Fnip1, down-regulated in IUGR-R adipose tissue compared to IUGR-L tissue, codes for a protein that interacts with folliculin, which in turn interacts with the AMPK and mTOR signaling pathways [32,33]. The known involvement of AMPK and mTOR signaling in cellular energy and nutrient sensing [34] may be relevant here although our results did not point specifically toward alterations to either pathway.

The reduced expression of Agtr2 in IUGR-R animals relative to macrosomic animals is also interesting since it is thought that an overactive renin-angiotensin system (RAS) is involved in metabolic syndrome [35]. Increased angiotensinogen (AGT) production by white adipose tissue has been related to both obesity and hypertension, and loss of Agtr2 expression is sufficient to rescue obesity induced by adipose tissue AGT overexpression [36]. We found additional effects on the RAS pathway by manual searching rather than through selection by the software, and these results generally distinguished between heavier (at week 12) and lighter animals. Although the overall direction of the impact on this pathway is not clear, the multiple hits do suggest a possible association between our fetal/postnatal growth paradigm and altered signaling within the renin-angiotensin system.

Recent work in cardiomyocytes has shown cross-talk between mTOR and RAS signaling, acting in part through changes to Pik3 expression [37,38]. It is possible that our results reflect this: although the link to the mTOR pathway is only by association with the effects of Fnip1, we did see up-regulation of Pik3 in IUGR-R compared to IUGR-L and a general indication of effects on the RAS pathway in Clusters C and D (heavier vs. lighter animals).

Although these are very different cell types, these findings may provide some mechanistic information.

The Klf family of genes are widely involved in diverse biological processes that include adipogenesis and adipocyte differentiation, many of which interact with Cebpa or Cebpb and/or Pparg [39–41]. Of the five members of this family identified in our dataset, Klf4, Klf6 and Klf9 promote adipogenesis, Klf2 suppresses adipogenesis, and Klf9 and Klf13 are pro-adipogenic transcription factors [40,42]. The general trend in our data was for the expression of these genes to be higher in the heavier (IUGR-R and macrosomic) animals. Klf6 promotes adipocyte differentiation via suppression of the pre-adipocyte differentiation factor Dlk1 [43], and since our data show reduced expression of Dlk1 in IUGR-R compared to IUGR-L, this is generally consistent and point to a role of this gene family in the observed growth effects.

Alterations to adipocyte morphology can be directly related to alterations in adipocyte function as well as to insulin resistance [44,45]. Repin1 expression is reported to be increased during adipogenesis and reduced expression is associated with reduced adipocyte size as well as altered glucose uptake in 3T3 cells [46]. The observed down-regulation of Repin1 in macrosomic animals compared to IUGR-L may be consistent with the reduced number of smaller and mid-size adipocytes that we observed in macrosomic animals. However, Repin1 expression was also down-regulated in IUGR-R animals, which had a very similar adipocyte size distribution to that of IUGR-L animals, so it is not clear if whether Repin1 expression directly relates to adipocyte size in these animals or whether another factor modulates the effect in IUGR-R animals.

The phenomenon of "centile crossing" is well documented for IUGR babies that experience a rapid increase in body weight and fat in early childhood (here, IUGR-R males), while other IUGR babies (here, IUGR-L males) do not show this (Figure 11). There are significant adverse consequences that occur as a result of rapid weight gain in childhood that creates an obese phenotype [6], referred to as a "thrifty phenotype", and leads to the spectrum of metabolic diseases related to obesity, including glucose intolerance, that are extremely difficult to reverse [1,15,47]. As a result of a dramatic increase in body weight immediately after weaning, IUGR-R males exhibit postweaning growth rate centile crossing, and thus the prenatal and postnatal nutrient mismatch that is theorized in the thrifty phenotype hypothesis. This could occur even if a child is subsequently provided with a nutritionally balanced diet, although the impact on obesity would be greatly exacerbated with a highly processed, high-calorie meal typical of the standard Western diet [48].

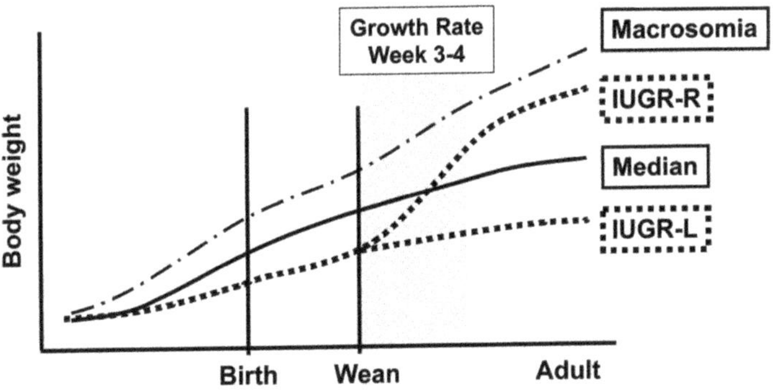

Figure 11. Centile crossing by IUGR-R male mice. IUGR-R mice show 120% increase in body weight during the first week (week 3–4) after weaning that by week four results in IUGR-R males having a body weight similar to that of males that were macrosomic at birth (see Figure 3).

Other developmental studies of IUGR have focused on exposures to environmental stressors, such as metabolic disrupting chemicals, physiological/psychological stress, and the consequences of consumption of processed foods instead of the nutrient (breast milk) that infants evolved to consume for far longer than is common in developed countries [1,49,50]. Not surprising is that our preliminary findings of different outcomes in males and females are, in fact, a common outcome [51].

While our prior findings suggest that the IUGR male mice were first malnourished due to decreased transplacental nutrient availability in utero, they were also possibly malnourished prior to weaning, potentially due to competition for resources against larger siblings (there were ~2 more pups per litter than nipples), although this remains to be investigated. It is also necessary to examine the neural control systems that regulate food hunger and satiety in IUGR-L and IUGR-R as well as macrosomic males to determine how they differ from median body weight at birth males. It was only in the period immediately after weaning when the IUGR mice had the ability to feed freely that differences between IUGR-L and IUGR-R males became apparent. It is during this brief post-weaning period that should be the focus of future studies, even though we were still able to identify differences in GTT and gene activity in the adipose tissue when the mice were examined much later in adulthood.

4. Materials and Methods

4.1. Animal Husbandry

CD-1 mice (*Mus domesticus*) were purchased from Charles River Breeding Laboratories (Wilmington, MA, USA) and maintained as an outbred colony with periodic replacement. The mice used in this study were housed in 18 × 29 × 13 cm polypropylene cages on corncob bedding. Pregnant and lactating mice were fed Purina mouse breeder chow 5008 (soy-based, Purina-Mills, St. Louis, MO, USA). After weaning, offspring were fed Purina standard laboratory chow 5001 (soy-based). Water was provided ad libitum in glass bottles and was purified by ion exchange followed by a series of carbon filters. Rooms were maintained at 25 ± 2 °C under a 12-h light:dark cycle, with the lights on at 1030 h. All animal procedures were approved by the University of Missouri Animal Care and Use Committee and conformed to the Guide for the Care and Use of Laboratory Animals of the National Institutes of Health. The animal facility is accredited by the Association for Assessment and Accreditation of Laboratory Animal Care, International (AAA-LAC).

4.2. Hemi-Ovariectomy Procedure

To examine the consequences of developing in a crowded uterine horn, offspring from two blocks of hemi-ovariectomized pregnant female CD-1 mice were examined. In Block 1, 52 postpartum dams were hemi-ovariectomized under ketamine-based anesthesia. In Block 2 an additional 21 postpartum dams were hemi-ovariectomized. We used postpartum dams so that we could examine offspring from the second litter, which has more pups than the first litter [23]; our objective was to increase the crowding of fetuses within one uterine horn. The left ovary was removed with a small incision, and the ovarian artery was cauterized. The left ovary was removed rather than the right ovary because of differences in the anatomy of the blood vessels associated with the left and right uterine horns.

Figure 1A depicts the normal female mouse reproductive tract with a duplex uterus. We have reported that the cranial end of the right utero-ovarian artery consistently inserts into the descending aorta, whereas the left utero-ovarian artery has a variable insertion into the descending aorta or the renal artery, including in some cases immediately adjacent to the left kidney. The variability in the left uterine horn vasculature at the cranial end of the uterus impacts placental blood flow at the cranial end of the uterine horn in both mice and rats [20,21]. Since placental blood flow impacts fetal growth [19,22], we chose to only examine offspring that developed in the right uterine horn to avoid variability due to the vascular anatomy of the left uterine horn. A schematic of the crowded uterine horn is shown in Figure 1B. This model has been shown to lead to differences in placental blood flow between fetuses located in the middle vs. the cranial or caudal ends of the uterus [22].

4.3. *Mating of Females and Determination of Offspring Body Weights*

Beginning five days after surgery, the hemi-ovariectomized female mice were singly housed with a stud male for up to 14 days. The day of natural birth was designated as postnatal day (PND) 1. At approximately 1400 h on the day of birth (PND 1), all pups were weighed and toe-clipped for identification, so that each individual could be categorized by birth weight percentile and then followed for the postnatal rate of growth and other outcomes. Toe-clipping was performed on the back paws of the animal using a set numbering system. Mouse pups were then not handled again between the day of birth and weaning at week 3 (postnatal day 21), since handling multiple times prior to weaning reduces pre-pubertal growth rate (unpublished observation). After the pups were weaned, the dams were euthanized with CO_2 and then necropsied to confirm the location of implantation sites as well as the vascular anatomy of the right uterine horn. After weaning at week 3, all male and female offspring from all of the birth weight percentile groups were housed 3–4 siblings of the same sex per cage and were weighed once per week until they were 12 weeks old. Experimental animals were generated from two separate breedings, referred to below as Block 1 (52 litters) and Block 2 (21 litters). As stated in the Introduction, we defined animals below the bottom 5th percentile for birth weights as IUGR and animals above the 95th percentile as macrosomic. We defined animals in the median range of birth weights as median-at-birth.

4.4. *Experimental Animals*

Summarized information of the experimental animals, group designations and experiments are in Table S1 for reference.

4.4.1. Block 1 Animals

All 605 animals (308 males and 297 females at birth) from 52 litters were used for the analysis of body weight between birth and postnatal week 12. After week 12, only males in the IUGR, median and macrosomic ranges were retained for further study.

The group of male mice selected from Block 1 for further study consisted of groups of IUGR, macrosomic and median-at-birth animals that were selected, respectively, from the bottom 9.9%, above the 93rd%, and close to the median (47.4–51.2%) of all birth weights (n = 29, 21 and 16, respectively). Based on the percent increase in body weight between postnatal week 3–4, we divided the IUGR males into low (IUGR-L, up to 65% weight gain) and rapid (IUGR-R, over 65% weight gain) subgroups, since there is evidence that IUGR interacts with the rate of postnatal growth in terms of risk for developing metabolic diseases [15]. Some of these male mice (IUGR-L n = 6, IUGR-R n = 10, median n = 16, and macrosomic n = 16) were used for glucose tolerance tests (GTT), fat pad (gonadal, renal and inguinal) and organ (liver, kidney, heart, spleen, testes, epididymides) collections, as well as analysis of gonadal adipocyte number and size; other males (IUGR-L n = 5, IUGR-R n = 5, and macrosomic n = 5) were used for further fat pad collections and adipocyte analysis, and also for analysis of gene expression by qPCR and microarray. While we followed the body weights of female siblings through postnatal week 12, due to an absence of body weight differences by week 12 in females (Figure S1), we focused here on males.

4.4.2. Block 2 Animals

A second block of 21 hemi-ovariectomized pregnant females produced pups solely for a follow-up GTT test. These IUGR-L (n = 7), IUGR-R (n = 19), median (n = 24) and macrosomic (n = 25) males were selected using the above criteria (IUGR < 10th percentile of birth weight, and macrosomic > 93rd percentile for birth weight). The body weights of these animals were recorded to establish the post-weaning weight gain and again at 24 weeks old.

4.5. Glucose Tolerance Test (GTT)

For the first glucose tolerance test (GTT), we examined Block 1 IUGR-L, IUGR-R, median and macrosomic males (n = 5, 9, 16 and 16, respectively). The GTT was conducted prior to sacrifice when animals were approximately 6 months old. Animals were fasted for 14 h prior to the test, but water was available ad libitum. Following the fast, animals were weighed and injected i.p. with a single bolus of glucose at a dose of 2 g glucose per kg body weight in an average volume of 176 μL of saline. Serum glucose measurements were taken by tail nick just prior to glucose administration and again at 30, 60 and 120 min after glucose administration. Gonadal, renal and inguinal fat pads were collected and weighed from these males at sacrifice. The area under the concentration-time curve (AUC) for the 0–120 min prior to and after dosing was calculated using the linear trapezoidal rule.

We conducted a second GTT test on Block 2 males, using the same methods as described for the first GTT, except that in this study the animals only experienced a 4-h fast prior to the GTT. The 4 h commenced at the start of the light phase; mice do not eat much food during the early light phase of the light:dark cycle [52] and this was thus a minimal fasting period. Bodyweights were recorded prior to the GTT, and gonadal fat-pad weights were recorded at sacrifice. Animals were sacrificed at least 7 days after the GTT rather than immediately afterwards as for Block 1.

4.6. Collection of Gonadal Fat

Gonadal fat was collected from all 14-h fasted and non-fasted Block 1 males, for analysis of adipocyte number and size distribution and gene expression analysis. For all animals, the weight of both gonadal fat pads was recorded. A 50–60 mg portion of the fat was placed in 5.2 mL of 2.9% OsO_4 in 0.05 M acidic collidine and retained for cell counting, and the remainder of the fat was snap-frozen in liquid nitrogen for later analysis of gene expression by qPCR and microarray analysis.

4.7. Analysis of Fat Cell Number and Volume

Fat from all Block 1 males was used in this study. The reserved 50–60 mg portion of the fat pad (above) was prepared for cell counting using a modification of the method of Hirsch and Gallian [53] as described by Kump and Booth [54], and cell number and size distribution were measured by the Coulter method [55]. Briefly, the fat was fixed and adipocytes were separated from other tissue in the OsO_4-collidine solution for 3–4 weeks; this procedure results in free cells. The cells were washed with isotonic saline solution and left in saline for 24 h, and then washed with 8 M urea in saline and left in that solution for 3–4 days. The cells were finally rinsed with 0.1% Triton X-100 and filtered through a 250 μm filter onto a 10 μm filter. The collected cells were suspended in Isoton II (Beckman-Coulter, Fullerton, CA, USA) containing 10% glycerol. Cells were counted and cell size was determined on a Coulter Multi-Sizer II, with the particle counting window set to 8.03–271.1 μm. Cells were counted in three 15-s bursts, and the three counts were summed to give a single measure per 45 s period. Cell counts reported here are the total number of cells contained in the combined fat pads.

4.8. Analysis of Gonadal Fat Gene Expression

We used both qPCR and microarray analysis to examine gene expression in gonadal fat. For qPCR we used samples from both the 14-h fasted and non-fasted males; we selected 4–5 animals per IUGR and macrosomic group, but because of the low numbers we did not stratify the IUGR animals further into L and R categories. For microarray analysis we used samples from selected non-fasted animals only, using 4 of the five macrosomic males, 3 IUGR-L males, and 3 IUGR-R males.

RNA was prepared from 50–80 mg of gonadal fat pad samples. Fat was homogenized in TRI Reagent (Sigma, St. Louis, MO, USA), and RNA was isolated according to the manufacturer's protocol, with a final precipitation using lithium chloride. The integrity of the RNA samples was verified by electrophoresis in 0.75% agarose gel. The RNA concentrations were measured spectrophotometrically. For microarray analysis, the samples went through a further cleanup on Qiagen RNeasy kit.

4.8.1. qPCR Assay

Expression of specific mRNAs was measured by one-step qPCR as described by Bustin [56] with the TaqMan EZ RT-PCR kit (Applied Biosystems, Foster City, CA, USA) on the ABI PRISM 7700 Sequence Detection System (Applied Biosystems, Waltham, MA, USA). The concentrations of $Mn2^{+}$, probe, and primers were optimized for each primer/probe set. Six target genes were selected based on their known roles in adipose tissue. Of these 6 genes, Pparg2 is a well-known master regulator of adipogenesis [57] that is also important for the regulation of glucose and lipid homeostasis and insulin sensitivity [58]. Cebpa interacts with Pparg2 to promote adipocyte differentiation as well as the transcription of glucose transporters, such as Glut4 (Slc2a4) [59]. These three genes are thus early-stage adipogenic genes [60]. Lpl is a central enzyme of lipoprotein metabolism that is important to the development of obesity, metabolizing lipoproteins and triglyceride particles, and producing fatty acids that are taken up by the adipocyte [61]. Hsd11b1 and Cyp19 are both candidate obesity genes [62].

Primer/probe sets for Hsd11b1 (11ß-hydroxysteroid dehydrogenase; GenBank accession no. NM-001044751.1) and Cyp19 (aromatase; GenBank accession no. NM-007810.3) were designed using Primer Express software (Applied Biosystems) and are shown in Table 2. Primers were designed to span exon boundaries in order to prevent the amplification of genomic DNA. Primers were synthesized by Invitrogen (Carlsbad, CA, USA), and probes were synthesized by Applied Biosystems. Other primer/probe sets, for Pparg2 (peroxisome proliferator receptor gamma 2), Ceba (CCAAT enhancer-binding protein α), Lpl (lipoprotein lipase) and Glut4 (glucose transporter type 4), were TaqMan Gene Expression Assays (pre-optimized and validated) obtained from Applied Biosystems; details are given in Table 3. The relative concentrations of specific mRNAs in each sample were normalized to total RNA per sample, as previously described [56,63,64].

Table 2. Sequences of primers and probes synthesized for real-time qPCR assays.

Gene Name		Sequence (5′–3′)
Hsd11b1	Forward	GCAGCATTGCCGTCATCTC
	Reverse	GAACCCATCCAGAGCAAACTTG
	Probe	TGGCTGGGAAAATGACCCAGCCTATG
Cyp19	Forward	CCGAGCCTTTGGAGAACAATT
	Reverse	TCCACACAAACTTCCACCATTC
	Probe	TTTCTTTATGAAAGCTCTGACGGGCCCT

Table 3. Primer/probe gene expression assays purchased for real time qPCR from Applied Biosystems.

Gene Name	Assay ID	Context Sequence
Pparg	Mm00440945_m1	TCAGTGGAGACCGCCCAGGCTTGCT
Cebpa	Mm00514283_s1	ACCAGCCACCGCCGCCACCGCCACC
Lpl	Mm00434764_m1	ATGGATGGACGGTAACGGGAATGTA
Glut4 (Slc2a4)	Mm00436615_m1	CTGCTGCTGCTGGAACGGGTTCCAG

4.8.2. Microarray Analysis

Total RNA was isolated from gonadal fat as described above and purified with the RNeasy Mini kit (Qiagen, Valencia, CA, USA) according to the manufacturers' instructions, and RNA quality was determined on an Agilent Bioanalyzer (Agilent, Palo Alto, CA, USA). The transcriptomal profiles were determined using Affymetrix mouse 430 2.0 microarrays, which assess over 39,000 gene transcripts. Scanned image data were converted into numerical tables using Affymetrix GeneChip Operating Software and Gene

Expression Console. Data analysis and mining, including gene ontology enrichment analysis, were performed using GeneSifter (Giospiza Inc., Seattle, WA, USA) and Partek Genomics Suite (Partek Inc., St. Louis, MO, USA) Further functional characterization was performed using DAVID (The Database for Annotation, Visualization and Integrated Discovery; http://david.abcc.ncifcrf.gov (accessed on 12 January 2020) [65,66]. Microarray data were deposited in NCBI Gene Expression Omnibus (accession number GSE33761).

4.9. Statistics

Data (other than microarray data, described above) were analyzed by analysis of variance or analysis of covariance (organ weights with body weight as the covariate) using SAS (v9.2, SAS Institute, Cary, NC, USA). In a few cases, data were log-transformed. Comparisons of overall statistically significant findings were made using Fisher's LSD with Bonferroni correction for multiple comparisons. qPCR data were compared using *t*-tests. Statistical significance was set at $p < 0.05$).

5. Conclusions

We identified here the effects of both prenatal and postweaning growth on factors related to adult metabolic disease. While the effects of pre- and post-natal growth are complex, we have clearly identified two groups in this study that have characteristics of metabolic syndrome in adulthood: the first being the result of overgrowth during fetal life (macrosomic males), and the second resulting from an interaction of prenatal growth restriction followed by a very rapid rate of postnatal growth over a short period of time after weaning, which occurred in IUGR-R males. We describe an interesting set of similarities between macrosomic and IUGR-R males, which experienced very different pathways to elevated weight by the end of the first week of free-feeding after weaning, but also outcomes that revealed similarities between IUGR-L and IUGR-R males in comparison to macrosomic males. Our studies here provide an initial profile of these three groups of males, but future research will be required to move beyond these preliminary findings.

In summary, with the increasing incidence of obesity and its related co-morbidities throughout the developed world, further research is needed on any promising method to identify those at risk for obesity and to develop customized, realistic therapies for those individuals. Our crowded uterus model presents a novel method to study the progression of three different phenotypes, IUGR with rapid postweaning growth, IUGR without rapid postweaning growth and macrosomic animals. The objective of future experiments would be to better understand the different characteristics of these three groups (in comparison to median bodyweight at birth animals). A long-term objective would be the potential to develop focused treatments to not only mitigate the abnormal metabolic consequences for these different sub-groups, but to intervene in the processes that lead to abnormal adult phenotypes.

Supplementary Materials: The following supporting information can be downloaded at: https://www.mdpi.com/article/10.3390/metabo12020102/s1. Figure S1: Birth weight and growth of female mice from Block 1, Figure S2: Glucose tolerance test in 4-h fasted females, Figure S3: qPCR quantification of gene expression in gonadal fat in 14-h fasted IUGR and macrosomic male mice, Figure S4: Expression of selected genes of interest from the microarray data set, Table S1: Outline of Block 1 and Block 2 experiments conducted on male offspring. Table S2: Organ weights of 14-h fasted Block 1, Set 1 males. Table S3: Body weights and fat pad weights in 14-h fasted and non-fasted males, Table S4: Growth and fat weights of the sub-group of 14-h fasted males and of non-fasted males used in qPCR analyses, Table S5: Growth and fat weights of animals used in Microarray analyses, Table S6: Gene pathways manually identified as being potentially impacted by either prenatal or postweaning development, Table S7A, Microarray data: T-test results, Table S7B, Microarray Data: Genes identified by clustering analysis.

Author Contributions: Conceptualization, F.S.v.S.; methodology, F.S.v.S., J.A.T. and T.S.; validation, F.S.v.S., T.S., B.L.C. and J.A.T.; formal analysis, F.S.v.S., T.S. and J.A.T.; investigation, B.L.C. and J.A.T.; resources, F.S.v.S. and T.S.; data curation, B.L.C. and J.A.T.; writing—original draft preparation, B.L.C., F.S.v.S. and J.A.T.; writing—review and editing, F.S.v.S., J.A.T., T.S. and B.L.C.; visualization, F.S.v.S., T.S. and J.A.T.; supervision, F.S.v.S.; project administration, F.S.v.S.; funding acquisition, F.S.v.S. and T.S. All authors have read and agreed to the published version of the manuscript.

Funding: This research was funded by NIEHS grant numbers ES11283 and ES01876 to F.S.v.S. and grant numbers R21ES024861, R01ES023316, and R01ES020454 to TS. The APC was funded by an internal research account at the University of Missouri.

Institutional Review Board Statement: All animal procedures were approved by the University of Missouri Animal Care and Use Committee and conformed to the Guide for the Care and Use of Laboratory Animals of the National Institutes of Health. The program is fully accredited by the Association for Assessment & Accreditation of Laboratory Animal Care, International (AAA-LAC).

Informed Consent Statement: Not applicable.

Data Availability Statement: The microarray data have been deposited in the NCBI GEO database, accession number GSE33761. Other data presented in this study are contained within the article and Supplementary Materials and may be obtained from the authors.

Acknowledgments: The authors are grateful for assistance from Jiude Mao, James Kirkpatrick, Maren Bell Jones and Brittany Angle for their technical help; Mark Ellersieck for assistance with statistical analysis; Boyd O'Dell for guidance on adipocyte counting and generous use of his laboratory equipment, and Linda Bennett and Keiko Shioda for their work and help with the initial microarray analysis.

Conflicts of Interest: The authors declare no conflict of interest.

References

1. Heindel, J.J.; Blumberg, B.; Cave, M.; Machtinger, R.; Mantovani, A.; Mendez, M.A.; Nadal, A.; Palanza, P.; Panzica, G.; Sargis, R.; et al. Metabolism disrupting chemicals and metabolic disorders. *Reprod. Toxicol.* **2017**, *68*, 3–33. [CrossRef] [PubMed]
2. Longo, S.; Bollani, L.; Decembrino, L.; Di Comite, A.; Angelini, M.; Stronati, M. Short-term and long-term sequelae in intrauterine growth retardation (IUGR). *J. Matern. Fetal. Neonatal. Med.* **2013**, *26*, 222–225. [CrossRef] [PubMed]
3. Heindel, J.J.; vom Saal, F.S. Role of nutrition and environmental endocrine disrupting chemicals during the perinatal period on the aetiology of obesity. *Mol. Cell. Endocrinol.* **2009**, *304*, 90–96. [CrossRef] [PubMed]
4. Barker, D.J. The malnourished baby and infant. *Br. Med. Bull.* **2001**, *60*, 69–88. [CrossRef] [PubMed]
5. Fall, C.H. The Genesis of 'Fetal Origins of Adult Disease'. *Int. J. Diab. Dev. Ctries.* **2001**, *21*, 5.
6. Baird, J.; Fisher, D.; Lucas, P.; Kleijnen, J.; Roberts, H.; Law, C. Being big or growing fast: Systematic review of size and growth in infancy and later obesity. *Bmj* **2005**, *331*, 929. [CrossRef]
7. Styrud, J.; Eriksson, U.J.; Grill, V.; Swenne, I. Experimental intrauterine growth retardation in the rat causes a reduction of pancreatic B-cell mass, which persists into adulthood. *Biol. Neonate* **2005**, *88*, 122–128. [CrossRef]
8. Jones, A.P.; Friedman, M.I. Obesity and adipocyte abnormalities in offspring of rats undernourished during pregnancy. *Science* **1982**, *215*, 1518–1519. [CrossRef]
9. Vuguin, P.M. Animal Models for Small for Gestational Age and Fetal Programing of Adult Disease. *Horm. Res.* **2007**, *68*, 113–123. [CrossRef]
10. Woodward-Lopez, G.; Davis, R.L.; Gerstein, D.E.; Crawford, P.B. *Obesity Dietary and Developmental Influences;* CRC Press: Boca Raton, FL, USA, 2006; p. 336.
11. Ravelli, G.P.; Stein, Z.A.; Susser, M.W. Obesity in young men after famine exposure in utero and early infancy. *N. Engl. J. Med.* **1976**, *295*, 349–353. [CrossRef]
12. Jansson, T.; Ylven, K.; Wennergren, M.; Powell, T.L. Glucose transport and system A activity in syncytiotrophoblast microvillous and basal plasma membranes in intrauterine growth restriction. *Placenta* **2002**, *23*, 392–399. [CrossRef] [PubMed]
13. Ogata, E.S.; Finley, S.L. Selective ligation of uterine artery branches accelerates fetal growth in the rat. *Pediatr. Res.* **1988**, *24*, 384–390. [CrossRef] [PubMed]
14. Baumann, H.; Gauldie, J. The acute phase response. *Immunol. Today* **1994**, *15*, 74–80. [CrossRef]
15. Priante, E.; Verlato, G.; Giordano, G.; Stocchero, M.; Visentin, S.; Mardegan, V.; Baraldi, E. Intrauterine Growth Restriction: New Insight from the Metabolomic Approach. *Metabolites* **2019**, *9*, 267. [CrossRef] [PubMed]
16. Oken, E.; Gillman, M.W. Fetal origins of obesity. *Obes. Res.* **2003**, *11*, 496–506. [CrossRef]
17. Taveras, E.M.; Rifas-Shiman, S.L.; Belfort, M.B.; Kleinman, K.P.; Oken, E.; Gillman, M.W. Weight status in the first 6 months of life and obesity at 3 years of age. *Pediatrics* **2009**, *123*, 1177–1183. [CrossRef]

18. Overview. Elsevier Point of Care Editorial Board. Intrauterine Growth Restriction. Available online: https://www-clinicalkey-com.proxy.mul.missouri.edu/#!/content/clinical_overview/67-s2.0-e1056319-3828-4963-b385-e379fb2f33a9 (accessed on 16 November 2021).
19. McLaren, A.; Michie, D. Control of pre-natal growth in mammals. *Nature* **1960**, *187*, 363–365. [CrossRef]
20. vom Saal, F.S.; Dhar, M.D. Blood flow in the uterine loop artery and loop vein is bi-directional in the mouse: Implications for intrauterine transport of steroids. *Physiol. Behav.* **1992**, *52*, 163–171. [CrossRef]
21. Even, M.D.; Laughlin, M.H.; Krause, G.F.; vom Saal, F.S. Differences in blood flow to uterine segments and placentae in relation to sex, intrauterine location and side in pregnant rats. *J. Reprod. Fertil.* **1994**, *102*, 245–252. [CrossRef]
22. Coe, B.L.; Kirkpatrick, J.R.; Taylor, J.A.; vom Saal, F.S. A new 'crowded uterine horn' mouse model for examining the relationship between foetal growth and adult obesity. *Basic Clin. Pharm. Toxicol.* **2008**, *102*, 162–167. [CrossRef]
23. vom Saal, F.S.; Finch, C.E.; Nelson, J.F. Natural history and mechanisms of reproductive aging in humans, laboratory rodents and other selected vertebrates. In *The Physiology of Reproduction*, 2nd ed.; Knobil, E., Neil, J.D., Eds.; Raven Press: New York, NY, USA, 1994; Volume 2, pp. 1213–1314.
24. vom Saal, F.S. Variation in phenotype due to random intrauterine positioning of male and female fetuses in rodents. *J. Reprod. Fertil.* **1981**, *62*, 633–650. [CrossRef]
25. Jensen, T.L.; Kiersgaard, M.K.; Sorensen, D.B.; Mikkelsen, L.F. Fasting of mice: A review. *Lab. Anim.* **2013**, *47*, 225–240. [CrossRef]
26. Pan, D.Z.; Miao, Z.; Comenho, C.; Rajkumar, S.; Koka, A.; Lee, S.H.T.; Alvarez, M.; Kaminska, D.; Ko, A.; Sinsheimer, J.S.; et al. Identification of TBX15 as an adipose master trans regulator of abdominal obesity genes. *Genome Med.* **2021**, *13*, 123. [CrossRef] [PubMed]
27. Andrikopoulos, S.; Blair, A.R.; Deluca, N.; Fam, B.C.; Proietto, J. Evaluating the glucose tolerance test in mice. *Am. J. Endocrinol. Metab.* **2008**, *295*, E1323–E1332. [CrossRef] [PubMed]
28. Taylor, J.A.; Sommerfeld-Sager, J.M.; Meng, C.X.; Nagel, S.C.; Shioda, T.; Vom Saal, F.S. Reduced body weight at weaning followed by increased post-weaning growth rate interacts with part-per-trillion fetal serum concentrations of bisphenol A (BPA) to impair glucose tolerance in male mice. *PLoS ONE* **2018**, *13*, e0208846. [CrossRef]
29. vom Saal, F.S.; Vandenberg, L.N. Update on the Health Effects of Bisphenol A: Overwhelming Evidence of Harm. *Endocrinology* **2021**, *162*, bqaa171. [CrossRef]
30. Xue, B.; Sukumaran, S.; Nie, J.; Jusko, W.J.; DuBois, D.C.; Almon, R.R. Adipose Tissue Deficiency and Chronic Inflammation in Diabetic Goto-Kakizaki Rats. *PLoS ONE* **2011**, *6*, e17386. [CrossRef] [PubMed]
31. Ailhaud, G.; Grimaldi, P.; Negrel, R. Cellular and molecular aspects of adipose tissue development. *Annu. Rev. Nutr.* **1992**, *12*, 207–233. [CrossRef]
32. Baba, M.; Hong, S.-B.; Sharma, N.; Warren, M.B.; Nickerson, M.L.; Iwamatsu, A.; Zhen, W.; Burke, T.R.J.; Linehan, W.M.; Schmidt, L.S.; et al. Folliculin encoded by the BHD gene interacts with a binding protein, FNIP1, and AMPK, and is involved in AMPK and mTOR signaling. *Proc. Natl. Acad. Sci. USA* **2006**, *103*, 15552–15557. [CrossRef]
33. Linehan, W.M.; Srinivasan, R.; Schmidt, L.S. The genetic basis of kidney cancer: A metabolic disease. *Nat. Rev. Urol.* **2010**, *7*, 277–285. [CrossRef]
34. Blanchard, P.-G.; Festuccia, W.T.; Houde, V.P.; St-Pierre, P.; Brule, S.; Turcotte, V.; Cote, M.; Bellmann, K.; Marette, A.; Deshaies, Y. Major involvement of mTOR in the PPARγ-induced stimulation of adipose tissue lipid uptake and fat accretion. *J. Lipid Res.* **2012**, *53*, 1117–1125. [CrossRef] [PubMed]
35. Wang, C.-H.; Li, F.; Takahashi, N. The renin angiotensin system and the metabolic syndrome. *Open Hypertens. J.* **2010**, *3*, 1–13. [CrossRef] [PubMed]
36. Yvan-Charvet, L.; Masseira, F.; Lamande, N.; Ailhaud, G.; Teboul, M.; Moustaid-Moussa, N.; Gasc, J.M.; Quignard-Boulange, A. Deficiency of angiotensisn type 2 receptor rescues obesity but not hypertension induced by overexpression of angiotensinogen in adipose tissue. *Endocrinology* **2009**, *150*, 1421–1428. [CrossRef] [PubMed]
37. Hafizi, S.; Wang, X.; Chester, A.; Yacoub, M.H.; Proud, C.G. ANG II activates effectors of mTOR via PI3-K signaling in human coronary smooth muscle cells. *Am. J. Physiol.—Heart Circ. Physiol.* **2004**, *287*, H1232–H1238. [CrossRef]
38. Pulakat, L.; DeMarco, V.G.; Whaley-Connell, A.; Sowers, J.R. The Impact of Overnutrition on Insulin Metabolic Signaling in the Heart and the Kidney. *Cardiorenal. Med.* **2011**, *1*, 102–112. [CrossRef]
39. Brey, C.W.; Nelder, M.P.; Hailemariam, T.; Gaugler, R.; Hashmi, S. Krüppel-like family of transcription factors: An emerging new frontier in fat biology. *Int. J. Biol. Sci.* **2009**, *5*, 622–636. [CrossRef]
40. Pei, H.; Yao, Y.; Yang, Y.; Liao, K.; Wu, J.-R. Kruppel-like factor KLF9 regulates PPARg transactivation at the middle stage of adipogenesis. *Cell Death Differ.* **2011**, *18*, 315–327. [CrossRef]
41. Wu, Z.; Wang, S. Role of kruppel-like transcription factors in adipogenesis. *Dev. Biol.* **2013**, *474*, 235–243. [CrossRef]
42. Jiang, S.; Wei, H.; Song, T.; Yang, Y.; Zhang, F.; Zhou, Y.; Peng, J.; Jiang, S. KLF13 promotes porcine adipocyte differentiation through PPARγ activation. *Cell Biosci.* **2015**, *5*, 28. [CrossRef]
43. Li, D.; Yea, S.; Li, S.; Chen, Z.; Narla, G.; Banck, M.; Laborda, J.; Tan, S.; Friedman, J.M.; Friedman, S.L.; et al. Kruppel-like Factor-6 Promotes Preadipocyte Differentiation through Histone Deacetylase 3-dependent Repression of DLK1. *J. Biol. Chem.* **2005**, *280*, 26941–26952. [CrossRef]

44. Verboven, K.; Wouters, K.; Gaens, K.; Hansen, D.; Bijnen, M.; Wetzels, S.; Stehouwer, C.D.; Goossens, G.H.; Schalkwijk, C.G.; Blaak, E.E.; et al. Abdominal subcutaneous and visceral adipocyte size, lipolysis and inflammation relate to insulin resistance in male obese humans. *Sci. Rep.* **2018**, *8*, 4677. [CrossRef]
45. Liu, F.; Wang, H.; Zhu, D.; Bi, Y. Adipose Morphology: A Critical Factor in Regulation of Human Metabolic Diseases and Adipose Tissue Dysfunction. *Obes. Surg.* **2020**, *30*, 5086–5100. [CrossRef]
46. Ruschke, K.; Illes, M.; Kern, M.; Kloting, I.; Fasshauer, M.; Schon, M.R.; Kosacka, H.; Fitzl, G.; Kovacs, P.; Stumvoll, M.; et al. Repin1 maybe involved in the regulation of cell size and glucose transport in adipocytes. *Biochem. Biophys. Res. Commun.* **2010**, *400*, 246–251. [CrossRef]
47. Hales, C.N.; Barker, D.J. The thrifty phenotype hypothesis. *Br. Med. Bull.* **2001**, *60*, 5–20. [CrossRef]
48. Gluckman, P.D.; Hanson, M.A.; Beedle, A.S. Early life events and their consequences for later disease: A life history and evolutionary perspective. *Am. J. Hum. Biol.* **2007**, *19*, 1–19. [CrossRef]
49. Martinez, M.A.; Castro, I.; Rovira, J.; Ares, S.; Rodriguez, J.M.; Cunha, S.C.; Casal, S.; Fernandes, J.O.; Schuhmacher, M.; Nadal, M. Early-life intake of major trace elements, bisphenol A, tetrabromobisphenol A and fatty acids: Comparing human milk and commercial infant formulas. *Environ. Res.* **2019**, *169*, 246–255. [CrossRef]
50. Ruiz, D.; Padmanabhan, V.; Sargis, R.M. Stress, Sex, and Sugar: Glucocorticoids and Sex-Steroid Crosstalk in the Sex-Specific Misprogramming of Metabolism. *J. Endocr. Soc.* **2020**, *4*, bvaa087. [CrossRef]
51. Palanza, P.; Paterlini, S.; Brambilla, M.M.; Ramundo, G.; Caviola, G.; Gioiosa, L.; Parmigiani, S.; Vom Saal, F.S.; Ponzi, D. Sex-biased impact of endocrine disrupting chemicals on behavioral development and vulnerability to disease: Of mice and children. *Neurosci. Biobehav. Rev.* **2021**, *121*, 29–46. [CrossRef]
52. Sieli, P.T.; Jasarevic, E.; Warzak, D.A.; Mao, J.; Ellersieck, M.R.; Liao, C.; Kannan, K.; Collet, S.H.; Toutain, P.L.; vom Saal, F.S.; et al. Comparison of serum bisphenol A concentrations in mice exposed to bisphenol A through the diet versus oral bolus exposure. *Env. Health Perspect.* **2011**, *119*, 1260–1265. [CrossRef]
53. Hirsch, J.; Gallian, E. Methods for the determination of adipose cell size in man and animals. *J. Lipid Res.* **1968**, *9*, 110–119. [CrossRef]
54. Kump, D.S.; Booth, F.W. Sustained rise in triacylglycerol synthesis and increased epididymal fat mass when rats cease voluntary wheel running. *J. Physiol.* **2005**, *565*, 911–925. [CrossRef] [PubMed]
55. Graham, M.D. The Coulter Principle: Foundation for an Industry. *JALA* **2003**, *8*, 10. [CrossRef]
56. Bustin, S.A. Absolute quantification of mRNA using real-time reverse transcription polymerase chain reaction assays. *J. Mol. Endocrinol.* **2000**, *25*, 169–193. [CrossRef] [PubMed]
57. Rosen, E.D.; Sarraf, P.; Troy, A.E.; Bradwin, G.; Moore, K.; Milstone, D.S.; Spiegelman, B.M.; Mortensen, R.M. PPAR gamma is required for the differentiation of adipose tissue in vivo and in vitro. *Mol. Cell.* **1999**, *4*, 611–617. [CrossRef]
58. Jones, J.R.; Barrick, C.; Kim, K.-A.; Lindner, J.; Blondeau, B.; Fuimoto, Y.; Shiota, M.; Kesterson, R.A.; Kahn, B.B.; Magnuson, M.A. Deletion of PPAR in adipose tissues of mice protects against high fat diet-induced obesity and insulin resistance. *Proc. Natl. Acad. Sci. USA* **2005**, *102*, 6207–6212. [CrossRef] [PubMed]
59. Kaestner, K.H.; Christy, R.J.; Lane, M.D. Mouse insulin-responsive glucose transporter gene: Characterization of the gene and trans-activation by the CCAAT/enhancer binding protein. *Proc. Natl. Acad. Sci. USA* **1990**, *87*, 251–255. [CrossRef]
60. Moseti, D.; Regassa, A.; Kim, W.-K. Molecular Regulation of Adipogenesis and Potential Anti-Adipogenic Bioactive Molecules. *Int. J. Mol. Sci.* **2016**, *17*, 124–147. [CrossRef]
61. Gonzales, A.M.; Orlando, R.A. Role of adipocyte-derived lipoprotein lipase in adipocyte hypertrophy. *Nutr. Metabolism* **2007**, *4*, 22. [CrossRef]
62. Rosmond, R. Association studies of genetic polymorphisms in central obesity: A critical review. *Int. J. Obes.* **2003**, *27*, 1141–1151. [CrossRef]
63. Latil, A.; Bieche, I.; Vidaud, D.; Lidereau, R.; Berthon, P.; Cussenot, O.; Vidaud, M. Evaluation of androgen, estrogen (ER alpha and ER beta), and progesterone receptor expression in human prostate cancer by real-time quantitative reverse transcription-polymerase chain reaction assays. *Cancer Res.* **2001**, *61*, 1919–1926.
64. Richter, C.A.; Taylor, J.A.; Ruhlen, R.R.; Welshons, W.V.; vom Saal, F.S. Estradiol and bisphenol A stimulate androgen receptor and estrogen receptor gene expression in fetal mouse prostate cells. *Environ. Health Perspect.* **2007**, *115*, 902–908. [CrossRef] [PubMed]
65. Huang, D.W.; Sherman, B.T.; Lempicki, R.A. Systematic and integrative analysis of large gene lists using DAVID bioinformatics resources. *Nat. Protoc.* **2009**, *4*, 44–57. [CrossRef] [PubMed]
66. Huang, D.W.; Sherman, B.T.; Lempicki, R.A. Bioinformatics enrichment tools: Paths toward the comprehensive functional analysis of large gene lists. *Nucleic Acids Res.* **2009**, *37*, 1–13. [CrossRef] [PubMed]